Lecture Notes in Physics

The Lecture Notes in Physics

The series Lecture Notes in Physics (LNP), founded in 1969, reports new developments in physics research and teaching – quickly and informally, but with a high quality and the explicit aim to summarize and communicate current knowledge in an accessible way. Books published in this series are conceived as bridging material between advanced graduate textbooks and the forefront of research and to serve three purposes:

- to be a compact and modern up-to-date source of reference on a well-defined topic
- to serve as an accessible introduction to the field to postgraduate students and nonspecialist researchers from related areas
- to be a source of advanced teaching material for specialized seminars, courses and schools

Both monographs and multi-author volumes will be considered for publication. Edited volumes should, however, consist of a very limited number of contributions only. Proceedings will not be considered for LNP.

Volumes published in LNP are disseminated both in print and in electronic formats, the electronic archive being available at springerlink.com. The series content is indexed, abstracted and referenced by many abstracting and information services, bibliographic networks, subscription agencies, library networks, and consortia.

Proposals should be sent to a member of the Editorial Board, or directly to the managing editor at Springer:

Christian Caron
Springer Heidelberg
Physics Editorial Department I
Tiergartenstrasse 17
69121 Heidelberg / Germany
christian.caron@springer.com

T. Sato
T. Takahashi
K. Yoshimura (Eds.)

Particle and Nuclear Physics at J-PARC

Takahiro Sato
High Energy Accelerator
Research
Organization, KEK
1-1 Oho
Tsukuba, Ibaraki
305-0801 Japan

Toshiyuki Takahashi
High Energy Accelerator
Research
Organization, KEK
1-1 Oho
Tsukuba, Ibaraki
305-0801 Japan

Koji Yoshimura
High Energy Accelerator
Research
Organization, KEK
1-1 Oho
Tsukuba, Ibaraki
305-0801 Japan

Sato, T. et al. (Eds.), *Particle and Nuclear Physics at J-PARC*, Lect. Notes Phys. 781 (Springer, Berlin Heidelberg 2009), DOI 10.1007/978-3-642-00961-7

Lecture Notes in Physics ISSN 0075-8450
e-ISSN 1616-6361
ISBN 978-3-642-00960-0
e-ISBN 978-3-642-00961-7
DOI 10.1007/978-3-642-00961-7
Springer Dordrecht Heidelberg London New York

Library of Congress Control Number: 2009926956

Cover design: Integra Software Services Pvt. Ltd., Pondicherry

Printed on acid-free paper

Springer is part of Springer Science+Business Media (www.springer.com)

Preface

An accelerator complex which gives extremely high-intensity proton beams is being constructed in Tokai, Japan. The project is operated by JAEA (Japan Atomic Energy Agency) and KEK (High Energy Accelerator Research Organization) and called J-PARC (Japan Proton Accelerator Research Complex). J-PARC accelerator complex consists of 200 MeV linac, 3 GeV rapid cycling synchrotron, and 30 GeV main synchrotron. The energy of linac will be extended to 400 MeV and the energy of the main ring will be increased to 50 GeV in the near future.

J-PARC aims to perform various researches of life and material sciences by using neutron beams from the 3 GeV rapid cycling synchrotron. J-PARC also aims to perform various particle and nuclear physics experiments by using the 50 GeV main synchrotron. In this book we collected several proposals of particle and nuclear physics experiments to be performed by using 50 GeV main synchrotron.

Prof. Nagamiya gives a brief introduction of J-PARC. He describes the purpose of the project, the aims of the various facilities, and the researches to be done by using these facilities.

Prof. Ichikawa discusses about the long baseline nutrino oscillation experiment. This proposal is called T2K (Tokai to Kamioka) and it aims to measure mixing angles in the lepton sector. They try to perform a precise measurement of θ_{23} by measuring the ν_μ disappearance. Then they go to determine θ_{13} by measuring ν_μ–ν_e appearance signal. They also search for sterile components by measuring NC events.

Prof. Lim discusses about the experiment which searches a very rare decay of the neutral kaon: ${K^0}_L \to \pi^0 \nu\bar{\nu}$. This decay occurs via a direct CP violation. He will search this decay mode with higher sensitivity than the standard model expectation level.

Prof. Imazato presents his plan to measure T-violation effect in $K^+ \to \pi^0\mu^+\nu$ decay. He performed a precise measurement of this T-violation effect by measuring the transverse polarization of the decay muon using the TOROIDAL spectrometer system at KEK 12-GeV PS. He will use the same detector system with some modifications at J-PARC. His goal is to measure the muon transverse polarization with an accuracy of 10^{-4}.

Prof. Tamura describes the gamma-ray spectroscopy of Λ-hypernuclei. It aims to study hyperon–nucleon interactions, impurity effect in the nuclear structure, and medium effect in baryon properties through the precise measurements of the hypernuclear level structure using Ge detectors. With the development of a large-acceptance Ge detector array with fast readout electronics, he succeeded in the high-precision spectroscopy of Λ-hypernuclei of p-shell region. He will extend such studies in the wide range of hypernuclei from ${}^{4}_{\Lambda}$He to ${}^{208}_{\Lambda}$Pb at J-PARC.

Dr. Naruki reports their experiment to observe a pentaquark state Θ^{+}. Since the LEPS collaboration observed a narrow resonance at 1540 MeV/c^2 in 2003, many positive and negative results have been reported all over the world. They tried to observe this exotic state at KEK 12-GeV PS and obtained a hint to the possible existence of this exotic state. She is planning to search for this resonance by measuring $\pi^- + p \rightarrow K^- + \Theta^+$ with much higher sensitivity.

Dr. Yokkaichi discusses about mass modification of vector mesons in nuclear medium. This subject is related to the spontaneous breaking of the chiral symmetry in QCD, which is considered to play the main role in the mass generation mechanism of light quarks and hadrons. He reports on the vector meson mass modification, in particular, on the ϕ meson modification, observed in KEK-PS E325 experiment. At J-PARC he is going to collect 100 times statistics as that of the previous KEK experiment and investigate the in-medium meson properties in detail.

Prof. Iwasaki describes experimental mesonic bound states in nuclei. He reviews past experimental results which were not conclusive. These facts brought motivation for new experiments at J-PARC. He introduces three approaches: (1) study of kaonic atom, (2) search for the kaon bound states, and (3) search for the ϕ-meson bound states. The first two use different modes from the previous experiments, i.e., (1) ^{3}He instead of ^{4}He and (2) K^-pp instead of K^-ppn. Both modes will provide decisive tests for the Akaishi–Yamazaki prediction of deeply bound kaonic nuclei. The third one is a new experimental approach to study vector meson in medium.

Prof. Kuno reviews nuclear and particle physics by using intense muon source produced at J-PARC MR. There are a variety of potential projects which could be realized with a high-intensity muon beam, 10^{10}–10^{12} muons/s. Among them, there are three important particle physics programs: (1) precision measurement of muon $g-2$ anomalous magnetic moment, (2) search for muon electric dipole moment (EDM), and (3) search for lepton flavor violation of charged leptons (LFV). All three are very important to search for new physics. Here he focuses mainly on one of LFV search projects, i.e., search for coherent muon electron conversion (COMET) and its future extension (PRISM).

In this book, eight representative experimental proposals are presented. Many more experiments are actually planned at J-PARC which will explore the high-intensity frontier of particle and nuclear physics.

Tsukuba,
August 2008

Takahiro Sato
Toshiyuki Takahashi
Koji Yoshimura

Contents

Overview of J-PARC

Shoji Nagamiya

J-PARC Center, 2-4 Shirakata, Tokai-Mura, 319-1195, Japan
shoji.nagamiya@kek.jp

In the Japanese fiscal year JFY01, which started on April 1, 2001, a new accelerator project to provide high-intensity proton beams proceeded into its construction phase. This project, which is called the J-PARC (Japan Proton Accelerator Research Complex), is a joint project between two institutions, KEK and JAEA. We set a goal to achieve 1 MW proton beams at 3 GeV and 0.75 MW beams at 50 GeV. The construction period is 8 years, with anticipated first beams from the entire facility in the late fall of 2008, although the beams from the linac and 3 GeV were already accelerated and extracted in January and October 2007, respectively. In this chapter I describe the present status of the J-PARC.

1 What Is J-PARC?

J-PARC is the acronym of Japan Proton Accelerator Research Complex, which is under construction jointly by KEK (National High Energy Accelerator Research Organization) and JAEA (Japan Atomic Energy Agency). The facility is located in the Tokai campus of JAEA, which is about 60 km northeast of the Tsukuba campus of KEK.

This new proton accelerator is targeted at a wide range of fields, particle and nuclear physics, materials science, life science, and nuclear engineering. Figure 1 illustrates a one-page summary of sciences to be conducted at J-PARC. The atomic nucleus is made of protons and neutrons. When a low-energy proton (typically, at the energy of 1 GeV) hits the nucleus, the constituents of the nucleus, unstable nuclei, neutrons, and protons, will be ejected with the proton-induced spallation reaction. Among those, neutrons will be used at J-PARC for materials and life sciences. In addition, a copious production of pions is expected, so that the research on sciences with low-energy muons, for example, μSR or muonium science will be conducted. At higher beam energies, typically 50 GeV, the proton–nucleus collisions will produce kaons, high-energy neutrinos, anti-protons, etc. The usage of kaons

Nagamiya, S.: *Overview of J-PARC*. Lect. Notes Phys. **781**, 1–16 (2009)
DOI 10.1007/978-3-642-00961-7_1

and neutrinos is planned at J-PARC. The entire view of the J-PARC facility is illustrated in Fig. 2.

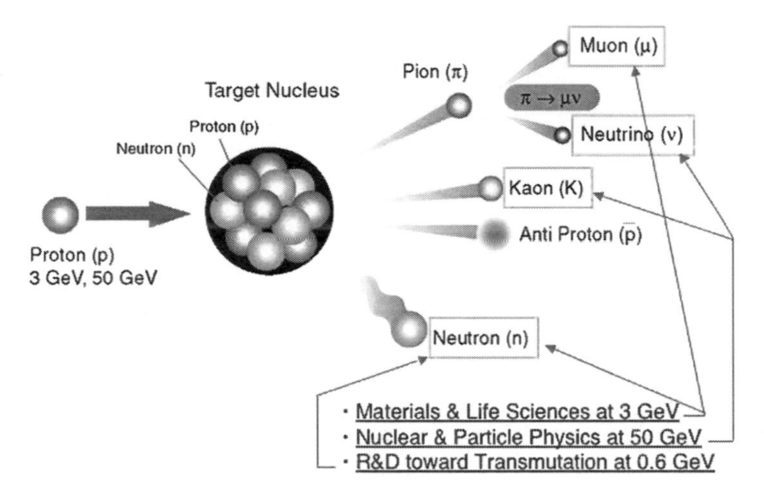

Fig. 1. Production of primary, secondary, and tertiary beams at J-PARC to be used for a variety of scientific programs.

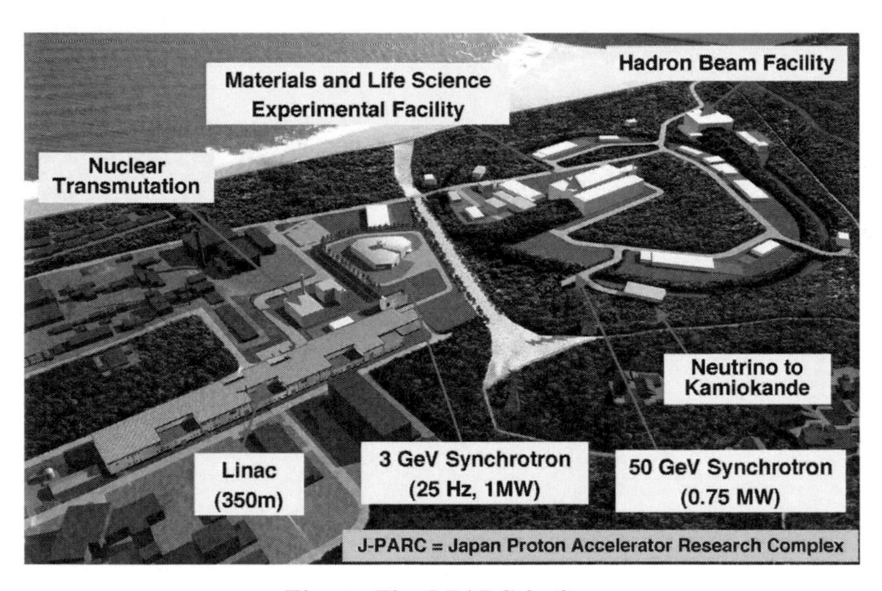

Fig. 2. The J-PARC facility.

J-PARC accelerator comprises the following components:

– 400 MeV (181 MeV on day-1) proton linac as an injector.

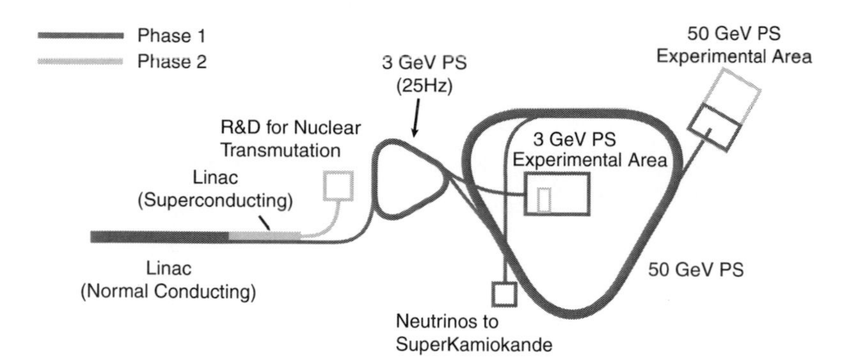

Fig. 3. Schematic diagram of the J-PARC facility. The project is divided into Phase 1 and Phase 2.

- 25 Hz 3 GeV proton synchrotron with 1 MW power, called the 3 GeV RCS (RCS = rapid cycle synchrotron).
- 50 GeV MR (MR = main ring) proton synchrotron with slow and fast extraction capabilities for nuclear–particle physics experiments.

The schematic diagram of the facility is shown in Fig. 3. The project is approved in two phases. In the first phase of the project, Phase 1, the linac up to 181 MeV, the 3 GeV RCS, and the 50 GeV MR (with an energy up to 40 GeV), together with three experimental halls, (a) an experimental hall for materials and life science, (b) a hadron experimental facility, and (c) a neutrino beamline, are included. Immediately after the completion of Phase 1, the linac energy will be recovered to the original design value of 400 MeV. In the second phase, Phase 2, the construction of a facility for R&D of nuclear transmutation, an expansion of the Hadron Experimental Hall, and an addition of beamlines and experimental devices in the Materials and Life Experimental Hall, etc., are planned. The upgrade plan is described later.

2 Accelerator and Facility Construction

The construction of the J-PARC facility started in 2001 and the provision of beams is set to commence in the late 2008. In January 2007 we succeeded in acceleration of proton beams through the linac. At the end of October 2007 the beam acceleration and extraction from the 3 GeV RCS was successfully achieved. The beam injection to the 50 GeV MR is planned in the spring of 2008. The schedule together with the entire view of the J-PARC facility is illustrated in Fig. 4. The present construction schedule, which was created in February 2006, has not been changed since then.

Concerning the facility construction, the groundbreaking ceremony was held in June 2002. Subsequently, construction work has started at a rapid

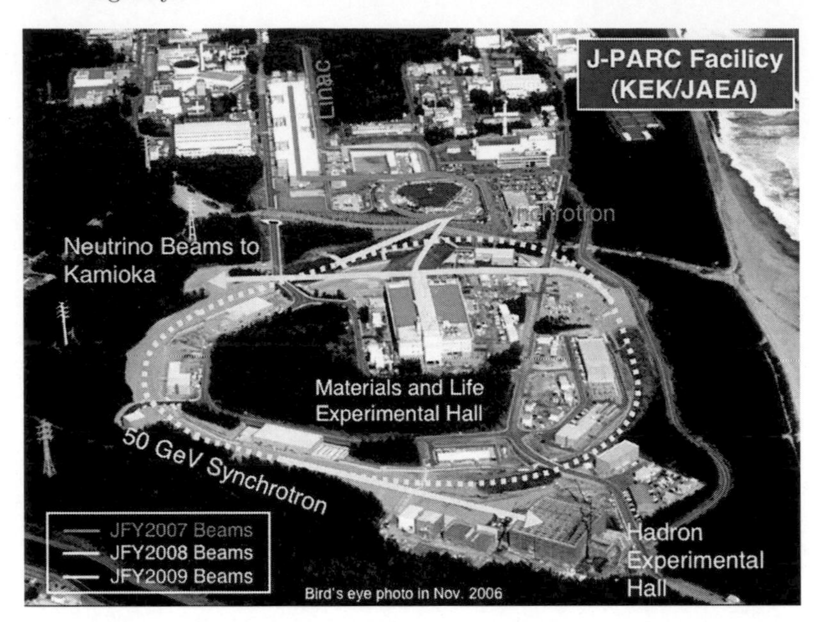

Fig. 4. An entire view of the J-PARC and the present construction plan.

speed. By now (February 2008), almost all the facility buildings, except for those for neutrinos, were completed.

2.1 Accelerator Components

At J-PARC an H^- ion source is connected to RFQ and then to a series of DTLs (drift tube linacs). At the end of October 2003, the first test for the DTLs (the first 20 MeV portion) was done at KEK. On the first day, 6 mA was successfully accelerated and in a week later, the acceleration of 30 mA was achieved.

The DTL is then connected to a series of SDTLs (separated drift tuber linacs) in order to accelerate to 181 MeV. As described earlier, the 181 MeV beam through the linac was obtained in January 2007.

In regard to the linac, a unique idea of the π-mode stabilizing loop, which is one of the devices invented at KEK to stabilize the fields in an RFQ, was adopted, which allows acceleration energy above 3 MeV with frequency range from 300 to 400 MHz. Also, a new drift tube linac (DTL) has newly developed coils for the electro-quadrupole magnets in order to improve a packing factor. The entire scenery of the linac is shown in Fig. 5.

The linac beam will, then, be injected to the 3-GeV RCS (Fig. 6) and to the 50-GeV MR (Fig. 7). About 96% of the beams at the 3 GeV will be extracted to the Materials and Life Experimental Hall and the remaining 4% of the beam will be sent to the 50-GeV ring. The J-PARC synchrotron

Fig. 5. Linac for J-PARC. The beam is already available.

Fig. 6. 3-GeV synchrotron for J-PARC (*left*) and a new RF sector (*right*).

has many unique features. For example, an arrangement of synchrotron magnets was designed so as to have no transition energies in the ring. A novel idea of the RF cavity to attain a high-voltage gradient was implemented by using a specific and new magnetic alloy, called the Finemet. At the 3 GeV, 20 different types of ceramic vacuum pipes are prepared in order to avoid "eddy currents" induced by a rapid cycle operation of the synchrotron magnet (25 Hz).

Due to the budget constraint, the initial beam energy for the 50 GeV PS will be 40 GeV. Later, when an additional budget to allow the storage of an electricity power is funded, a full 50-GeV operation is possible. In addition, due to budget constraint, the initial beam energy of the linac is 181 MeV. The recovery of the beam energy to 400 MeV will be made as soon after the completion of Phase 1.

Fig. 7. 50-GeV synchrotron for J-PARC.

2.2 Experimental Halls

At J-PARC three major experimental halls will be prepared. The first one is an experimental hall to use 3-GeV beams, called the Materials and Life Experimental Hall. One of the most useful aspects in the research at J-PARC is a broad application by using neutron and muon beams. The 3-GeV RCS provides 25 Hz proton beams. This means that the pulsed neutron/muon beams can be obtained 25 pulses per second. The proton beam hits the first production target to produce muons. There, 5% of the full beam power will be used. See Fig. 8 (left) for the muon production target. The remaining 95% power will be sent to the final target where neutron beams will be produced. An expected total beam power is 1 MW. The present world frontier is at the level of 0.1 MW, so that the most powerful neutron beams are available at J-PARC. For pulsed neutron beams, 23 beamlines will be prepared, as shown in Fig. 8 (right). About 10 beamlines will be available at the time when the first beams are available.

On the other hand, the 50 GeV MR has two beam extraction lines. The slow extracted beam will be delivered to the Hadron Experimental Hall, where kaon beams will be prepared for variety of experiments, in addition to primary proton beams. This area is called the Kaon Factory. Three major beamlines, (a) charged kaon beams with 1.8 GeV/c, (b) neutral kaon beams, and (c) kaon beams with 1.1 GeV/c, are under construction. Requests from users to use this hall are rapidly increasing, since many experimental proposals including the usage of primary beams, muon beams, etc., have been sent to the J-PARC. In order to accommodate all these requests, the Phase 1 experimental hall is already too small. An expansion of the hall is planned as one of Phase 2 projects. Figure 9 shows the present view of this Hadron Experimental Hall.

Fig. 8. *Left*: Muon source. The beam comes from the back toward the front direction. Four beamlines for pulsed muons will be prepared. *Right*: Pulsed neutron experimental hall. Both pictures were taken in August 2007.

Another beamline from the 50 GeV MR is a fast extraction line which will be used for the production of neutrino beams. Here, muon neutrino beams, ν_μ, will be created and they are sent to the Super-Kamiokande neutrino detector which is located at 295 km west of J-PARC.

This beamline is illustrated in Fig. 10. Extracted proton beams are bent sharply by a series of superconducting magnets toward the west direction.

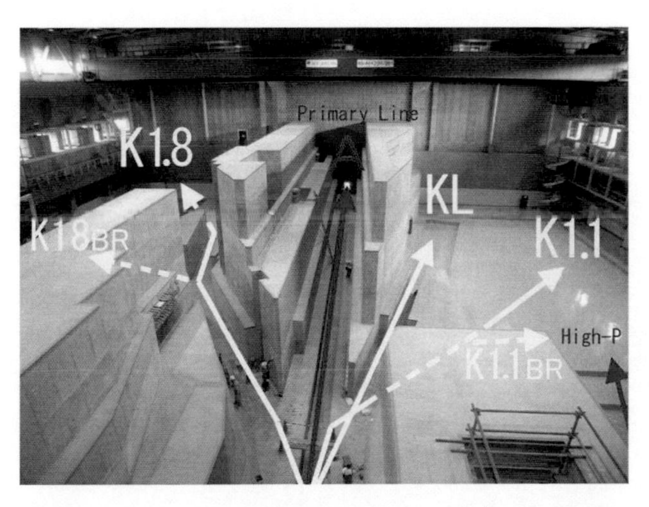

Fig. 9. Hadron Experimental Hall as of August 2007.

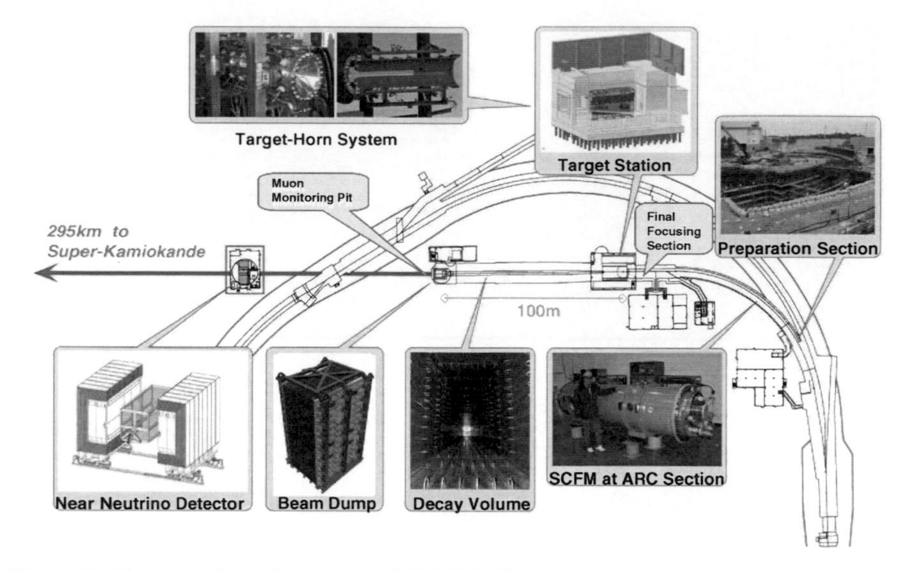

Fig. 10. Neutrino beamline at the J-PARC. The neutrino beam will be sent toward the west direction where the Super-Kamiokande detector exists.

Then, at the target area the beam is sharply focused by the device called the "horn." A copious pions are produced there. Then, the pion decays into a muon and muon neutrino ($\pi \rightarrow \mu + \nu_\mu$) when the pion traverses through the area called the "decay volume." The muon neutrino will be detected immediately after the tunnel of the neutrino beamline by the detector called the "Near Neutrino Detector." After that, the neutrinos will be measured by the second detector which is placed at 295 km west of the J-PARC location. This second detector is the Super-Kamiokande.

3 Examples of Sciences at J-PARC

The J-PARC is a multi-purpose facility in which scientific research in nuclear and particle physics, materials science, life science, nuclear engineering, etc., will be conducted. In addition, the usage by industries is highly encouraged at J-PARC. In this chapter, I show very limited examples only.

3.1 Neutrino Experiment, T2K

The first experiment is related to the neutrino mass and mixing. It is known that, from the most fundamental principle, there are no reasons to assume that the neutrino mass is zero, although it has been believed for many years that the neutrino carries zero mass. In a recent experiment at Super-Kamiokande [1, 2], where atmospheric neutrinos were detected, it was demonstrated that muon neutrinos ν_μ from the sky were converted to tau neutrinos ν_τ while traversing through the Earth. This phenomenon proved evidence on the existence of neutrino oscillation. Because this oscillation can occur only when the neutrino carries a finite mass, this measurement demonstrated the existence of a finite mass of neutrino.

A recent K2K experiment [3] using ν_μ beams from the KEK 12-GeV accelerator to detect ν_μ at Super-Kamiokande showed an additional evidence that ν_μ would oscillate while traversing from KEK to Super-Kamiokande. Furthermore, a later SNO result [4] and, independently, a KamLAND experiment [5, 6] show that neutrinos from the Sun (primarily ν_e) also oscillate due to a finite mass of neutrinos.

At the forthcoming J-PARC an anticipated ν_μ neutrino flux is by 100 times stronger than the flux obtained at the 12-GeV PS at KEK. Therefore, precise measurements of the neutrino mass can be expected at the 50-GeV facility of J-PARC. Experimental group also expects to observe $\nu_\mu \rightarrow \nu_e$ oscillation by observing ν_e appearance at Super-Kamiokande (see Fig. 11). By observing this new mode, a new mixing angle, θ_{13}, can be determined. This T2K experiment, thus, will determine a completely new and unknown mixing parameter, as illustrated in Fig. 12.

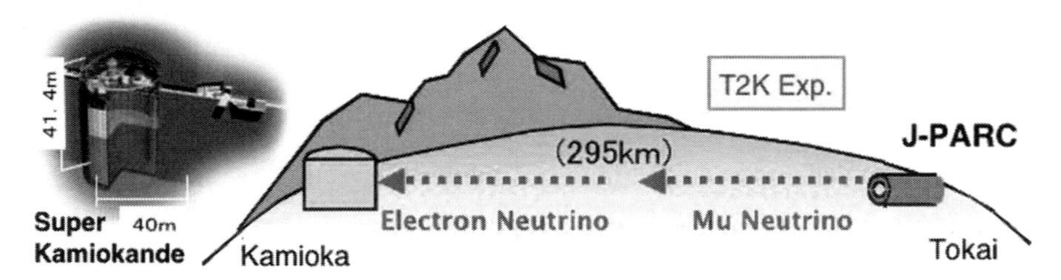

Fig. 11. T2K (Tokai to Kaimioka) experiment using ν_μ beams from J-PARC.

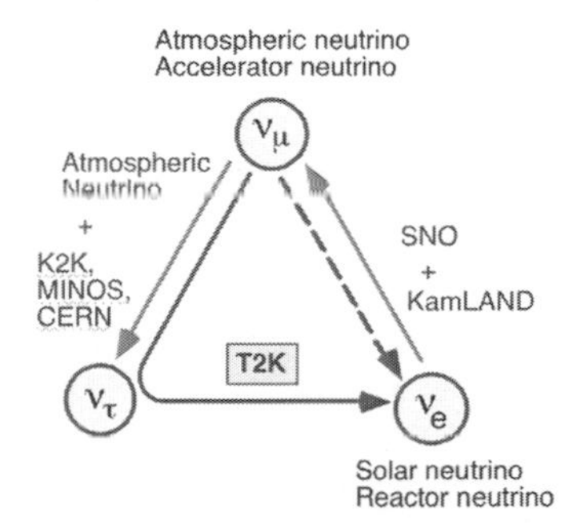

Fig. 12. Three neutrinos and the goal of the T2K experiment. The mixing between the first and the third neutrinos will be measured.

3.2 Hadron Experiments with Kaon Beams

At the Hadron Experimental Hall, many different pieces of experiments are planned. Details will be described by other author(s) in this book. Here, one or two examples are given.

One experiment is related to the mass of matter. It is known that over 99% mass of the visible matter of the universe is carried by atomic nuclei and, thus, by protons and neutrons. Each proton or neutron is made of three quarks. One puzzle, which has not been solved quantitatively until now, is an exact reason that the mass of proton (or neutron) is as heavy as $1\,\mathrm{GeV}/c^2$, whereas the quark mass is much lighter, being less than 1/100 of the proton mass. It is believed that the creation of a large proton mass is connected with spontaneous chiral symmetry breaking. The quantitative nature of this symmetry breaking needs to be further explored.

Theoretically, it is expected that this symmetry breaking can be studied by implanting a meson (which is made of quark and anti-quark) in the interior of matter under extreme conditions and, then, by measuring the change of properties such as the chiral order parameter [7, 8] in these conditions. Implanting mesons or baryons in the interior of nuclear matter and studying their properties in nuclear matter are, thus, extremely interesting.

Figure 13 illustrates the current status of properties of hadrons in nuclei. In a recent experiment [9] a hint of the existence of a meta-stable bound state of negative kaon (K^-) in the nucleus was discovered. If a kaon is implanted inside the nucleus, it was hypothetically predicted that this kaon might play a role as a catalyzer to induce the formation of an extremely high-density

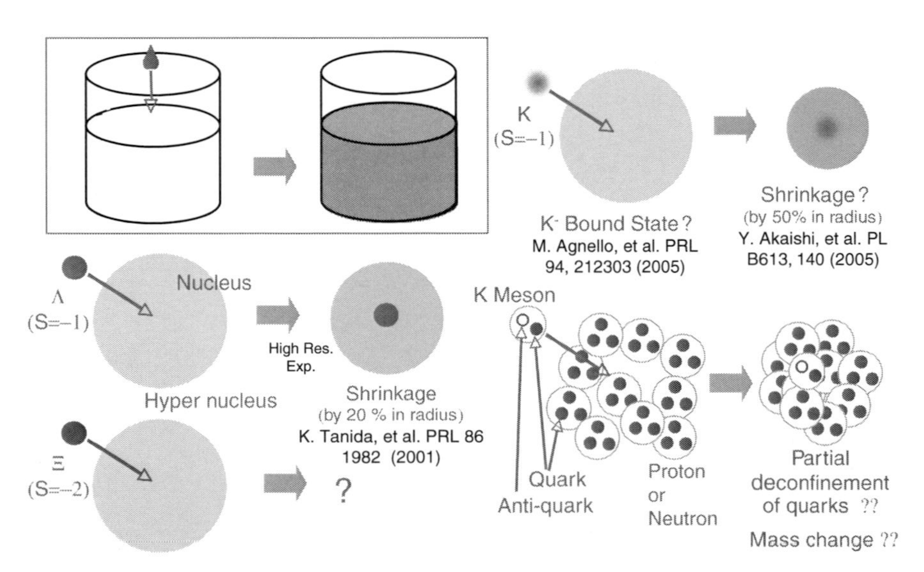

Fig. 13. Implantation of strange baryon or meson inside the nucleus, to be studied with kaon beams at J-PARC.

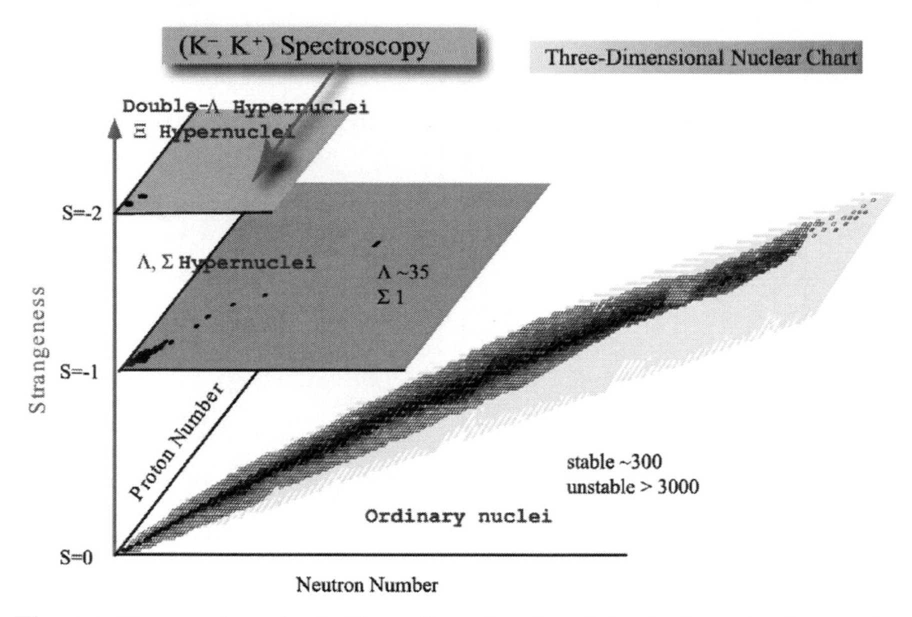

Fig. 14. Hypernuclear chart. Three-dimensional nuclear chart can be drawn when a strangeness freedom is incorporated.

system [10]. If the matter were compressed, the property of K^- could also change, as illustrated schematically in the lower right in Fig. 13. The change of meson properties inside nuclear matter would lead to important conclusions concerning the experimental study of chiral symmetry breaking, as hinted in Refs. [7, 8, 11].

Another example of hadron implantation inside the nucleus is a strange baryon implantation, called the hypernucleus. As shown in Fig. 14, three-dimensional nuclear chart can be drawn when the baryon freedom includes strangeness. The spectroscopy itself is interesting. In addition, in an earlier experiment [12] it was demonstrated that an implantation of Λ hyperon inside the nucleus induces a shrinkage of the nucleus. Detailed studies on strange baryons and strange mesons in nuclei are planned as day-1 experiments at J-PARC by using high-flux kaon beams.

3.3 Materials and Life Sciences

The J-PARC covers broad sciences other than nuclear and particle physics. The largest area is materials and life sciences. There, experimental studies will be carried by using pulsed neutrons and muons.

The neutron beam carries many unique features, as illustrated in Fig. 15. First, the neutron penetrates through the matter. This feature is unique, as compared to X-rays, since X-rays are easily reflected by a metallic substance, whereas neutrons go through any metallic materials. For example, the movement of automobile engine has already been studied with neutron beams. In

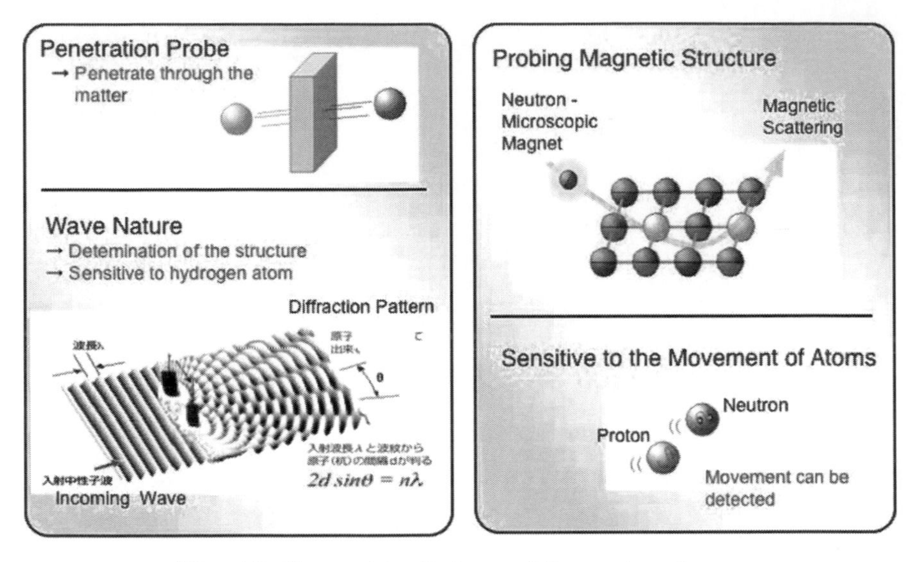

Fig. 15. Four unique features of the neutron beam.

the second, the neutron is a wave, so that it produces a diffraction pattern through a crystal. In particular, since neutrons interact with atomic nuclei instead of electrons, the neutron is very sensitive to atoms with small atomic number Z. In order to measure hydrogen, water, etc., with small Z, therefore the neutron beam is extremely powerful. In the third, the neutron carries a

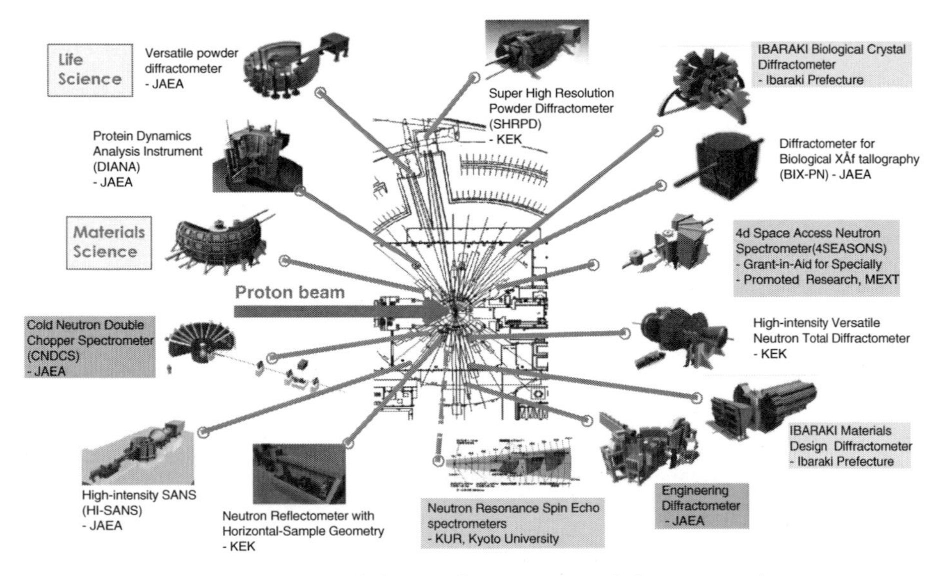

Fig. 16. Experimental devices being prepared for neutron beams.

magnetic moment. The magnetic structure of a crystal or any material can be studied with neutron beams. Finally, high-intensity neutron beams are very powerful for the study of the movement of atoms. This unique feature allows us to study a "function" of a protein in addition to the "structure" of the protein. Many experimental devices, as illustrated in Fig. 16, are being prepared for the study of materials and life sciences with neutrons.

4 Future of the J-PARC

4.1 J-PARC in the World

New facilities that are competing with, and/or comparable to, the J-PARC are being constructed in the world, as illustrated in Fig. 17. In the area of pulsed neutron beams a new facility called the SNS in the USA has been constructed, and the neutron beams have started to be used by experimental groups. This SNS will provide beams at the MW level, and this power is comparable to the power expected at the J-PARC. The ISIS at the Rutherford Laboratory in the UK is already sending pulsed neutron beams at 0.1–0.2 MW. The J-PARC together with the SNS and ISIS will form three regional centers for neutron sciences.

In the area of neutrino physics, the J-PARC, FNAL, and CERN will compete with each other in the area of accelerator-driven neutrino beams. The J-PARC has the advantage that it already has a working detector, the Super-Kamiokande. On the other hand, FNAL and CERN have the advantage that

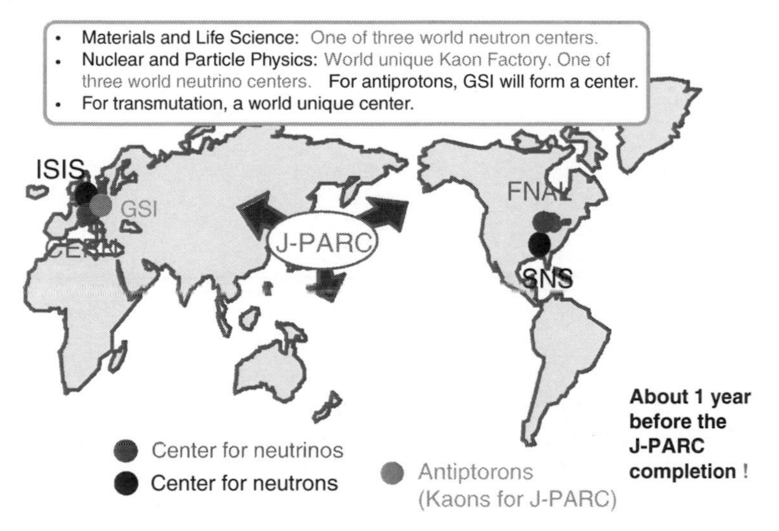

Fig. 17. J-PARC when compared with other competing and comparable accelerators in the world.

they already have neutrino beams. A strong competition exists among these three facilities in the future neutrino sciences.

Concerning the hadron physics, the J-PARC is unique in terms of kaon beams. This is the reason why J-PARC is called a unique Kaon Factory in the field of hadron physics in the world. On the other hand, a future facility called FAIR in Germany will produce intense anti-proton beams. Although there is a capability at J-PARC to produce anti-proton beams, the present arrangement is such that the anti-proton physics will be conducted at FAIR, whereas the kaon physics will be done at J-PARC. Other programs, such as programs using intense muon beams for flavor violation experiment and/or g-2 experiment, are being studied at J-PARC. In this area, a good coordination with FNAL fixed target program might have to be made.

4.2 Upgrades of the J-PARC

The present construction of the J-PARC will be completed in the spring of 2009. Many possible options after 2009 for upgrades are discussed. Described below is a list of major items that are currently under discussion:

- Neutrons and muons: How to fill the 23 beamlines for neutrons? So far, about 10 were funded. Concerning muons, one among four beamlines will be in operation in 2008. Others have to be filled after 2008.
- Hadrons: Unfinished beamlines, including neutral kaons, 1.1 GeV/c beamlines, and a primary beamline have to be completed. Also, in order to accommodate many users an expansion of the present hadron hall has to be done.
- Neutrinos: The power upgrade of the 50-GeV MR to above 1 MW must be done. Also, the detector at a distance of 2 km from the J-PARC is under discussion.
- Nuclear transmutation: This is the major item for Phase 2 in J-PARC, as illustrated in Fig. 3.
- Others: An energy upgrade of the main ring to 50 GeV must be done. Also, an issue on the table is the third extraction line from the 50-GeV MR. Other proposals, such as polarized proton beams at the MR, have to be considered as well.

All the above items are now under discussion at the Users Steering Committee at the J-PARC, where this committee is composed of representatives of user communities for the J-PARC. Among the items listed above, the program of nuclear transmutation is discussed at the Atomic Energy Committee of Japan, since this program is related to the policy of the waste disposal of long-lived radioactive materials produced in nuclear fuel plants. It is believed that the proton power required for the practical treatment of nuclear waste transmutation is on the order of 20–50 MW. Our J-PARC has a much smaller power when compared to these numbers. Thus, we plan to perform R&D studies to establish a concept of the accelerator-driven nuclear transmutation

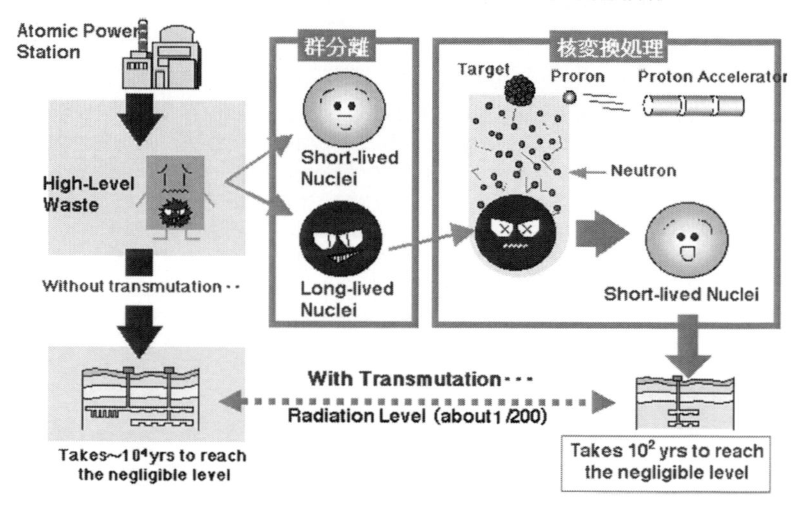

Fig. 18. Need of the nuclear transmutation program at J-PARC. The site for this program is reserved for the J-PARC, as illustrated in Fig. 2.

(ADT). The concept of the transmutation is illustrated in Fig. 18. Development of mechanical and nuclear engineering together with collection of nuclear reaction data under critical conditions will be the goal in our project.

4.3 Operation of the J-PARC

From the end of 2008 the J-PARC will start its operation. The J-PARC Center was created in 2006 to take a responsibility of the operation of J-PARC. This center is located under two institutions, KEK and JAEA.

Many infrastructures are being prepared: lodging, road access, users' office, visas, technical support, proposal handling, safety, etc. In addition, the project team is now working hard for the completion of the accelerators and the experimental halls and facilities.

Acknowledgments

At this opportunity, I would like to thank all the J-PARC members for their patient and energetic work on the completion of J-PARC. Also, I would like to thank all the administrative members of KEK and JAEA for their strong support of J-PARC.

References

1. Fukuda, Y., et al.: Phys. Rev. Lett. **81**, 1562 (1998)
2. Fukuda, Y., et al.: Phys. Rev. Lett. **82**, 2644 (1999) and references therein

3. K2K Collaboration: Phys. Lett. **B511**, 178 (2001); Also see http://neutrino.kek.jp/
4. SNO Collaboration: Phys. Rev. Lett. **87**, 8707301 (2001)
5. KamLAND Collaboration: Phys. Rev. Lett. **90**, 021802 (2003)
6. KamLAND Collaboration: Phys. Rev. Lett. **94**, 081801 (2005)
7. Hatsuda, T., Kunihiro, T.: Phys. Rev. Lett. **55**, 158 (1985)
8. Weise, W.: Nucl. Phys. **A44**, 59c (1993)
9. Agnello, M., et al.: Phys. Rev. Lett. **94**, 212303 (2005)
10. Akaishi, Y., Yamazaki, T.: Phys. Rev. **C65**, 044005 (2002)
11. Suzuki, K., et al.: Phys. Rev. Lett. **92**, 072302 (2004)
12. Tanida, K., et al.: Phys. Rev. Lett. **86**, 1982 (2001)

The T2K Long-Baseline Neutrino Experiment

A.K. Ichikawa

Department of Physics, Kyoto University, Kitashirakawa, Sakyo, Kyoto 606-8502, Japan
ichikawa@scphys.kyoto-u.ac.jp

1 Introduction – The Current Situation in Neutrino Oscillation Physics

The neutrino oscillation phenomenon and the existence of the finite neutrino mass had been discovered through the observation of the neutrinos from the Sun and atmosphere. Initial indication came from the observation of the deficit of the solar neutrino flux by the pioneering experiments of Ray Davis and his collaborators [1] and by the Kamiokande experiment [2, 3].

The neutrino oscillation phenomenon occurs when neutrino being observed is a superposition of more than two different mass eigenstates and when the mass difference is so small that the waves interfere. Usually neutrino is produced as a flavor eigenstate. After traveling long distance, that flavor changes as a consequence of the interference. In a simple case of two types of neutrino, flavor eigenstates, $|\nu_\alpha>$ and $|\nu_\beta>$, can be written as follows:

$$|\nu_\alpha >= e^{-i(E_1 t - p_1 L)} \cos\theta + e^{-i(E_2 t - p_2 L)} \sin\theta, \quad (1)$$

$$|\nu_\beta >= e^{-i(E_1 t - p_1 L)} \sin\theta - e^{-i(E_2 t - p_2 L)} \cos\theta, \quad (2)$$

where t and L are time and position, respectively, at observation; E_i and p_i are the energy and momentum of mass eigenstates with mass m_i, respectively; and θ is the mixing angle. Since the neutrino mass is so small t can be approximated as L and $E_i \approx p + m_i^2/2E$ with the average energy E. Then, the phase factor is approximated to $e^{-i(m_i^2/2E)L}$.

With this approximation and with the initial condition that a neutrino is produced as an eigenstate $|\nu_\alpha>$ with energy E (GeV), the probability that it has another flavor β after traveling distance L (km) can be calculated as follows:

$$P(\nu_\alpha \rightarrow \nu_\beta) = \sin^2 2\theta \sin^2(1.27\Delta m^2 L/E), \quad (3)$$

where $\Delta m^2 = m_1^2 - m_2^2$ in eV^2. In the actual observation, flavor of neutrino is identified by the charged current interaction. The charged current interaction

Ichikawa, A.K.: *The T2K Long-Baseline Neutrino Experiment.*
Lect. Notes Phys. **781**, 17–43 (2009)
DOI 10.1007/978-3-642-00961-7_2

is not allowed when the energy of the neutrino is lower than the threshold energy to produce the corresponding charged lepton. Therefore, if the energy of the oscillating neutrino is lower than the threshold for the appearing flavor, then the neutrino oscillation is observed as "disappearance."

In reality, it has been confirmed that, at least, three flavors exist in neutrino, that is, ν_e, ν_μ, and ν_τ. Then the mixing among them with mass eigenstates can be naturally extended by a unitary 3×3 matrix (Maki–Nakagawa–Sakata [4] (MNS) matrix) defined by a product of three rotation matrices with three angles (θ_{12}, θ_{23}, and θ_{13}) and a complex phase (δ) as in the Cabibbo–Kobayashi–Maskawa matrix [5]:

$$\begin{pmatrix} \nu_e \\ \nu_\mu \\ \nu_\tau \end{pmatrix} = \begin{bmatrix} U_{\alpha i} \end{bmatrix} \begin{pmatrix} \nu_1 \\ \nu_2 \\ \nu_3 \end{pmatrix}, \tag{4}$$

$$U = \begin{pmatrix} 1 & 0 & 0 \\ 0 & C_{23} & S_{23} \\ 0 & -S_{23} & C_{23} \end{pmatrix} \begin{pmatrix} C_{13} & 0 & S_{13}e^{-i\delta} \\ 0 & 1 & 0 \\ -S_{13}e^{i\delta} & 0 & C_{13} \end{pmatrix} \begin{pmatrix} C_{12} & S_{12} & 0 \\ -S_{12} & C_{12} & 0 \\ 0 & 0 & 1 \end{pmatrix}, \tag{5}$$

where $\alpha = e,\ \mu,\ \tau$ are the flavor indices, $i = 1,\ 2,\ 3$ are the indices of the mass eigenstates, and S_{ij} (C_{ij}) represents $\sin\theta_{ij}$ ($\cos\theta_{ij}$). Neutrinos are produced as a flavor eigenstate and each component of mass eigenstate gets a different phase after traveling a certain distance. The detection of neutrinos by the charged current interactions projects these new states back onto flavor eigenstates. When only two of the mass states dominate the oscillations, the oscillation probability is same as Eq. (3). Note that these oscillations depend only on $|\Delta m^2|$, that is, they do not depend on either the absolute mass scale or ordering of the mass states. While this formula is valid in vacuum, interactions with electrons in matter can modify the oscillations [6, 7]. It is called the matter effect which depends on the sign of Δm^2 and provides a way to determine the ordering of the mass states. In other words, the vacuum oscillation cannot distinguish a mixing angle θ and $(\pi/2 - \theta)$ (see Eq. (3)). However, with significant matter effect, this degeneracy can be resolved.

After the experiments of Davis et al. and Kamiokande, their results were supported by measurements of the low-energy solar neutrino fluxes by the SAGE [8, 9] and GALLEX [10–12] experiments. While the solar neutrino anomaly was under investigation, the Super-Kamiokande (SK) experiment showed a zenith-angle-dependent suppression of the ν_μ flux (atmospheric neutrino flux, arising from cosmic ray interactions with the atmosphere), which indicated that neutrinos have mass [13, 14]. Recently the SNO [15–17] experiment has demonstrated that the solar neutrino anomaly is caused by neutrinos changing flavor, which was shown to be consistent with neutrino oscillations by the recent results from the KamLAND experiment [18, 19]. The K2K long-baseline experiment [20] has observed the disappearance of ν_μ with a controlled neutrino beam produced with an accelerator. The observed disappearance feature is very consistent with the atmospheric neutrino anomaly

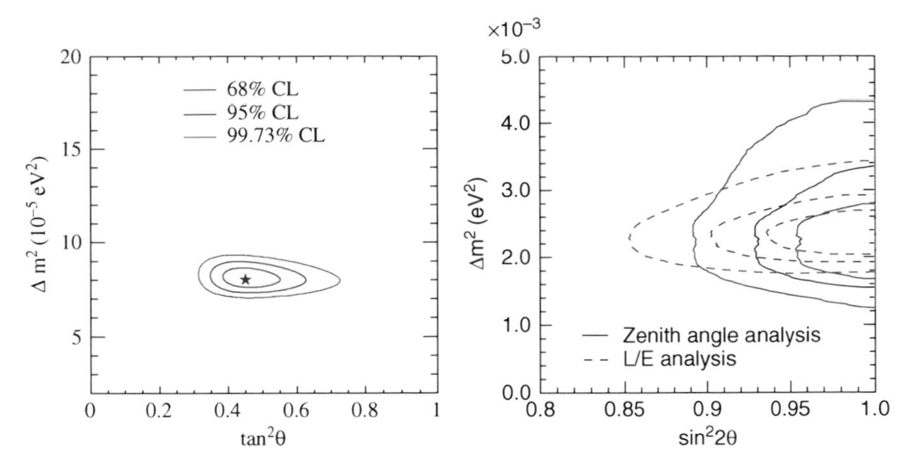

Fig. 1. Allowed regions for 2-neutrino mixing parameters for solar+KamLAND (*left figure*) and from the latest Super-Kamiokande analysis (*right figure*). Thanks to the matter effect, the degeneracy on a mixing angle θ and $\pi/2 - \theta$ is resolved and the plot on the $\tan^2\theta$ has been obtained in the case of the *left figure*.

observed by Super-Kamiokande. It has been established that the neutrino does oscillate and two Δm^2s exist.

The current results for oscillations from solar neutrino plus KamLAND data can be fit using two-neutrino oscillations with the matter effect, and the resulting parameters are shown in the left part of Fig. 1. Similarly, for the atmospheric oscillations the current best-fit parameters from Super-Kamiokande are shown in the right part of Fig. 1. These results constrain θ_{12}, θ_{23}, Δm^2_{21}, and $|\Delta m^2_{23}|$.

Two accelerator-based experiments are now running. The MINOS experiment is using the NuMI neutrino beam from Fermilab. Its far detector locates in the Soudan mine, 735 km away from Fermilab. It is making a sensitive measurement of the muon neutrino disappearance [21] and thereby is making a measurement of Δm^2_{23} accurate to about 10%. The CNGS neutrino beam from CERN is aimed at the Laboratori Nazionali del Gran Sasso (LNGS) in Italy, where the OPERA [22] experiment starts data-taking and the ICARUS [23] experiment is under construction. The baseline length is 730 km. These two experiments are aimed at making a conclusive demonstration of the mechanism for the atmospheric neutrino disappearance by actually observing the appearance of ν_τ from the $\nu_\mu \to \nu_\tau$ oscillation.

There are still three undetermined parameters in the MNS matrix – the angle θ_{13}, the sign of Δm^2_{23}, and the value of the CP-violating phase δ. These are fundamental parameters which should be predicted by a deeper theory of the particle physics, and hence intrinsically interesting in themselves.

Our current knowledge of θ_{13} comes from three sources. Atmospheric and solar neutrino oscillation results give some information, but the most useful

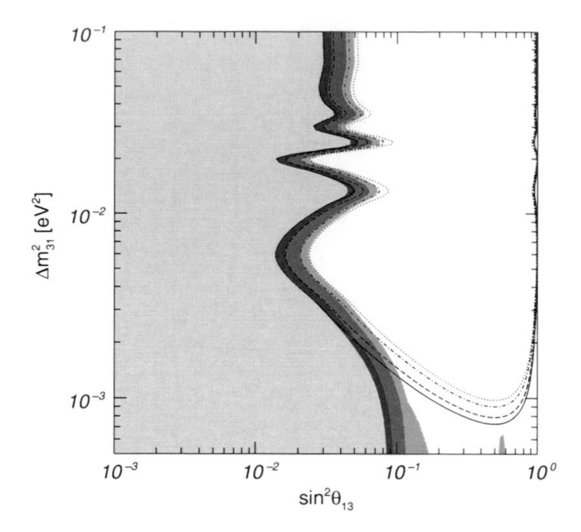

Fig. 2. Allowed region for θ_{13} from CHOOZ alone (*lines*) and also including data from solar neutrino experiments and KamLAND (*colored regions*) at 90, 95, 99%, and 3σ. From Maltoni et al. hep-ph/0309130.

data for constraining the value of θ_{13} comes from the CHOOZ reactor $\bar{\nu}_e$ oscillation experiment [24, 25]. This experiment measured the flux of $\bar{\nu}_e$ at a distance of 1 km from the CHOOZ reactor in France. The constraint from these measurements is summarized in Fig. 2.

The next goal would then be to improve our sensitivity to θ_{13} in the hope of a positive observation. Our current knowledge of neutrino oscillations tells us that Δm^2_{13} is approximately equal to Δm^2_{23} because $\Delta m^2_{12} \ll |\Delta m^2_{23}|$ and that oscillation originated by θ_{13} may be observable as the appearance of ν_e from ν_μ beam at the same distance and energy as the experiment optimized to observe the ν_μ disappearance by θ_{23}. In the three-generation frame, the oscillation probability for ν_μ goes to ν_e can be written as [26]

$$
\begin{aligned}
&P(\nu_\mu \to \nu_e) = \\
&4C^2_{13}S^2_{13}S^2_{23}\sin^2 \Phi_{31} \times \left(1 + \frac{2a}{\Delta m^2_{31}}(1 - 2S^2_{13})\right) \\
&+ 8C^2_{13}S_{12}S_{13}S_{23}(C_{12}C_{23}\cos\,\delta - S_{12}S_{13}S_{23})\cos\,\Phi_{32}\cdot\sin\,\Phi_{31}\cdot\sin\,\Phi_{21} \\
&- 8C^2_{13}C_{12}C_{23}S_{12}S_{13}S_{23}\sin\,\delta\cdot\sin\,\Phi_{32}\cdot\sin\,\Phi_{31}\cdot\sin\,\Phi_{21} \\
&+ 4S^2_{12}C^2_{13}\left(C^2_{12}C^2_{23} + S^2_{12}S^2_{23}S^2_{13} - 2C_{12}C_{23}S_{12}S_{23}S_{13}\cos\,\delta\right)\sin^2\,\Phi_{21} \\
&- 8C^2_{13}S^2_{13}S^2_{23}\left(1 - 2S^2_{13}\right)\frac{aL}{4E_\nu}\cos\,\Phi_{32}\cdot\sin\,\Phi_{31},
\end{aligned}
\tag{6}
$$

where $\Phi_{ij} \equiv \Delta m^2_{ij}L/4E$. The terms multiplied by the parameter a represent "matter effect." When neutrinos pass through inside the Earth, the Sun, or any dense object, they feel the potential by the weak interaction with matter.

Since the matter is made of quarks and electron, not of muon nor tau, the potential is different for ν_e and ν_μ/ν_τ. That potential difference causes another shift of the phase of propagation. In the above equation, the matter effect is simplified by approximating that the matter density ρ is constant across the path of the neutrino. Then,

$$a = 2\sqrt{2}G_F n_e E = 7.6 \times 10^{-5}\rho(g/cm^3)E(GeV)[(eV)^2]. \tag{7}$$

Given the already known limits on the parameters and under the condition that the matter effect is quite small, Eq. (6) can be approximated as

$$\begin{aligned} P(\nu_\mu \to \nu_e) \approx & \sin^2 2\theta_{13} \sin^2 \Phi_{31} \\ & + 1.9 S_{13} \cos\delta \cos\ \Phi_{32} \cdot \sin\ \Phi_{31} \cdot \sin\ \Phi_{21} \\ & - 1.9 S_{13} \sin\delta \sin\ \Phi_{32} \cdot \sin\ \Phi_{31} \cdot \sin\ \Phi_{21} \\ & + 0.44 \sin^2\ \Phi_{21}, \end{aligned} \tag{8}$$

where the values of $\theta_{12} = 33.9^\circ, \theta_{23} = 45^\circ$ are used.

If the value of $\sin^2 2\theta_{13}$ is within an order of magnitude lower than its current upper limit, the first term dominates and Eq. (8) becomes a relatively simple formula for the oscillatory appearance of ν_e in ν_μ beam.

The goal of the first phase of the T2K experiment, which is the main topic of this literature, is to discover finite θ_{13} with at least an order of magnitude better sensitivity than the current upper limit by searching for the ν_e appearance signal from the ν_μ beam. At the same time measuring θ_{23} and $|\Delta m^2_{23}|$ with greater precision by observing the disappearance of ν_μ is interesting in its own right. In particular, deviation of θ_{23} from $\pi/4$ is a matter of great interest.

Once the appearance of ν_e is observed, the goal of the experiment would be expanded dramatically. The new developments in the solar and reactor neutrino experiments confirmed that ν_μ oscillates to ν_e with a rather large mixing angle and Δm^2_{21} among the regions allowed before these measurements. Then it makes non-negligible contribution to Eq. (8) as the fourth term. This oscillation can compete with the first term of Eq. (8). The former oscillation is suppressed by a small Δm^2_{21} compared to $|\Delta m^2_{23}|$ and the latter is suppressed by a small mixing angle θ_{13}. Hence, the two processes can compete. This is one of the necessary conditions for a CP violation effect to be observable. The second and third terms of Eq. (8) are generated by the interference between these two oscillations. These features are illustrated in Fig. 3. Indeed, the third term has $\sin\delta$ and changes its sign depending on whether the observing neutrino is neutrino or anti-neutrino. Therefore, the CP violation phase δ may be observable through the comparison of the appearance signal from ν_μ beam and $\bar{\nu_\mu}$ beam. One possible source of the matter–antimatter asymmetry of the present-day universe is the CP violation in the neutrino sector via the mechanism known as leptogenesis [27]. Measuring the CP-violating phase δ in the MNS matrix should give valuable insights into the CP properties of

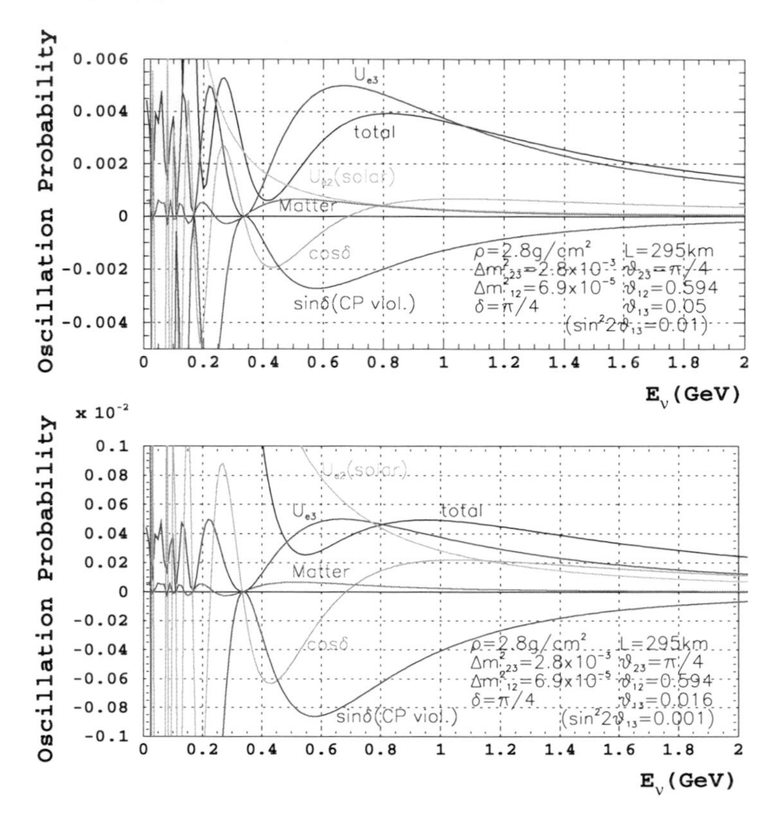

Fig. 3. Contribution of different components to $\nu_\mu \to \nu_e$ appearance at $\sin^2 2\theta_{13} = 0.01$ and 0.001.

neutrinos. Measurement of δ is only possible if θ_{13} is sufficiently large, and hence a measurement of θ_{13} is a critical step for future neutrino physics.

2 The T2K Long-Baseline Experiment

The T2K experiment is a long-baseline neutrino oscillation experiment. An almost pure muon neutrino beam is produced using a high-intensity proton accelerator complex, J-PARC at Tokai village, Japan. Then, after traveling 295 km, it is measured at a gigantic water Cerenkov detector, Super-Kamiokande, at Kamioka, Japan. Figure 4 shows the baseline of the experiment.

The main physics motivation of the T2K experiment comprises three goals.

1. Discovery of $\nu_\mu \to \nu_e$. More than an order of magnitude improvement in sensitivity over the present upper limit is possible in 5 years running with full operation of the J-PARC accelerator. The goal is to extend the search down to $\sin^2 2\theta_{13} \simeq 2\sin^2 2\theta_{\mu e} > 0.008$.

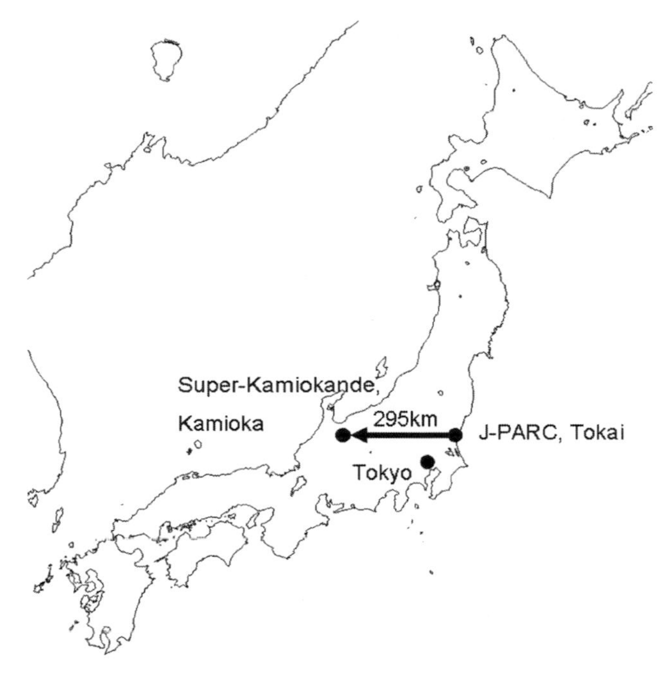

Fig. 4. Baseline of the T2K experiment.

2. Precision measurements of oscillation parameters in ν_μ disappearance. The goal is an observation of the oscillation minimum, a 1% measurement (about the same precision as Cabibbo angle in the quark sector) of the mixing angle and a 10% measurement of $|\Delta m^2|$ $(\delta(\Delta m^2_{23}) = 10^{-4}$ eV^2 and $\delta(\sin^2 2\theta_{23}) = 0.01)$. Deviation from or consistency with the maximal mixing at 1% accuracy may impose a constraint on the quark-lepton unification in future.
3. Search for sterile components in ν_μ disappearance by detecting the neutral-current events. If a non-zero sterile component is found, the physics of fermions will need modification to accommodate extra member(s) of leptons.

The J-PARC accelerator complex and neutrino facility is currently being constructed and the first beam is expected in April 2009.

With the successful achievement of these measurements, the construction of 1 Mt Hyper-Kamiokande detector at Kamioka, and a possible upgrade of the accelerator from 0.75 MW to a few MW in beam power, further experiments can be envisaged. These include another order of magnitude improvement in the $\nu_\mu \to \nu_e$ oscillation sensitivity, a sensitive search for the CP violation in the lepton sector (CP phase δ down to 10–20°), and an order of magnitude improvement in the proton decay sensitivity.

2.1 Neutrino Beam at J-PARC

A layout of the J-PARC facility is shown in Fig. 5.

A neutrino beam at J-PARC is produced with the "conventional method" where pions produced by hitting protons on a target decay into ν_μ and muons. The pions are focused by toroidal magnets called the electromagnetic horns and injected into a volume where they decay in flight (decay volume). The beam is almost pure ν_μ. ν_μ or $\bar{\nu}_\mu$ can be selected by flipping the polarity of the horns.

A key element of the design of the T2K facility is that the neutrino beam is directed so that the beam axis actually misses Super-Kamiokande as shown in Fig. 7. This method is called off-axis neutrino beam [28]. This, rather surprisingly, actually results in a considerable improvement in the quality of the beam for the ν_e appearance experiment. This arises from the kinematics of the π decay. As shown in Fig. 6, the neutrino energy is proportional to the energy of the parent π meson when the off-axis (OA) angle is 0 (general wide-band beam), while it is almost independent of the π meson energy with non-zero OA angle. Since pions in a wide momentum range contribute to narrow energy range of neutrino, a narrow intense neutrino beam is obtained.

The peak neutrino energy can be adjusted by choosing the off-axis angle. Recent results from the SK [13, 14], K2K [20], and MINOS [29] experiments point at $\Delta m^2_{23} = 2 \sim 3 \times 10^{-3}$ eV^2. With this parameter, the oscillation maximum is at around the neutrino energy of 500–700 MeV at the SK, i.e., 295 km away from the source. The off-axis beam has three major advantages

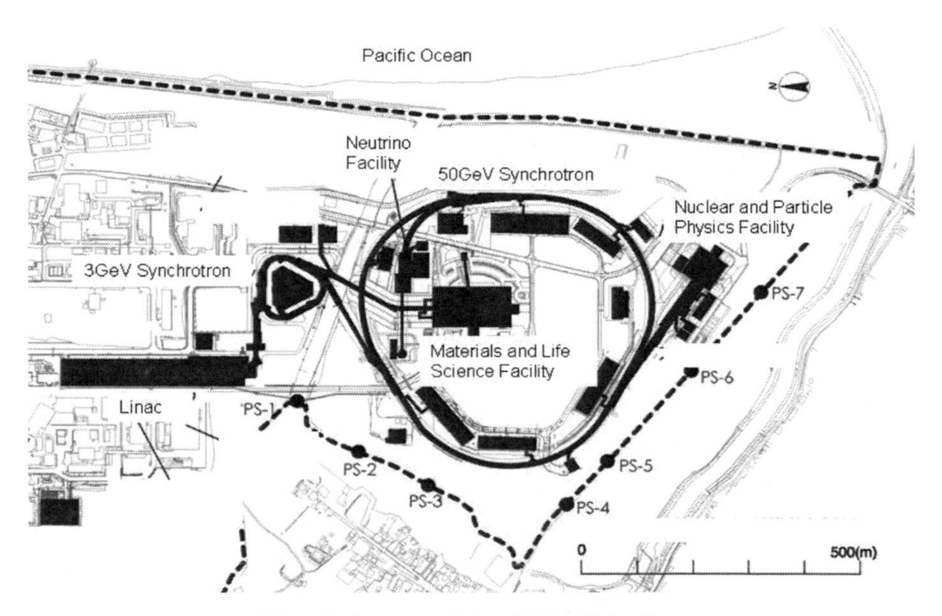

Fig. 5. Layout of the J-PARC facility.

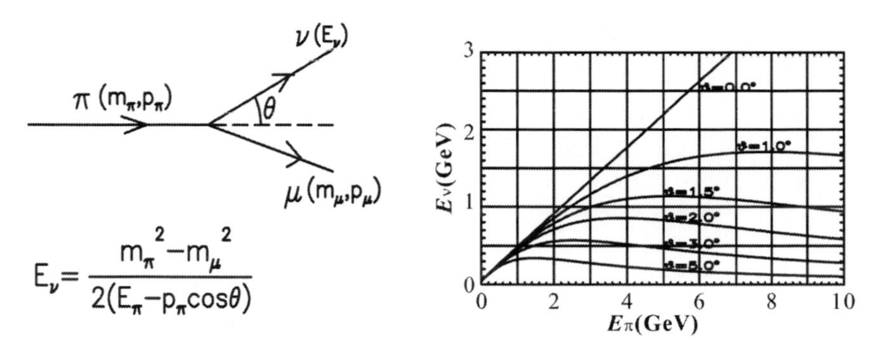

Fig. 6. Kinematics of the off-axis beam.

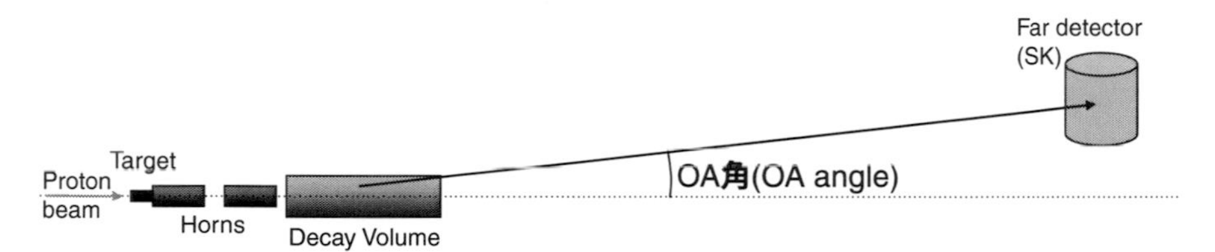

Fig. 7. Schematic of the off-axis beam.

over the conventional on-axis beam. First, the off-axis neutrino flux at the desired energy (near the oscillation maximum) is actually higher than that with on-axis. Second, there are fewer high-energy neutrinos, which do not contribute to the appearance signal but do contribute to its backgrounds in particular through the neutral current production of π^0s. The daughter γs from the π^0 decay are sometimes misidentified for the single electron signal. Third, the background due to the intrinsic contamination of the beam ν_e is actually less at the off-axis position due to the different kinematics of the decays that lead to ν_e.

The very kinematics that gives a useful selection in energy, however, means that the characteristics of the beam changes rapidly with the off-axis angle. Detailed measurement of the beam properties are therefore required to minimize the systematic uncertainties in any neutrino oscillation measurements. These detailed measurements will be made with a set of near detectors at 280 m downstream of the target.

Monte Carlo (MC) simulations using GEANT3 [30] have been used to optimize the beamline and to estimate the expected neutrino spectra and the number of events. Figure 8 shows the expected neutrino flux spectra. The J-PARC neutrino facility is capable to produce a beam whose off-axis angle is from 2.0–2.5°. The on-axis beam spectrum is shown as a reference in Fig. 8(a). The energy spectra of the contaminated ν_es are plotted in Fig. 8(b). The ν_e contamination in the beam is expected to be 1% at the off-axis angle of 2° (OA2°). The sources of ν_e are $\pi \to \mu \to e$ decay chain and K decay (K_{e3}). Their fractions are 37% for the μ decay and 63% for the K decay at OA2°.

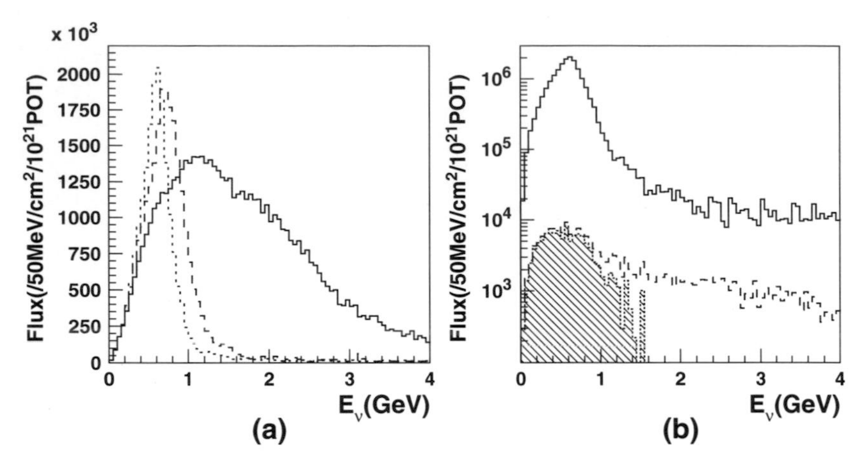

Fig. 8. (**a**) Energy spectra of the on-axis and off-axis ν_μ beams: *solid line* (on-axis), *dashed line* (2.0°), and *dotted line* (2.5°). As the off-axis angle increases, the energy peak narrows and moves lower in energy. (**b**) Energy spectra of the off-axis 2.5° beams: *solid line* (ν_μ) and *dashed line* (ν_e). The *hatched area* shows contribution to ν_e from the muon decay. Another main contributor is the K decay. Both histograms are normalized to 10^{21} 30-GeV protons on target, which roughly corresponds to 0.6-year operation at the designed intensity.

At the peak energy of the ν_μ spectrum, the ν_e/ν_μ ratio is as small as 0.2%. This indicates that the beam ν_e background is greatly suppressed (factor $\sim$4) by applying an energy cut on the reconstructed neutrino energy.

Figure 9 shows the expected energy spectra for ν_μs interacted in the Super-Kamiokande fiducial volume. The expected number of the total and CC interactions in 22.5 kton fiducial volume of SK is 2,300 events/yr and 1,700 events/yr, respectively, at 2.5° off-axis in the case of no oscillation with the 30-GeV, 0.75-MW, 3000-h/yr operation.

Neutrino Facility in J-PARC

A layout of the neutrino facility in J-PARC is illustrated in Fig. 10.

The neutrino facility consists of the following:

- primary proton beamline,
- target station,
- decay volume,
- beam dump,
- muon monitor,
- near neutrino detector,

and several utility buildings.

Special features of the T2K beamline are (1) the first use of superconducting combined function magnets [31] in the primary proton beamline, (2) the first application of the off-axis (OA) beam [28], and (3) capability for

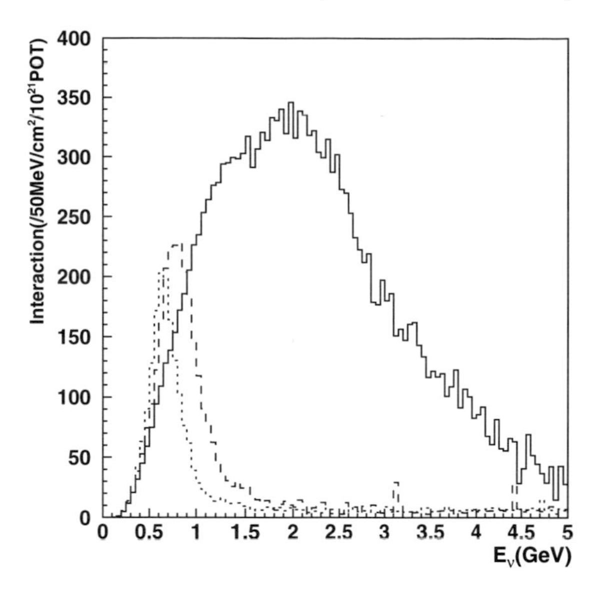

Fig. 9. Energy spectra for ν_μs interacted in the Super-Kamiokande fiducial volume: *solid* (on-axis), *dash* (2.0°), and *dot* (2.5°). The histogram is normalized to 10^{21} 30-GeV protons on target, which roughly corresponds to 0.6-year operation at the designed intensity.

radiation heating and shielding. In order to have tunability of the neutrino energy spectrum, the neutrino beamline is designed to cover OA angle from 2° to 2.5°.

Proton beam is extracted from the 50-GeV PS in a single turn (fast extraction). The design intensity of the beam is 3.3×10^{14} protons/pulse and repetition period is 3.5 s, resulting in the beam power of 0.75 MW at 50-GeV operation. The "50-GeV" PS will be operated at 30 or 40 GeV in early stages. One pulse (spill) consists of eight bunches each with 58-ns full width and separated by 598 ns, and the total spill width is ~4.2 μs.

The primary proton beamline transfers the proton beam from the 50-GeV PS to the neutrino production target. It consists of the preparation, arc, and final focusing sections.

Target station (TS) accommodates a production target and electromagnetic horns. The target is a graphite rod of 26 mm diameter and 90 cm length. About 85% of the incoming protons interact with the target material. When pulsed beam hits the target, big heat is generated in a very short period. This causes dynamical shock to the material. Hence graphite, which is rather light and strong against the heat shock, is adopted. Heat load from the beam is about 60 kJ/spill (~20 kW). This heat is removed by forced-flow Helium gas. The target is followed by three electromagnetic horns which focus generated pions in the forward direction. They are driven by pulsed current of 320 kA

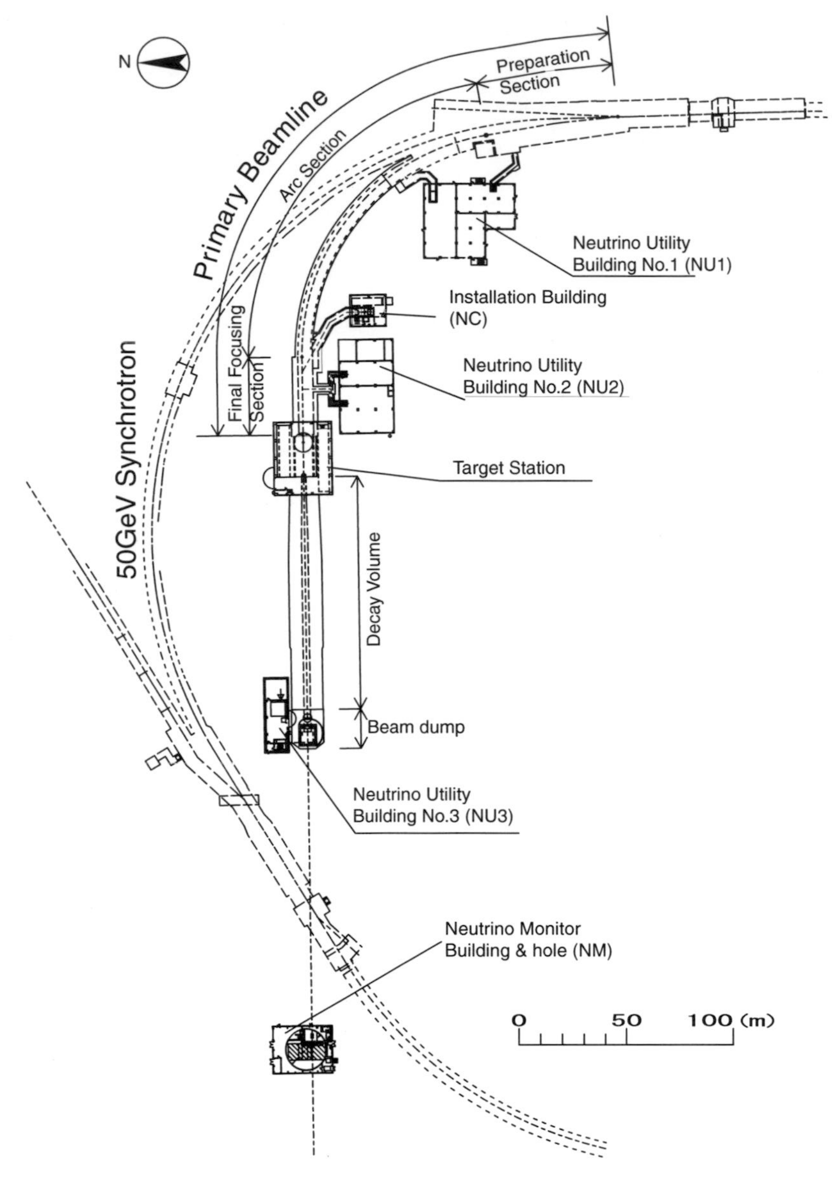

Fig. 10. Layout of the neutrino facility complex.

at peak synchronized with the proton beam timing. Strong force is applied on the horns by both the heat load and the Lorenz force, which is produced by the interaction of electrical current and magnetic field. The target station is filled with the helium gas in order to reduce tritium and NO_x production. The separation of the target station helium gas and the primary beamline vacuum is done by a beam window at the upstream entrance of the target station. It is made of Ti alloy and is cooled by helium gas. Everything inside the target station will be highly radio-activated and will never be accessible by hands once beam operation starts. All the beamline components in TS will be maintained remotely.

The decay volume (DV) is a free space downstream of the horn where pions decay in flight into ν_μs and muons. The length between the target and the end of the decay volume is about 110 m. DV is filled with helium gas to reduce pion absorption and is surrounded by concrete shields of about 6 m thick. The wall of helium vessel is water cooled to remove the heat load by the beam.

The beam dump is placed at the end of DV and stops hadrons such as remaining protons and secondary mesons. The dump consists of graphite blocks of 3.2 m thickness, 2.5 m thick iron, and 1 m thick concrete. The graphite blocks are cooled by aluminum blocks in which cooling water pipes are embedded.

The muon monitor is placed just behind the beam dump to measure the muon flux. While almost all hadrons are absorbed by the beam dump, muons of energy >5 GeV can penetrate the beam dump. Expected muon flux is $\sim 10^8/\mathrm{cm}^2/\mathrm{spill}$. The muon monitor is designed to provide pulse-by-pulse information on the intensity and profile (direction) of the beam. By monitoring the muon flux, the proton beam direction can be monitored with an accuracy of better than 1 mrad for each spill by segmented ionization chambers and/or semi-conductor detectors. The muon monitor also tracks the stability of the neutrino yield.

2.2 Near Detectors

The neutrino beam properties just after its production and interaction cross section will be measured precisely at the near detector experimental hall at 280 m from the target point. The location of the ND280 hall is shown as NM in Fig. 10.

The ND280 hall has a pit with a diameter of 17.5 m and a depth of 37 m, which incorporates both the on-axis and off-axis detectors. The B1 floor, which is about 24 m deep, is for the off-axis detector. The off-axis detector is nearly located on the line between the target point and the SK position. The B2 floor, which is about 33 m deep, is for the horizontal part of the on-axis detector. The B3 floor, which is 37 m deep, is for the deepest part of the vertical on-axis detector.

The On-Axis Neutrino Monitor (INGRID)

The muon monitor observes only the small fraction of muons because it observes muons whose energy is greater than 5 GeV. Therefore, it is also important to directly measure the direction of the neutrino beam. The on-axis neutrino monitor (INGRID) will be installed in the near detector hall to measure the neutrino beam direction and profile using neutrinos directly. It is required to measure the neutrino beam direction with precision much better than 1 mrad as the muon monitor.

The neutrino event rate around the beam center is estimated to be ~ 0.3 events/ton/spill. In order to monitor the stability of the neutrino event rate (direction) at the 1% (1 mrad) precision within a single day ($\simeq 2.5 \times 10^4$ spills), $\mathcal{O}(1)$ ton fiducial mass is necessary for the detector. The variation of the efficiency of the detector across the beam profile should be calibrated at a precision of 1% during the experiment.

The INGRID detector is designed to consist of $7 + 7 + 2$ identical units, arranged to form a grid which samples the beam on $8 \times 8\,\mathrm{m}^2$ ($\pm 4\,\mathrm{m} \times \pm 4\,\mathrm{m}$) area, as shown in Fig. 11. The design of one unit is shown in Fig. 12. The target of neutrino interaction is iron. Scintillators are used to detect a muon from the interaction.

Off-Axis Detector, ND280

Measurements of the flux of ν_μ/ν_e and their spectra are important for oscillation analysis. In addition, cross sections, which are quite unknown for neutrinos, need to be measured both for the reconstruction of spectra and background estimation at the far detector.

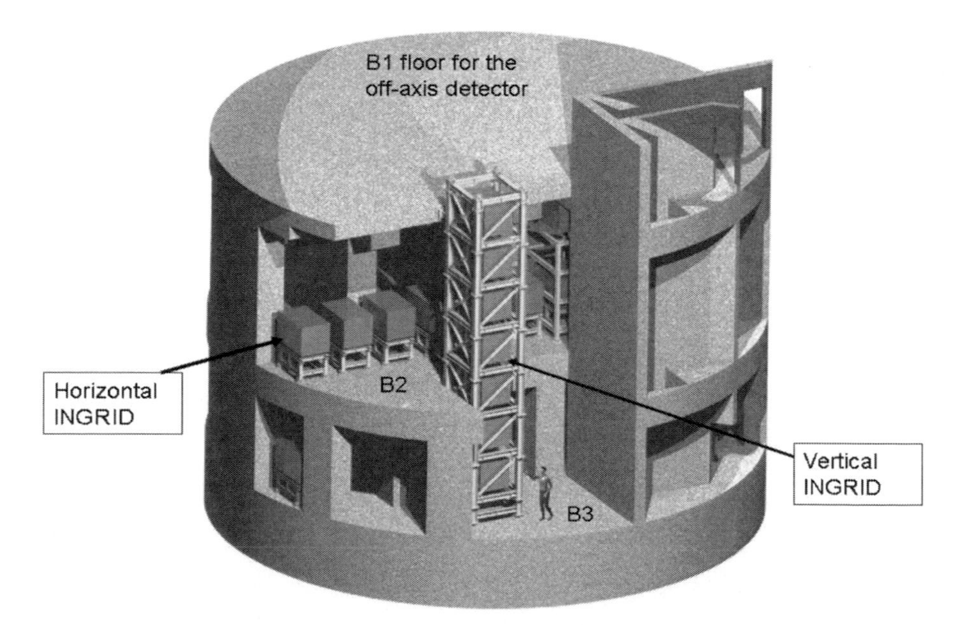

Fig. 11. The arrangement of the INGRID detectors at the near detector hall.

Fig. 12. One unit of the INGRID detector. Veto scintillators will surround this unit.

To make these measurements to the required precision, a highly segmented large volume detector, capable of charged and neutral particle energy measurements and particle identification will be constructed.

The reconstruction of the neutrino energy at the near detector is based on the charged current quasi-elastic (CC-QE) interaction, where the neutrino energy E_ν is reconstructed measuring the muon or electron energy E_l and its angle θ_l against the neutrino beam direction according to a formula

$$E_\nu = \frac{m_N E_l - m_l^2/2}{m_N - E_l + p_l cos\theta_l}, \tag{9}$$

where m_N and m_l are the masses of the neutron and lepton (= e or μ), respectively. The inelastic reactions of high-energy neutrinos constitute the background to E_ν measurement with CC-QE. In addition, the inelastic reactions produce π^0s which contribute to the main background for ν_e appearance search. Then the detection of γs and reconstruction of π^0s are also required for the background estimation.

The near detector complex (ND280) at 280 m from the target contains a fine-resolution magnetized detector designed to measure the neutrino beam's energy spectrum, flux, flavor content, and interaction cross sections before the neutrino beam has a chance to oscillate. This detector sits off-axis in the neutrino beam along a line between the average pion decay point in the decay volume and the Super-Kamiokande detector, at a distance of 280 m from the hadron production target.

The ND280 detector consists of the following elements, illustrated in Fig. 13:

- Magnet: ND280 uses the UA1 magnet from CERN. It will be operated with a magnetic field of 0.2 T to measure the momenta of penetrating charged particles produced by neutrino interactions in the near detector. The inner dimensions of the magnet are 3.5 m × 3.6 m × 7.0 m.

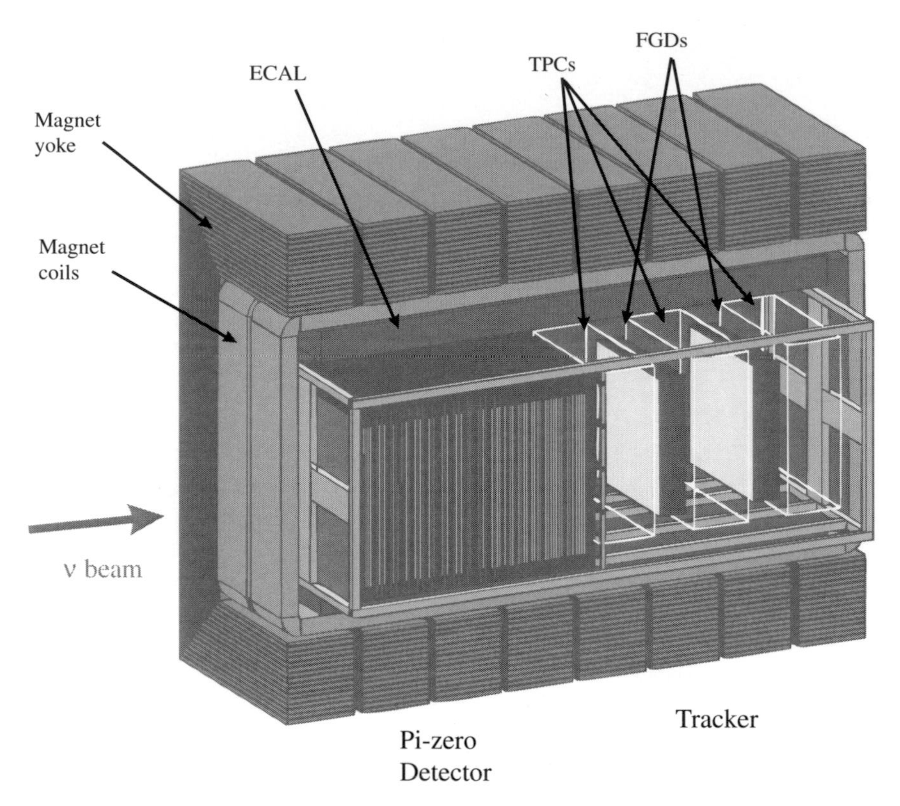

Fig. 13. Cutaway view of the T2K 280 m near detector. The neutrino beam enters from the *left*.

- Pi-zero detector (P0D): The P0D sits at the upstream end of ND280 and is optimized for measuring the rate of the neutral current π^0 production. The P0D consists of tracking planes composed of scintillating bars alternating with lead foil. Inactive layers of passive water in sections of the P0D provide a water target for measuring interactions on oxygen. The P0D is approximately cubical and is covered by the electromagnetic calorimeter.
- Tracker: Downstream of the P0D is a tracking detector optimized for measuring the momenta of charged particles, particularly muons and pions produced by the CC interactions, and for measuring the ν_e background in the beam.
  1. Time projection chambers (TPCs): Three time projection chambers will measure the momentum vector of muons produced by charged current interactions in the detector and will provide the most accurate measurement of the neutrino energy spectrum. The 3D tracking and dE/dx measurements in the TPC will also determine the sign of charged particles and identify muons, pions, and electrons.

2. Fine-grained detectors (FGDs): Two FGD modules, placed after the first and second TPCs, consist of layers of finely segmented scintillating tracker bars. The FGDs provide the target mass for neutrino interactions that will be measured by the TPCs and also measure the direction and ranges of recoil protons produced by the CC interactions in the FGDs, giving clean identification of the CC-QE and non-CC-QE interactions. One FGD module will consist entirely of plastic scintillator, while the second will consist of plastic scintillator and water to allow the separate determination of exclusive neutrino cross sections on carbon and on water.

- Electromagnetic calorimeter (ECAL): Surrounding the P0D and the tracker is an electromagnetic calorimeter. ECAL is a segmented Pb-scintillator detector whose main purpose is to measure γ-rays produced in ND280. It is critical for the reconstruction of π^0 decays.
- Side muon range detector (SMRD): Air gaps in the sides of the UA1 magnet are instrumented with plastic scintillator bars to measure the ranges of muons that exit the sides of the ND280. SMRD can also provide a veto for events entering the detector from the outside and a trigger useful for calibration.

2.3 Far Detector: Super-Kamiokande

The far detector, Super-Kamiokande, is located in the Kamioka Observatory, Institute for Cosmic Ray Research (ICRR), University of Tokyo, which has been successfully taking data since 1996. The detector was also used as a far detector for the K2K experiment, which was the first accelerator-based long-baseline experiment. In 2001, more than half of the photomultiplier tubes were destroyed by an accident. However, with the slogan "We will rebuild the detector. There is no question," the collaborators and volunteers have fully rebuilt it.

It is a 50,000 ton water Čerenkov detector located at a depth of 2,700 m water equivalent in the Kamioka mine in Japan. Its performance and results in atmospheric neutrinos and solar neutrinos have been well documented in [32–35].

A schematic view of the detector is shown in Fig. 14. The detector cavity is 42 m in height and 39 m in diameter, filled with 50,000 tons of pure water. There is an inner detector (ID), 33.8 m diameter and 36.2 m high, surrounded by an outer detector (OD) approximately 2 m thick. The inner detector has 11,146, 50 cm ϕ PMTs, instrumented on a 70.7 cm grid spacing on all surfaces of the inner detector. The outer detector is instrumented with 1,885, 20 cm ϕ PMTs and is used as an anti-counter to identify entering/exiting particles to/from ID. The fiducial volume is defined as 2 m away from the ID wall and the total fiducial mass is 22,500 ton.

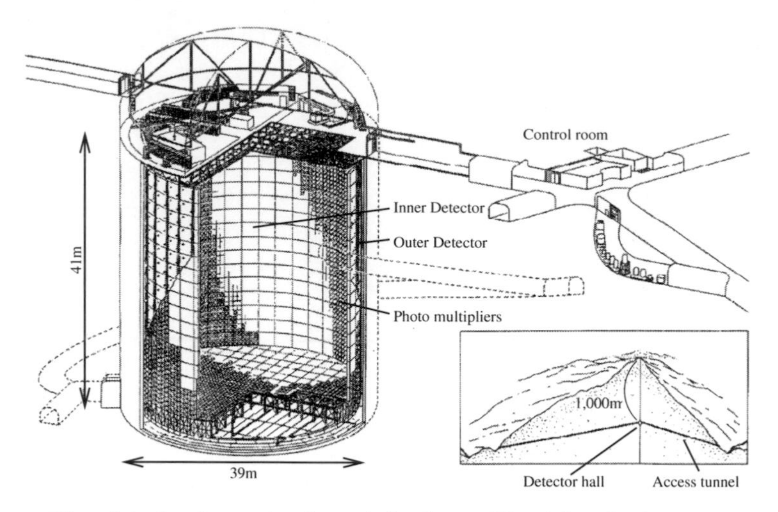

Fig. 14. A schematic view of the Super-Kamiokande detector.

The ID PMTs detect Čerenkov rings produced by relativistic charged particles. The trigger threshold was achieved to be as low as 4.3 MeV. The pulse height and timing information of the PMTs are fitted to reconstruct the vertex, direction, energy, and particle identification of the Čerenkov rings. A typical vertex, angle, and energy resolution for a 1-GeV μ is 30 cm, 3°, and 3%, respectively. The Čerenkov ring shape, clear ring for muons and fuzzy ring for electrons, provides good e/μ identification. A typical rejection factor to separate μs from es (or vice versa) is about 100 for single Čerenkov ring events at 1 GeV. The es and μs are further separated by detecting decay electrons from the μ decays. A typical detection efficiency of decay electrons from cosmic stopping muons is roughly 80%. A 4π coverage around the interaction vertex provides an efficient π^0 detection and e/π^0 separation.

Interactions of the neutrinos from an accelerator are identified by synchronizing the timing between the beam extraction time at the accelerator and the trigger time at Super-Kamiokande using the global positioning system (GPS). The synchronization accuracy of the two sites is demonstrated to be less than 200 ns in the K2K experiment. Because of this stringent time constraint, and the quiet environment of the deep Kamioka mine, chance coincidence of any entering background is negligibly low. A typical chance coincidence rate of the atmospheric neutrino events is 10^{-10}/spill, which is much smaller than the signal rate of about 10^{-3}/spill.

3 T2K Physics Goals

The main physics goals of the T2K experiment are as follows:

- A precision measurement of the neutrino oscillation parameters, $(\sin^2 2\theta_{23}, \Delta m^2_{23})$, by measuring the ν_μ disappearance

- The determination of $\sin^2 2\theta_{13}$ via the measurement of the $\nu_\mu \to \nu_e$ appearance signal
- Search for sterile components in the ν_μ disappearance or confirmation of $\nu_\mu \to \nu_\tau$ oscillation by detecting the neutral current events

3.1 ν_μ Disappearance

The neutrino oscillation parameters $(\sin^2 2\theta_{23}, \Delta m^2_{23})$ will be determined by measuring the survival probability of ν_μ after traveling 295 km. Neutrino events in SK are selected as fully contained events with visible energy ($E_{vis.}$) greater than 30 MeV in the fiducial volume of 22.5 kton. The events are further selected by requiring the presence of a single muon-like ring. The expected number of events without oscillation for an off-axis angle of 2.5° and 5×10^{21} protons on target (POT) are summarized in Table 1. The numbers of events after oscillation as a function of Δm^2_{23} are shown in Table 2 for the values of oscillation parameters $\sin^2 2\theta_{23} = 1.0$ and $\sin^2 2\theta_{13} = 0.0$.

The neutrino energy is reconstructed for the fully contained single muon-like ring events with the formula

$$E_\nu = \frac{m_N E_l - m_l^2/2}{m_N - E_l + p_l \cos\theta_l}, \tag{10}$$

Table 1. The expected number of neutrino events for 5×10^{21} POT for ν_μ disappearance analysis without oscillation. CC-QE refers to charged current quasi-elastic events and CC-non-QE to other charged current events, and NC refers to neutral current events.

	CC-QE	CC-non-QE	NC	All ν_μ
Generated in FV	4,114	3,737	3,149	11,000
(1) FCFV	3,885	3,011	1,369	8,265
(2) $E_{vis.} \geq 30$ MeV	3,788	2,820	945	7,553
(3) Single ring μ-like	3,620	1,089	96	4,805

Table 2. The expected number of neutrino events with neutrino oscillation at different values of Δm^2_{23}. The 5×10^{21} POT, $\sin^2 2\theta_{23} = 1.0$, and $\sin^2 2\theta_{13} = 0.0$ are assumed.

Δm^2 (eV^2)	CC-QE	CC-non-QE	NC	All ν_μ
No oscillation	3,620	1,089	96	4,805
2.0×10^{-3}	933	607	96	1,636
2.3×10^{-3}	723	525	96	1,344
2.7×10^{-3}	681	446	96	1,223
3.0×10^{-3}	800	414	96	1,310

where E_l, p_l, and $\cos\theta_l$ are energy, momentum, and scattered angle, respectively, and m_N and m_l are the masses of the neutron and muon with the assumption that the event is caused by the charged current quasi-elastic (CC-QE) interaction. In order to measure the oscillation parameters, the reconstructed energy distribution is fit assuming neutrino oscillation with an extended-maximum likelihood method. Figure 15 was generated by an MC simulation and shows the reconstructed neutrino energy distribution and the ratio of the distribution with oscillation to one without oscillation. The expected statistical uncertainty on the measurements is shown as a function of the true Δm^2_{23} in Fig. 16. It is 0.009 for $\sin^2 2\theta_{23}$ and $5 \times 10^{-5}\,\mathrm{eV}^2$ for Δm^2_{23} at $(\sin^2 2\theta_{23}, \Delta m^2_{23}) = (1, 2.7 \times 10^{3}\,\mathrm{eV}^2)$.

Systematic uncertainties have to be kept way down to achieve these sensitivities. Following systematic uncertainties have been studied:

1. prediction for the number of the fully contained single ring μ-like events (flux)
2. non-QE/QE ratio (non-QE/QE)
3. energy scale (energy scale)
4. spectrum shape (shape)
5. spectrum width (width)

Figure 17 shows the effect of systematic errors on $\sin^2 2\theta$ and Δm^2 measurement with the uncertainty of 10% for flux, 20% for non-QE/QE, 4% for energy scale, and 10% for width. To see the effect of the shape, the neutrino spectrum is multiplied with a weighting factor of $1 + 0.2(1 - E_\nu) + const.$ This change roughly represents the difference in spectra predicted by MARS

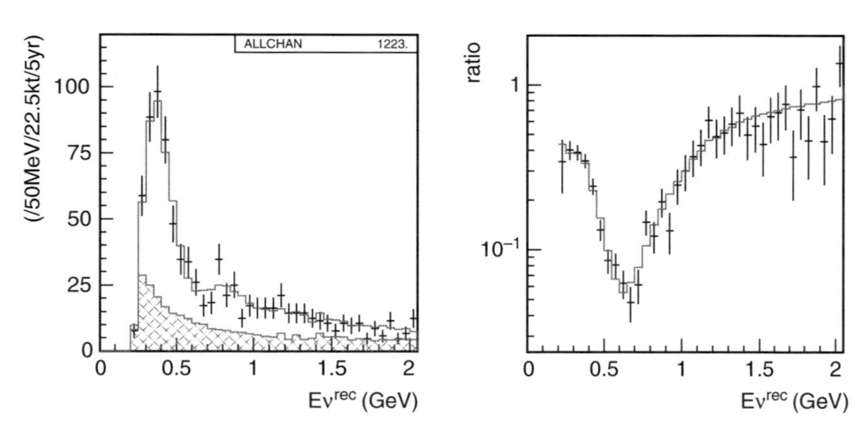

Fig. 15. Results of an MC simulation for the ν_μ disappearance. The $(\sin^2 2\theta_{23}, \Delta m^2_{23}) = (1.0, 2.7 \times 10^{-3}\,\mathrm{eV}^2)$ is assumed. (*Left*) The reconstructed neutrino energy distribution. The *line* is a prediction with the best-fit oscillation parameters. The *hatched area* shows the non-QE component. (*Right*) The ratio of the reconstructed neutrino energy distribution against the prediction without oscillation. The *line* is that of a prediction with the best-fit oscillation parameters.

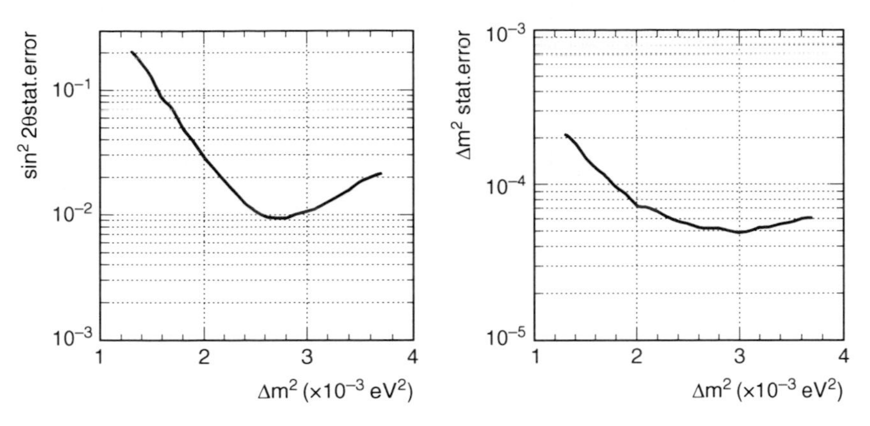

Fig. 16. The statistical uncertainty of the oscillation parameters as a function of true Δm^2_{23}. The value of $\sin^2 2\theta_{23}$ is assumed to be 1.

[36–38] and FLUKA [39] hadron production models. In order to keep the systematic uncertainties below the statistical error, the uncertainties should be less than about 5% for the prediction of the number of events, 2% for the energy scale, 5–10% for the non-QE/QE ratio, and 10% for the spectrum width. The spectrum shape uncertainty should be less than the difference of the hadron production models MARS and FLUKA. This should be achieved through the measurement at the near detector, calibration of the far detector, and measurement of the hadron production.

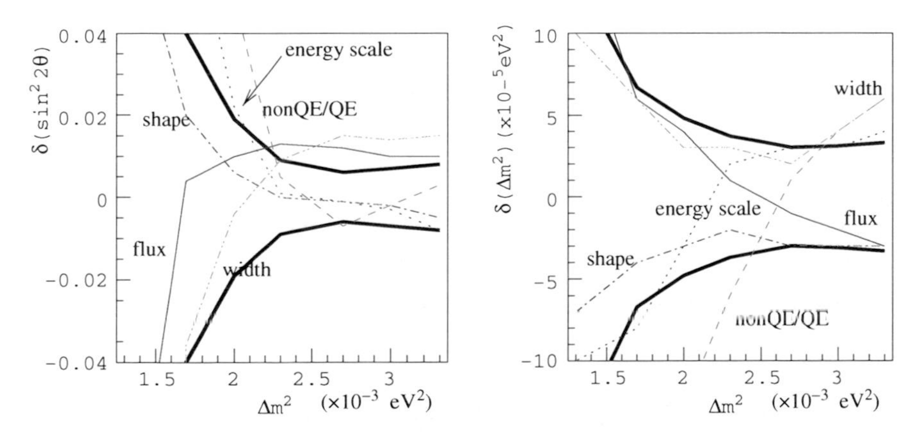

Fig. 17. Effects of systematic uncertainties on determination of oscillation parameters (OA2.5°). *Red curve*: the effect when flux is changed by 10%; *green*: non-QE/QE ratio changed by 20%; *blue*: energy scale changed by 4%; *pink*: spectrum shape (see text); *light blue*: width of spectrum changed by 10%. The *bold curves* indicate 1σ statistical error.

3.2 ν_e Appearance

The ν_e selection cuts are based on the SK-1 atmospheric neutrino analysis. Events are required to be fully contained within the 22.5 kton fiducial volume, and to have visible energy ($E_{vis.}$) greater than 100 MeV, a single electron like (e-like) ring, and no decay electrons. The electron identification eliminates most of the muon background events and the decay electron cut further reduces events from the inelastic charged current (CC) processes associated with the π^0 production. The dominant source of background events (see Table 3) at this stage is single π^0 production in the neutral current (NC) interactions. The backgrounds can be further reduced by requiring the reconstructed neutrino energy to be around the oscillation maximum: $0.35\,\text{GeV} \leq E_\nu^{rec.} \leq 0.85\,\text{GeV}$.

The remaining background from π^0 is further reduced with specific "e/π^0 separation" cuts. The π^0 background has a steep forward peak toward the neutrino direction due to the coherent π^0 production. Thus events in the extreme forward direction ($\cos\theta_{\nu_e} \geq 0.9$) are rejected. Then events with only one high-energy gamma detected in the asymmetric decay of the π^0 are the dominant backgrounds. In order to find the hidden lower energy gamma ring, the photomultiplier hit pattern, including scattered light, is fit under the hypothesis of two gamma rings. The energy and direction of each gamma ring are reconstructed, and the invariant mass of the two rings is calculated. The π^0 background is further suppressed by rejecting events with the mass at around the π^0 mass ($m_{2\gamma}^{rec.} \geq 100\,\text{MeV/c}^2$). These further "e/$\pi^0$' separation" cuts significantly reduce the π^0 background. The background level due to NC-π^0 is expected to be comparable to the predicted background from intrinsic ν_e in the beam.

Table 3 summarizes the expected number of events after the event selections for the 5×10^{21} POT exposure at $\Delta m^2 = 2.5 \times 10^{-3}\text{eV}^2$ and $\sin^2 2\theta_{13} = 0.1$. Figure 18 is the expected reconstructed energy distribution for the selected event.

Figure 19 shows the 90% CL ν_e sensitivity for 5×10^{21} POT exposure and for $\sin^2 2\theta_{23} = 1$ and $\delta = 0, \pi/2, -\pi/2, \pi$, assuming a 10% systematic uncertainty in the background subtraction. Here, the sensitivity is defined as the value where the expected observation is consistent with background rate

Table 3. The number of events selected by the ν_e appearance analysis, as predicted by NEUT Monte Carlo for 5×10^{21} POT exposure. For the calculation of oscillated ν_e, $\Delta m^2 = 2.5 \times 10^{-3}\,\text{eV}^2$ and $\sin^2 2\theta_{13} = 0.1$ are assumed.

	ν_μCC BG	ν_μNC BG	Beam ν_e BG	ν_eCC signal
Fully contained, $E_{vis} \geq$100 MeV	2215	847	184	243
1 ring e-like, no decay-e	12	156	71	187
0.35$\leq E_\nu^{rec.} \leq$0.85 GeV	1.8	47	21	146
e/π^0 separations	0.7	9	13	103

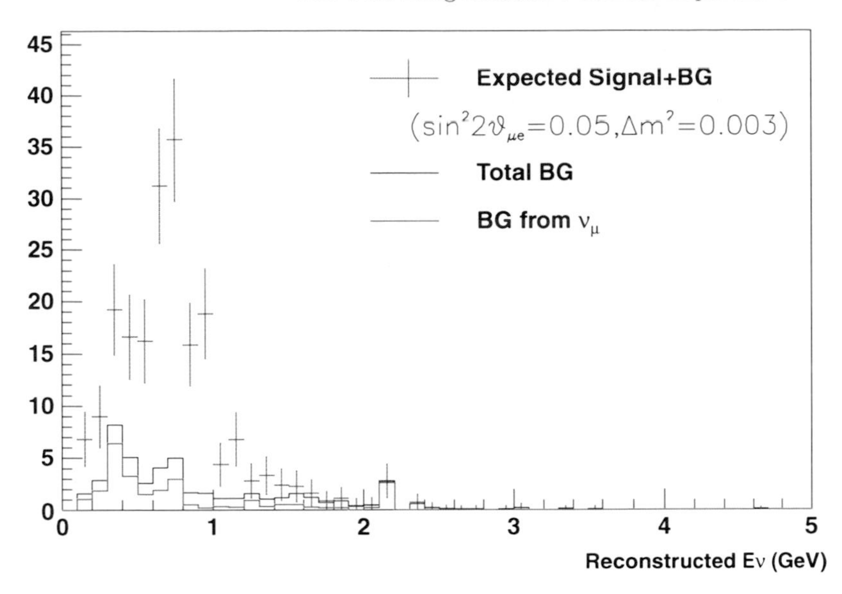

Fig. 18. Expected reconstructed energy distribution for the selected event as ν_e appearance signal.

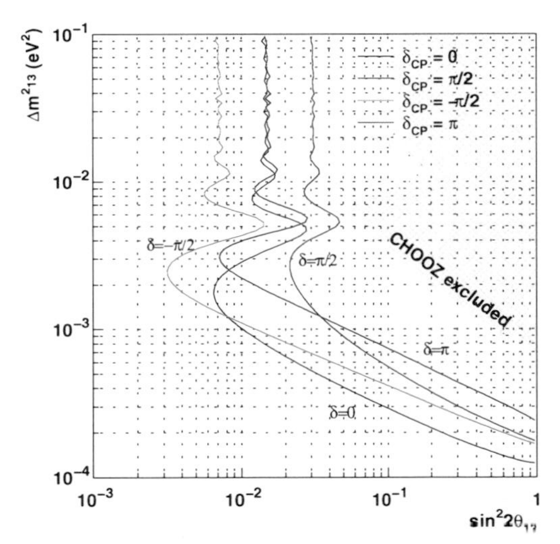

Fig. 19. The 90% CL sensitivity to $\sin^2 2\theta_{13}$ for an exposure of 5×10^{21} POT with the assumption of maximum mixing, $\sin^2 2\theta_{23} = 1.0$, and the CP violation phase $\delta = 0, \pi/2, -\pi/2, \pi$. The 90% excluded region of CHOOZ is overlaid for comparison to $\sin^2 2\theta_{23} = 1.0$.

within that confidence level. At $\Delta m^2 \sim 2.5 \times 10^{-3}\mathrm{eV}^2$, the sensitivity is $\sin^2 2\theta_{13} = 0.008$ at 90% CL.

3.3 ν_τ Versus $\nu_{sterile}$ Analysis

The analysis of the ν_τ appearance has been performed by the Super-Kamiokande collaboration [40]. This measurement disfavors a pure $\nu_\mu \rightarrow \nu_{sterile}$ solution to the atmospheric neutrino anomaly. The search for $\nu_\mu \rightarrow \nu_{sterile}$ with a small mixing angle is still an interesting topic that can be addressed by the T2K experiment. The analysis method is based on the neutral current measurement in the SK detector, a clear signature of which is single π^0 production. The analysis is performed in a similar way to the previous ones:

- 22.5 kton fiducial volume cut is applied.
- The visible energy is selected between 100 and 1500 MeV.
- Events are selected for two e-like rings.
- Events are selected with no decay electrons to further suppress the CC background.

The event suppression by the cuts is shown in Table 4. The purity of the final NC sample is 83%, with 255 total expected events for an exposure of 5×10^{21} POT.

The sensitivity to the fraction of ν_s in $\nu_\mu \rightarrow \nu_\tau$ oscillation is shown in Fig. 20 (left) as a function of Δm^2 and $\sin^2 2\theta_s$. The best limit, $\sin^2 2\theta_s \leq 0.2$, will be obtained for $\Delta m^2 \approx 3 \times 10^{-3}\,\mathrm{eV}^2$. The effect of systematic uncertainties is shown in Fig. 20 (right), where the sensitivity to $\sin^2 2\theta_s$ is plotted as a function of the exposure time for two values of systematic error on the NC prediction: 5 and 10%. A 10% error already dominates the measurement after the first year (10^{21} POT). The main systematic error for this analysis is a limit on the knowledge of the NC π^0 production, similar to the ν_e appearance analysis.

Table 4. The expected number of neutrino events in 5×10^{21} POT for ν_τ appearance analysis without oscillation. The third and fourth columns show the comparison for the two cases where the oscillation is purely due to ν_τ or due to ν_s. The numbers are computed assuming $\sin^2 2\theta_{23} = 1.0$ and $|\Delta m^2_{23}| = 2.7 \times 10^{-3}\,\mathrm{eV}^2$.

	ν_μ CC	$\nu_\mu + \nu_\tau$ NC $\nu_\mu -> \nu_\tau$	ν_μ NC $\nu_\mu -> \nu_s$	ν_e
Generated in fiducial volume	3,173	3,239	1,165	236
(1) 100 MeV $\leq E_{vis.} \leq$ 1500 MeV	184	724	429	109
(2) Two e-like rings	31	281	125	19
(3) No decay electron	9	255	104	14

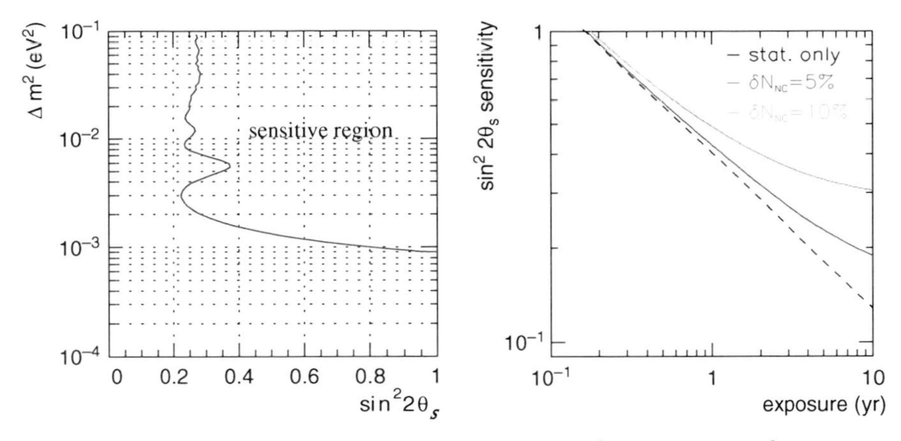

Fig. 20. (*Left*) Sensitivity to the ν_s fraction in the $\sin^2 2\theta_s$ versus Δm^2 plane with 5×10^{21} POT. (*Right*) The sensitivity as a function of exposure with different uncertainties in the background estimation.

3.4 Search for the CP Violation in the Lepton Sector – Future Project

Now that the large mixing angle solution of the solar neutrino deficit is confirmed by KamLAND, the chance of discovering CP violation is good providing that θ_{13} and δ_{CP} are not suppressed too much. The CP asymmetry is calculated as follows:

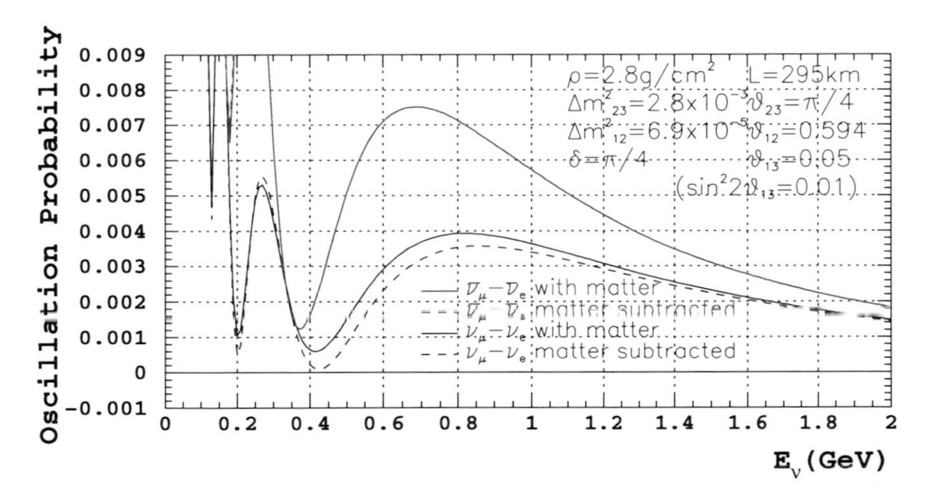

Fig. 21. Oscillation probabilities for $\nu_\mu \rightarrow \nu_e$ (*upper*) and $\bar{\nu}_\mu \rightarrow \bar{\nu}_e$ (*lower*). The *solid curves* include asymmetry due to matter effect. For the *dashed curves*, the matter effect is subtracted and the differences between $\nu_\mu \rightarrow \nu_e$ and $\bar{\nu}_\mu \rightarrow \bar{\nu}_e$ are all due to CP effect.

$$A_{CP} = \frac{P(\nu_\mu \to \nu_e) - P(\bar{\nu}_\mu \to \bar{\nu}_e)}{P(\nu_\mu \to \nu_e) + P(\bar{\nu}_\mu \to \bar{\nu}_e)} = \frac{\Delta m_{12}^2 L}{4E_\nu} \cdot \frac{\sin 2\theta_{12}}{\sin \theta_{13}} \cdot \sin\delta. \tag{11}$$

By choosing a low-energy neutrino beam at the oscillation maximum (E $\sim$ 0.75 GeV and L $\sim$ 295 km for T2K), the CP asymmetry is enhanced as 1/E. Taking the best-fit values of KamLAND, $\sin^2 2\theta_{12} = 0.91$ and $\Delta m_{12}^2 = 6.9 \times 10^{-5}$, 1/10 of the CHOOZ limit for $\sin^2 2\theta_{13} = 0.01$, and $\delta = \pi/4$ (half of the maximum CP angle), A_{CP} becomes as large as 40% as shown in Fig. 21. The matter effect, which creates fake CP asymmetry, increases linearly with the neutrino energy. Because of the use of low-energy neutrinos, the fake asymmetry due to the matter effect is small for the T2K experiment. Further upgrade of the J-PARC beam intensity up to 2–4 MW, and a mega-ton class neutrino detector would enable to explore the CP phase as small as $\sim$10°.

4 Summary

The neutrino beam produced at J-PARC will be the world's most intense neutrino beam in GeV region when the J-PARC accelerator complex succeeds to accelerate the protons at the designed intensity. Super-Kamiokande is already the world's largest neutrino detector. With this combination of the beam and far detector, the Tokai-to-Kamioka long-baseline neutrino oscillation experiment, T2K, is aiming for the discovery of finite θ_{13} and most precise measurement of θ_{23} and Δm_{23}^2. If the nature gives sufficiently large θ_{13}, discovery of the CP violation in the lepton sector may lie in the future of the project.

References

1. Cleveland, B., et al.: Astrophys. J. **496**, 505 (1998)
2. Fukuda, Y., et al.: Phys. Rev. Lett. **77**, 1683–1686 (1996)
3. Hirata, K.S., et al.: Phys. Rev. D **44**, 2241–2260 (1991)
4. Maki, Z., Nakagawa, N., Sakata, S.: Prog. Theor. Phys. **28**, 870 (1962)
5. Kobayashi, M., Maskawa, T.: Prog. Theor. Phys. **49**, 652 (1973)
6. Mikheyev, S.P., Yu Smirnov, A.: Sov. J. Nucl. Phys. **42**, 913 (1985)
7. Wolfenstein, L.: Phys. Rev. D **17**, 2369 (1978)
8. Gavrin, V.: Results from the Russian American Gallium Experiment (SAGE), VIIIth international conference on Topics in Astroparticle and Underground Physics (TAUP), Seattle, 5–9 Sept 2003
9. Abdurashitov, J.N., et al.: J. Exp. Theor. Phys. **95**, 181 (2002); the latest SAGE results were presented by C. Cattadori, results from radiochemical solar neutrino experiments, XXI international conference on Neutrino Physics and Astrophysics (Neutrino), Paris, 14–19 June 2004
10. Bellotti, E.: The Gallium Neutrino Observatory (GNO), VIIIth international conference on Topics in Astroparticle and Underground Physics (TAUP), Seattle, 5–9 Sept 2003

11. Altmann, M., et al.: Phys. Lett. B **490**, 16 (2000)
12. Hampel, W., et al.: Phys. Lett. B **447**, 127 (1999)
13. Ashie, Y., et al.: Phys. Rev. D **71**, 112005 (2005)
14. Fukuda, Y., et al.: Phys. Rev. Lett. **81**, 1562–1567 (1998)
15. Ahmad, Q.R., et al.: Phys. Rev. Lett. **87**, 071301 (2001)
16. Ahmad, Q.R., et al.: Phys. Rev. Lett. **89**, 011301 (2002)
17. Ahmed, S.N., et al.: Phys. Rev. Lett. **92**, 181301 (2004); nucl-ex/0502021, Phys. Rev. C72: 055502, 2005
18. Eguchi, K., et al.: Phys. Rev. Lett. **90**, 021802 (2003)
19. Eguchi, K., et al.: Phys. Rev. Lett. **94**, 081801 (2005)
20. Aliu, E., et al.: Phys. Rev. Lett. **94**, 081802 (2005)
21. Michael, D.G. et al. [MINOS Collaboration]: Phys. Rev. Lett. **97**, 191801 (2006) [arXiv:hep-ex/0607088]
22. The OPERA experiment, Pessard, H., et al.: (OPERA), hep-ex/0504033
23. The ICARUS experiment, a second-generation proton decay experiment and neutrino observatory at the Gran Sasso Laboratory, Arneodo, F., et al. (ICARUS), hep-ex/0103008
24. Apollonio, M., et al.: Eur. Phys. J. **C27**, 331–374 (2003)
25. Apollonio, M., et al.: Phys. Lett. B **466**, 415–430 (1999)
26. Richter, B.: SLAC-PUB-8587, hep-ph/0008222 (2000) and references there in
27. Fukugita, M., Yanagida, T.: Phys. Lett. B **174**, 45 (1986)
28. Beavis, D., Carroll, A., Chiang, I., et al.: Proposal of BNL AGS E-889 (1995)
29. First MINOS oscillation results, http://www-numi.fnal.gov/talks/results06.html
30. Brun, R., et al.: CERN DD/EE/84-1 (1987)
31. Ogitsu, T., et al.: IEEE Trans. Appl. Supercond. **14**, 604 (2004)
32. Super-Kamiokande collaboration: Phys. Rev. Lett. **81**, 1562 (1998)
33. Super-Kamiokande collaboration: Phys. Rev. Lett. **85**, 3999 (2000)
34. The Super-Kamiokande collaboration: Phys. Rev. Lett. **86**, 5651 (2001)
35. The Super-Kamiokande collaboration: Phys. Rev. Lett. **86**, 5656 (2001)
36. Mokhov, N.V.: The Mars code system user's guide, Fermilab-FN-628 (1995)
37. Krivosheev, O.E., Mokhov, N.V.: MARS code status, Proceedings of the Monte Carlo 2000 conference, Fermilab-Conf-00/181, p. 943, Lisbon, 23–26 Oct 2000
38. Mokhov, N.V.: Status of MARS code, Fermilab-Conf-03/053 (2003) http://www-ap.fnal.gov/MARS/
39. Fasso, A., Ferrari, A., Ranft, J., Sala, P.R.: http://pcfluka.mi.infn.it/
40. Fukuda, S., et al. (Super-Kamiokande collaboration): Phys. Rev. Lett. **85**, 3999–4003 (2000)
41. Athanassopoulos, C., et al.: Phys. Rev. C542685-27081996, nucl-ex/9605001
42. Aguilar, A., et al.: Phys. Rev. D641120072001
43. Armbruster, B., et al.: Phys. Rev. D **65**, 112001 (2002)
44. Church, E.D., Eitel, K., Mills, G.B., Steidl, M.: Phys. Rev. D **66**, 013001 (2002)
45. Church, E., et al.: nucl-ex/9706011
46. P-875: A long baseline neutrino oscillation experiment at Fermilab, Ayres, D., et al. MINOS Proposal, NuMI-L-63. http://www.hep.anl.gov/ndk/hypertext/numi notes.html
47. http://jkj.tokai.jaeri.go.jp/index-e.html
48. Fukuda, Y., et al.: Nucl. Instrum. Methods A **501**, 418–462 (2003)

$K_L \to \pi^0 \nu\bar{\nu}$

G.Y. Lim

IPNS, KEK, Tsukuba, Ibaraki 305-0801, Japan
gylim@post.kek.jp

1 Introduction

CP violation has been a central issue in particle physics since its discovery in 1964. It was one of the cornerstones of the standard model (SM) consisting of the three generations of quarks. Recent successful measurements by K- and B-meson decays [1–4] enabled us to understand how different types of quarks transform among them. The experimental results so far support the standard model based on the quark mixing expressed by the Cabibbo–Kobayashi–Maskawa (CKM) matrix [5].

However, there are still many questions about the CP violation, such as "Is there any underlying mechanism to determine the CKM parameters?", "Are there any other sources of CP violation to explain the baryon–antibaryon asymmetry in the universe?". In order to answer these questions, the CP violation studies have entered the era of precision in both theoretical calculations and experimental measurements to observe any deviation from the SM expectations.

The very rare decay $K_L^0 \to \pi^0 \nu\bar{\nu}$ [6] provides one of the best probes for understanding the origin of CP violation in the quark sector [7, 8]. The branching ratio, $Br(K_L \to \pi^0 \nu\bar{\nu})$, can be calculated with exceptionally small uncertainty, and any deviation from the SM expectation will be directly connected to a clear observation of the new physics. It is a flavor-changing neutral current (FCNC) process from strange to down quarks through the electroweak loop diagrams, which have a large potential to observe a new physics contribution.

Experimentally, the measurement of $Br(K_L \to \pi^0 \nu\bar{\nu})$ is very challenging as it is a three-body decay of a neutral particle (K_L) into all neutral particles including two neutrinos. Thus, there are only a small number of kinematic constraints to identify the signal events. In addition, we need a huge number of K_L decays as a consequence of its tiny branching ratio, expected to be 2.8×10^{-11} in the SM. A large number of background events from other K_L decay modes must also be rejected.

Lim, G.Y.: $K_L \to \pi^0 \nu\bar{\nu}$. Lect. Notes Phys. **781**, 45–74 (2009)
DOI 10.1007/978-3-642-00961-7_3

For a precise measurement of the decay, we take a step-by-step approach. We have performed the first trial, E391a, at the KEK 12-GeV proton synchrotron to study the detection method. With the experience at the E391a, we will proceed to a higher experimental sensitivity at the J-PARC. The J-PARC, newly constructed in Japan, will provide a 100-times-larger flux of protons to produce the K_L, compared to those from the KEK 12-GeV PS operated so far. After the first observation of the decay with an upgraded E391a detector, we will proceed to the next step by using an additionally upgraded and/or newly constructed detector and beamline to determine the $Br(K_L \to \pi^0\nu\bar{\nu})$ precisely.

1.1 $K_L \to \pi^0\nu\bar{\nu}$ in the Standard Model

In the standard model, the $K_L \to \pi^0\nu\bar{\nu}$ decay is expressed by the electroweak penguin and box diagrams as shown in Fig. 1. The hadronic matrix element can be factorized by the precisely measured branching ratio of the $K^+ \to \pi^0 e^+\nu$ decay [9]. The higher order QCD corrections to the diagrams with the virtual top quark are small due to the large mass of the top quark [10, 11]. Long-distance contributions are negligible [12–14] because neutrinos are weakly interacting particles. Consequently, the uncertainty in the theoretical calculations to determine the $Br(K_L \to \pi^0\nu\bar{\nu})$ is estimated to be only a few percent.

The decay amplitude of $K_L^0 \to \pi^0\nu\bar{\nu}$ is proportional to the imaginary part of a product of CKM matrix elements, $Im(V_{ts}^* V_{td})$, which corresponds to the height of the unitarity triangle as shown in Fig. 2. The unitarity of the CKM matrix has been considered as one of the most critical checks for new physics beyond the standard model. The pure and clean information obtained by the $K_L \to \pi^0\nu\bar{\nu}$ decay is crucial for the checks of the SM as well as those with the B decays. By using current estimates of the SM parameters, the branching ratio for the $K_L^0 \to \pi^0\nu\bar{\nu}$ decay is predicted to be

$$Br(K_L^0 \to \pi^0\nu\bar{\nu}) = (2.20 \pm 0.07) \times 10^{-10} \left(\frac{\lambda}{0.2248}\right)^8 \left[\frac{Im(V_{ts}^* V_{td})}{\lambda^5} X(x_t)\right]^2$$

and is expected to lie in the range $(2.8 \pm 0.4) \times 10^{-11}$ [15]. Here $\lambda \equiv |V_{us}|$, $X(x_t) = 1.464 \pm 0.041$ is the value of the Inami–Lim loop function [16, 17]

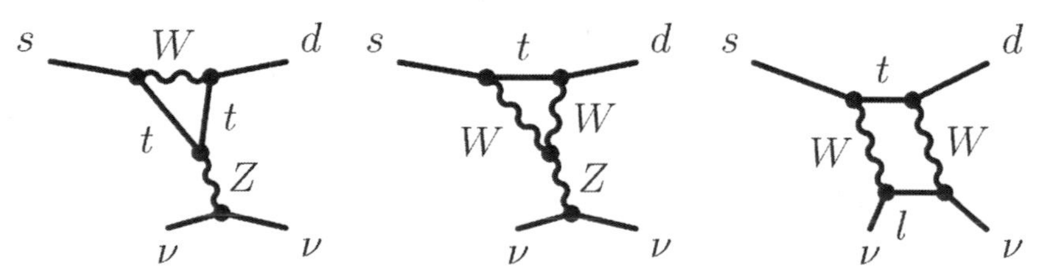

Fig. 1. SM diagrams for the $K_L^0 \to \pi^0\nu\bar{\nu}$ decay.

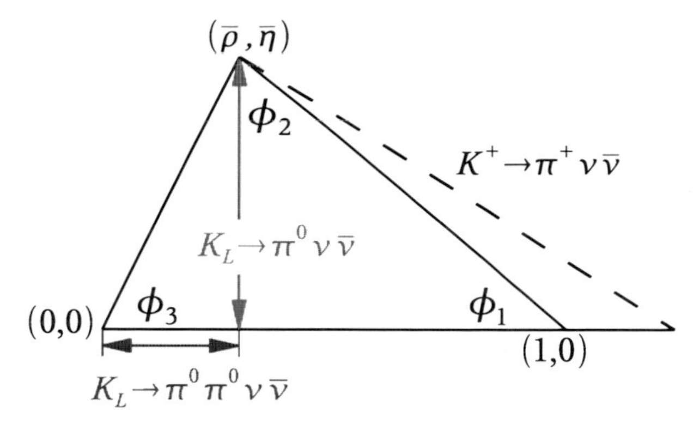

Fig. 2. Kaon unitarity triangle.

(including the QCD corrections), and the parameter x_t is the square of the ratio of the top-quark and W-boson masses.

1.2 $K_L \to \pi^0 \nu\bar{\nu}$ Beyond the Standard Model

Various models beyond the SM predict a sizable (new) contribution to the $K_L \to \pi^0 \nu\bar{\nu}$ decay [18]. As shown in Fig. 3, there is a wide parameter space where we can explore with improving experimental sensitivity (even before reaching the SM expectation). For example, in the minimal supersymmetric extensions of the SM (MSSM), the $Br(K_L \to \pi^0 \nu\bar{\nu})$ will be largely enhanced by a new source of flavor mixing. The Little Higgs model with T-parity (LHT) was recently suggested and a branching ratio 20 times larger than that of the SM is expected. Because the observables in B decays will be changed by only

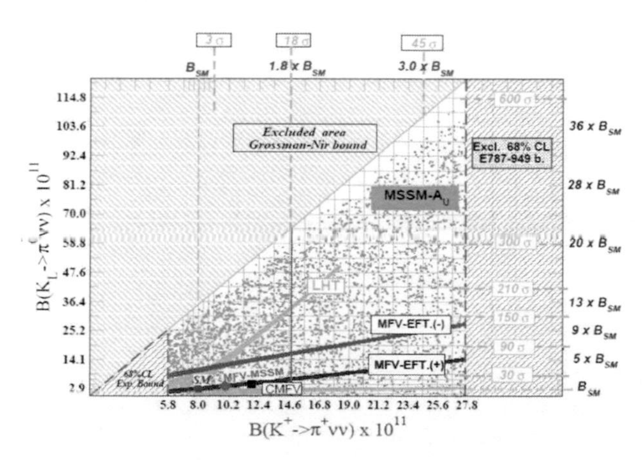

Fig. 3. Expected branching ratios of the $K_L \to \pi^0 \nu\bar{\nu}$ and $K^+ \to \pi^+ \nu\bar{\nu}$ decays in the models beyond the standard model. There is large parameter space for the new physics [19].

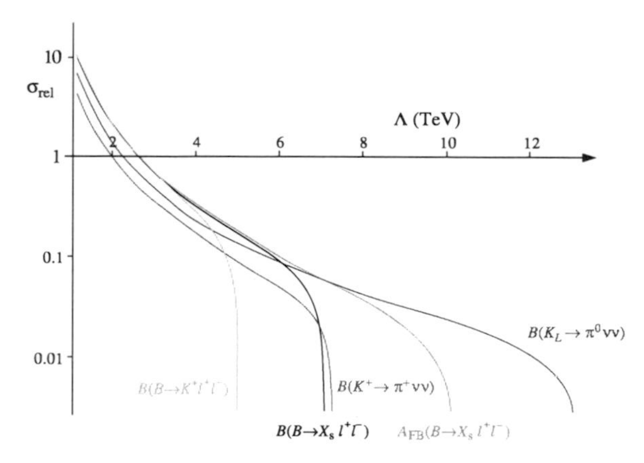

Fig. 4. Comparison of the effectiveness of different observables in rare K and B decays in setting future bounds on the energy scale of new physics operators [20].

a factor of 2 according to the model, the LHT will be critically tested by the $Br(K_L \to \pi^0\nu\bar{\nu})$ measurement in the early stage of J-PARC.

In the minimal flavor violation (MFV), a scenario assuming that there is no other source of the CP violation in the new physics, the $K_L \to \pi^0\nu\bar{\nu}$ decay will become more important. As shown in Fig. 4, a precise measurement of $Br(K_L \to \pi^0\nu\bar{\nu})$ will explore the physics in the highest energy scale.

2 KEK-PS E391a

E391a is the first dedicated experiment for the $K_L \to \pi^0\nu\bar{\nu}$ decay. Its primary goal was to establish an experimental method for precise measurement of the $Br(K_L \to \pi^0\nu\bar{\nu})$. The signal of the $K_L \to \pi^0\nu\bar{\nu}$ decay is a well-reconstructed single π^0, detected by the two photons from $\pi^0 \to \gamma\gamma$ decay without any accompanying particles.

Among the full list of K_L decay modes, only the $K_L \to \pi^0\nu\bar{\nu}$ and $K_L \to \gamma\gamma$ decays have two photons in the final states. Thus, the signal can be clearly identified if detection of two and only two photons can be ensured. A key issue for the $K_L \to \pi^0\nu\bar{\nu}$ decay measurement is thus to construct a detector system that has high detection efficiency for photons and charged particles (to veto background events). The $K_L \to \gamma\gamma$ decay can be easily removed by requiring two-body decay kinematics.

There is another important background source: π^0 production by neutron. When a neutron interacts with the residual gas in the decay volume, a π^0 will be produced via n+A→ π^0 +n+A reaction and can be misidentified as a signal event. Because there are lots of neutrons in the neutral beam, we have to evacuate the decay region (down to 10^{-5} Pa) in order to reduce the interactions. Neutrons that spread out from the center of the beam and enter detector components, *halo neutrons*, also produce a π^0. In order to

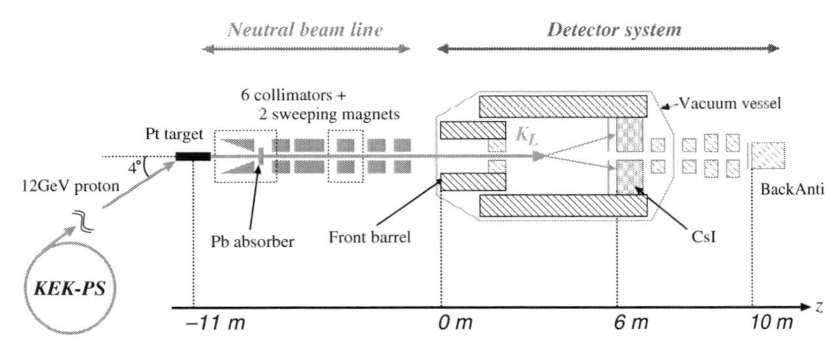

Fig. 5. Schematic view of the E391a experiment.

avoid the backgrounds, we have to collimate neutrons so that they will not hit the detector components. Because the positions of the π^0 production are determined, we can set a signal region to avoid the background.

Figure 5 shows a schematic view of the E391a experiment. The experiment was performed at the 12-GeV proton synchrotron (12-GeV PS) of High Energy Accelerator Research Organization (KEK) in Japan. The protons were directed onto a production target to produce K_L, and the well-collimated K_L beam entered the detector that consisted of a calorimeter and a hermetic veto system. The electromagnetic calorimeter consisted of undoped CsI crystals for detecting two photons. The veto system detected photons and charged particles from the K_L decay in order to remove backgrounds. Most of the detectors were put into a vacuum vessel to maintain a high detection efficiency without having a thick vacuum wall between the evacuated decay region and the detector components.

2.1 K0 Beamline

Protons were accelerated up to the kinetic energy of 12 GeV and slowly extracted to the experimental hall during 2 s with 4-s cycles through the EP2-C beamline; 2×10^{12} protons per spill hit the production target which was made of platinum (Pt), 60-mm long (0.68 λ_I, 20 X_0) and 8 mm diameter. The target was inclined by 4° horizontally with respect to the primary beam axis. The neutral beamline, *K0 beamline*, delivered the secondary particles into the detector which was located at 11 m downstream of the target.

The K0 beamline was designed to generate a *narrow* and *clean* beam, called a *pencil beam*. When the π^0 is reconstructed from energies and positions of two photons measured by the calorimeter, with a constraint of the π^0-mass, only the angle between the momentum vectors of the two photons can be determined. By assuming that the π^0 decays on the beam axis, the decay vertex along the beam can be determined. Thus, a beam with large transverse dimensions will introduce uncertainties in the reconstruction of the vertex

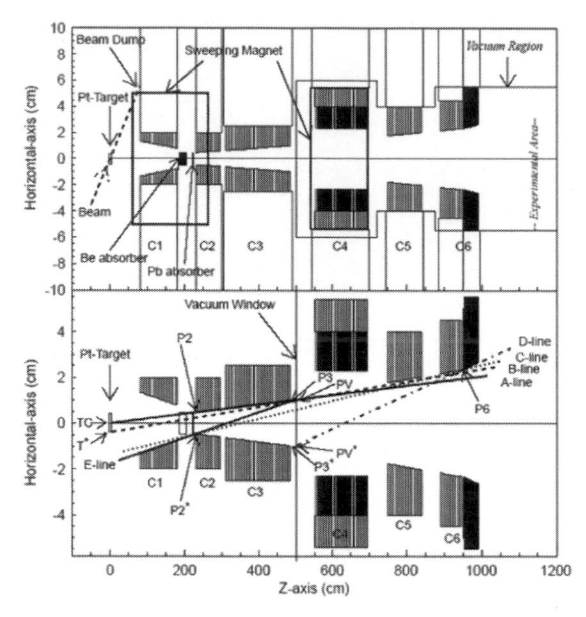

Fig. 6. Schematic views of the *K0 beamline*. The *top figure* shows an arrangement of the beamline components (note the different scales in the vertical and horizontal axes). The *bottom figure* shows the collimation scheme with various lines as explained in the text in detail.

and the transverse momentum (P_T). The clean beam means that it is well collimated and there is no halo component of neutrons that can enter the detector.

A schematic view of the K0 beamline is given in Fig. 6. Tungsten discs 5-cm thick were arranged to produce various inner lines indicated in the bottom figure. The two collimators in the upstream section (C2, C3) were aligned to the A line which defined the beam shape. The collimators in the downstream section (C5, C6) were aligned to the B and C lines; those were determined by the scattering points inside the collimators. The last half of C6, in which the line was determined by the last scattering point in the upstream section, was an active collimator consisting of alternating layers of tungsten and plastic scintillator.

Two dipole magnets were installed to sweep out charged particles. Inside the downstream magnets, thin Gd_2O_3 sheets were placed to reduce the thermal neutron flux (C4). In order to prevent halo generation through the scattering with air inside the collimators, the downstream half of the beamline (from C4) was evacuated down to the 1-Pa level, and it was directly connected to the vacuum vessel which contained the detector.

Figure 7 shows the performance of the K0 beamline obtained by a Monte Carlo calculation. The results were confirmed by the data obtained on a series of beam-survey experiments in April and December 2000 and December 2001

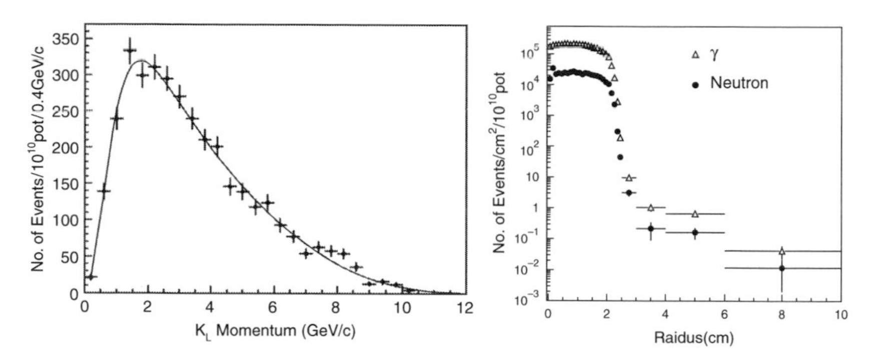

Fig. 7. Monte Carlo results for the K0 beamline. *Left*: K_L momentum distribution at the exit of the beamline. *Right*: Beam profile at the exit of the beamline. The *triangles* and *circles* indicate those for photon and neutron with energies above 1 MeV, respectively.

[21]. The momentum spectrum of K_L had an average value of 2.6 GeV/c, which was precisely determined later by using $K_L \to \pi^0\pi^0\pi^0$ data obtained at the physics run. The neutral beam was sharply collimated, and the flux ratio of the halo to the core was lower than 10^{-5}.

2.2 Detector System

Figure 8 shows the cross-sectional view of the E391a detector. K_L^0s entered from the left side, and the detector components were cylindrically assembled along the beam axis. Most of them were installed inside the vacuum vessel to minimize interactions of the particles before detection. The electromagnetic calorimeter was made of undoped CsI crystals, labeled "CsI," and was used to measure the energies and positions of the two photons from the π^0 decay [22]. The main barrel (MB) and front barrel (FB) consisted of alternating layers of lead and scintillator plates and surrounded the decay region. To

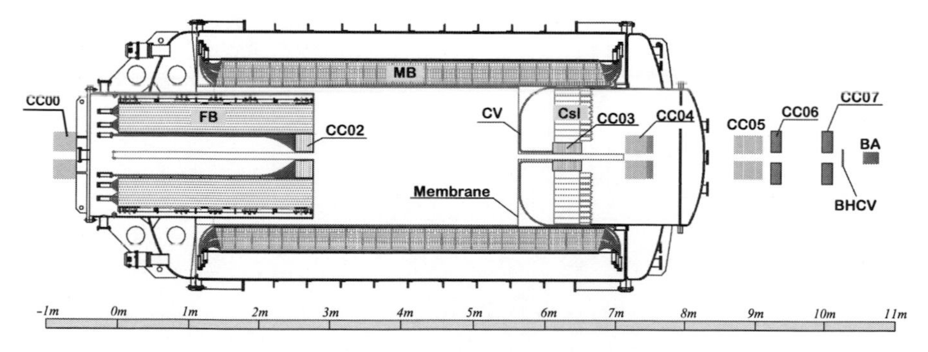

Fig. 8. Schematic cross-sectional view of the E391a detector; 0 m at the E391a coordinate corresponds to the entrance of the detector.

detect charged particles that entered the calorimeter, scintillator plates (CV) hermetically covered the region in front of the calorimeter. Multiple collar-shaped photon counters (CC00, CC02–07) were placed along the beam axis to detect particles escaping along the beam direction. In order to achieve a highly evacuated decay volume ($\sim 10^{-5}$Pa), we separated the decay region from the detector components by using a thin film called a *membrane.*

Calorimeter

Figure 9 shows an arrangement of the 496 undoped CsI crystals, whose size was $7 \times 7 \times 30$ cm^3 (normal CsI), in the electromagnetic calorimeter. A collar counter (CC03) surrounded the beam hole at the center of the calorimeter. Between the CC03 and normal CsI crystals, 24 CsI crystals of $5 \times 5 \times 50$ cm^3, which were used at the KTeV experiment (KTeV CsI), were located. At the periphery of the cylinder, there were 56 CsI modules with a trapezoidal shape and 24 modules of a sampling calorimeter made of alternating lead and scintillator plates (lead–scintillator sandwich).

Because the calorimeter was operated in vacuum, two kinds of special care were needed. One was to treat the heat generated at the high-voltage (HV) dividers for the PMTs and the other was HV failure due to the discharge between dynodes in the divider. We modified the HV divider so as to reduce heat dissipation and to reduce heat conduction from the divider to the PMT and the CsI crystals. The rear edge of the divider was connected to a copper pipe, in which temperature-controlled water was circulated. We measured the pressure dependence of the discharge voltage for several PMTs, whose results indicated that the vacuum pressure should be less than 1 Pa for the safe PMT operation.

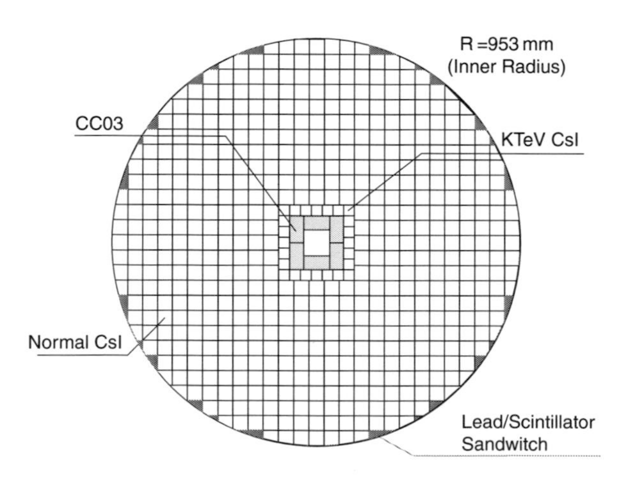

Fig. 9. Schematic view of the electromagnetic calorimeter.

Barrel Veto

There were two barrel veto counters, main barrel (MB) and front barrel (FB), made of alternating lead and scintillator plates. The MB surrounded the decay region to detect additional photons from K_L decays. The FB formed the upstream decay chamber in order to detect photons from K_L decay that occurred in front of the fiducial region.

We used wavelength-shifting (WLS) fibers for readout of the barrel counters in order to reduce the light attenuation in the long scintillator plates. We used a newly developed 2 in. PMT (Hamamatsu R329-EGP), which had a prism-shaped photocathode having 1.8 times larger sensitivity than that of the standard R329 in the range of emission light (500 nm of wavelength) of the WLS fiber [23].

- *Main barrel (MB)*: The MB was made of 32 trapezoidal modules, arranged around the beam axis without any holes or cracks to rays drawn from the beam axis outward. Each module was divided into inner (15 layers with 1-mm-thick lead plates) and outer parts (30 layers with 2-mm-thick lead plates) according to the thickness of the lead plates. The thickness of the scintillator plate was 5 mm and the total thickness of the module was 317.9 mm, which corresponds to 13.5 X_0, and its longitudinal length was 550 cm. Figure 10 shows the cross section of the MB module and the MB after assembling. A PMT was attached to each end of each module; the typical light output was 35 (10) photoelectrons per MeV energy deposit at the nearest (farthest) point from the PMT. The response of the MB down to a 1-MeV deposit was well reproduced by the Monte Carlo simulation as shown in Fig. 11.
- *Front barrel (FB)*: Figure 12 explains the role of the upstream decay chamber surrounded by the FB and CC02. When a $K_L \to \pi^0\pi^0$ decay occurs in front of the fiducial region, the calorimeter has an acceptance to two photons and the remaining two photons possibly move to the upstream direction. When the two photons detected at the calorimeter are not generated from the same π^0, the reconstructed vertex can be moved downstream and fake the signal. In order to suppress the background, we installed the upstream chamber to detect the photons escaping upstream.

Figure 13 shows a cross section of the FB module and the upstream decay chamber after assembling the 16 modules of the FB and the CC02. The thickness of scintillator and lead plates of the FB were 5 and 1.5 mm, respectively. The total thickness of the module was 413 mm (17.2 X_0). The FB signal was read out by only one end due to the geometrical constraint. At the other end, aluminized mylar sheets were attached for light reflection. The typical light yield was 20 (10) photoelectrons per MeV energy deposit in the scintillator at the nearest (farthest) point from the PMT.

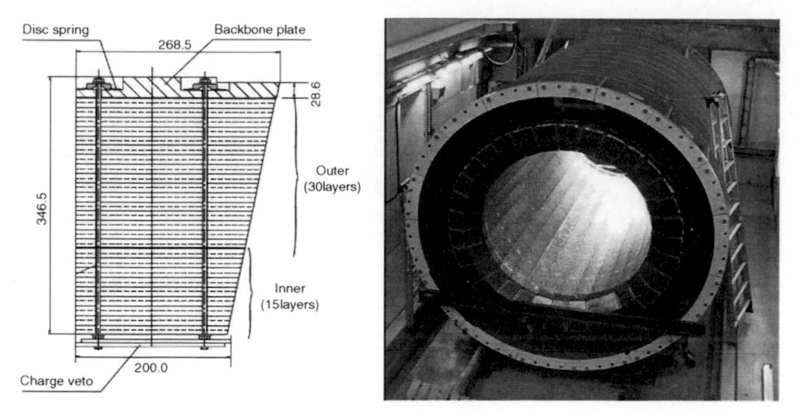

Fig. 10. *Left*: Cross section of the MB module. *Right*: Photograph of the MB after assembling.

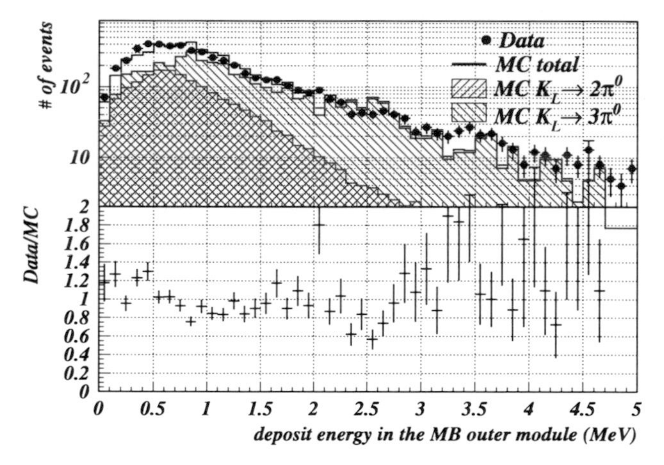

Fig. 11. *Top*: Distribution of the energy deposit in the main barrel for the data (*dots*) and Monte Carlo (MC) results to the $K_L \to \pi^0\pi^0$ decay (*blue hatched*) and the $K_L \to \pi^0\pi^0\pi^0$ decay (*red hatched*). *Bottom*: Ratio of the data to MC.

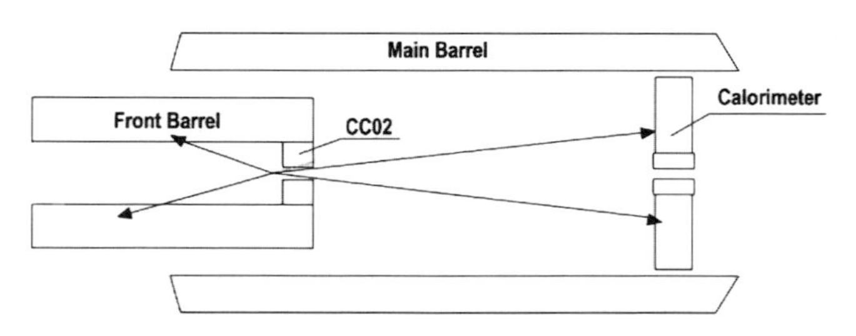

Fig. 12. Schematic view of the role of the upstream decay chamber.

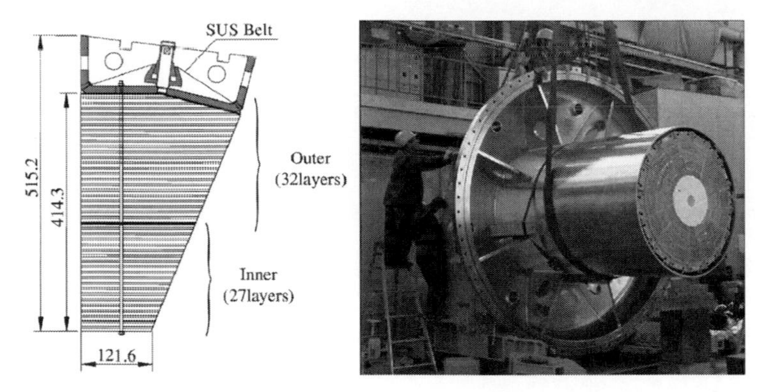

Fig. 13. *Left*: Cross section of the FB module. *Right*: Photograph of the upstream chamber after assembling.

Collar Counters

A series of counters surrounded the beam hole in order to detect photons escaping with a small angle from the beam direction. Four different types of detectors were prepared according to their position and function: CC02 (lead–scintillator shashlik sampling counter), CC03 (tungsten–scintillator sampling calorimeter), CC04 and CC05 (lead–scintillator sampling calorimeter), and CC06 and CC07 (lead glass crystal). All the collar counters satisfied the requirement that the light output be more than 10 photoelectrons per 1-MeV deposit, which was the typical energy threshold for vetoing.

- **CC02**: CC02 was located at the downstream edge and inside the cylinder formed by the FB. It consisted of eight modules that were *shashlik* sampling calorimeters and formed an octagonal shape. Each module consisted of alternating lead and scintillator plates, and the scintillating light was extracted up to the upstream edge of the FB by using optical fibers.
- **CC03**: CC03 was located at the center of the calorimeter and made of alternating tungsten and scintillator plates. The main role of CC03 was to prevent the beam particles from entering the calorimeter and producing a fake photon signal. Also, it rejected a fusion event in which two photons entered the calorimeter nearby and misidentified as one electromagnetic shower. The probability for there to be a fusion event increases when the K_L decays close to the calorimeter. In particular, an event in which a photon passed through the innermost crystal and combined with the shower of another photon was a harmful background source. In order to prevent the event, we used the tungsten–scintillator option to get enough radiation lengths in the limited space.
- **CC04 and CC05**: CC04 and CC05 were lead–scintillator sampling calorimeters positioned behind the calorimeter. CC04 was located inside the vacuum vessel, while the CC05 was outside. These counters detected photons escaping through a hole of the CC03.

- **CC06 and CC07**: The role of CC06 and CC07 was essentially identical to that of CC04 and CC05. CC06 and CC07 consisted of 10 blocks of the lead glass, sized $15 \times 15 \times 30\,\mathrm{cm}^3$, which were used in a previous KEK experiment. They were needed to form a 4π hermetic veto with optimization of the BA.

Back Anti and Beam Hole Charged Veto

The back anti (BA) detector was located at the downstream end of the neutral beam in order to detect photons going through the beam hole. Because there were many accidental hits due to neutrons in the beam, we suffered an acceptance loss because an event that has a sizable in-time energy deposit at the BA should be rejected regardless of whether the energy deposit was generated by a real photon or not. Thus, the BA was designed to separate the signal produced by photons and neutrons by using the alternating lead–scintillator sampling calorimeter modules and quartz layers (Fig. 14). The secondary particles produced by neutrons move slowly and do not emit Čerenkov light in the quartz, while the shower particles by photon emit the light. Thus, the photon was effectively rejected by requiring the energy deposit at both the calorimeter modules and the quartz layers, in the same time. In addition, we required a narrow time-window for vetoing with multi-hit TDC.

A PMT gain drift due to high counting rate was monitored by comparing the ADC spectrum of PMT outputs against the (same amount of) light from a Xe lamp between beam on and off. The difference was less than 10%, and it was applied to the calibration of the counter. Also, we took recorded data on the counting rate of the particles from the production target and used them to estimate the effect of accidental hits.

The $K_L \to \pi^+\pi^-\pi^0$ decay would effectively be a background process when π^+ and π^- are undetected. Because the selection criteria of the BA for photon detection were not tight enough to reject the charged particles, we placed another counter in front of BA, the beam hole charged veto (BHCV), which

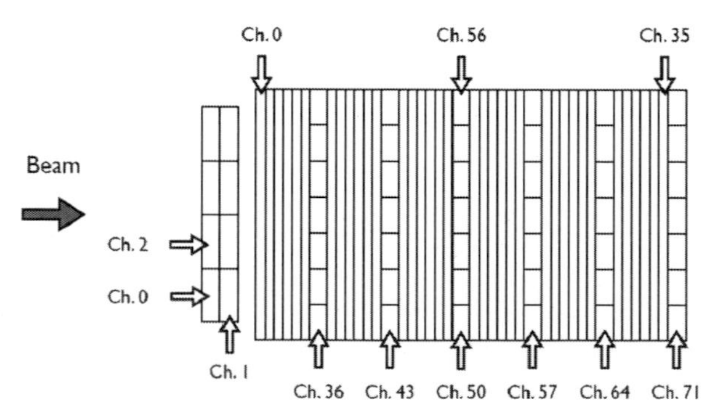

Fig. 14. The beam hole charged veto (BHCV) and back anti (BA). The BA was made of alternating lead–scintillator plates (Ch. 0–35) and quartz layers (Ch. 36–77).

consisted of eight scintillator plates for the charged particles. The effect of accidental hits on the counter was estimated with the same method as BA.

2.3 Data Taking and Analysis

The E391a experiment started taking data in February 2004. In the first period (called Run I), whose partial analysis was reported in [24], the membrane drooped into the neutral beam near the calorimeter and caused many neutron-induced backgrounds. After fixing this problem, we took data twice during February–April (Run II) and October–December (Run III) 2005. Recently, we completed data analysis for the Run II [25]. The Run III data are being analyzed. In this note, we will concentrate on the results from the Run II analysis.

The data analysis has five steps:

- **Clustering**: When a photon hits the calorimeter, it generates an electromagnetic shower and its energy is spread over multiple CsI crystals. Thus, the incident energy is deduced by collecting energies in the continuous group of the CsI crystals, called a cluster. To improve the accuracy of the energy measurement, clusters located at the boundary of the calorimeter were not used as the photon candidates. Also, the clusters should have transverse shower shape consistent with a shower generated by a single photon in order to remove neutron-induced clusters and fusion clusters.
- **Selection of two photon events**: We selected events that had exactly two photons in the calorimeter and no additional in-time hits in all the counters. The typical energy threshold for the veto was 1 MeV for veto counters except the charged veto for which it was 0.3 MeV. In order to reconstruct the π^0 properly, the photon clusters were required to have more than 150 and 250 MeV for lower and higher energy photons, respectively.
- **π^0 reconstruction**: By assuming that two photons came from a π^0 decay on the beam axis, we calculated the decay vertex along the beam axis (Z) and the transverse momentum of π^0 (P_T).
- **Kinematic requirements**: The reconstructed π^0 should be kinetically consistent with a π^0 decay within the proper K_L^0 momentum range. Also, each photon should satisfy the condition of the π^0 decay.
- **Signal box**. We defined the region for the candidate events (signal box) in the P_T vs. Z plot to be $0.12 < P_T < 0.24$ GeV/c and $340 < Z < 500$ cm as shown in Fig. 19. In the analysis of the Run II data, we masked the signal box so that all the selection criteria (cuts) were determined without examining the candidate events.

2.4 Backgrounds

There were two types of background events. One was the event from K_L decays and the other was the event due to *halo neutrons*.

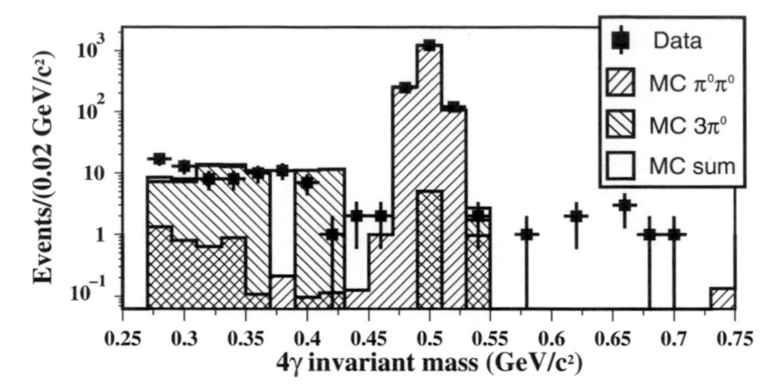

Fig. 15. Reconstructed invariant mass distribution of the events with four photons in the calorimeter. The points show the data and the histograms indicate the contribution of $K_L \to \pi^0\pi^0$ and $K_L \to \pi^0\pi^0\pi^0$ decays (and their sum) estimated from the simulation, normalized the number of events in the $K_L \to \pi^0\pi^0$ peak.

K_L Decay Backgrounds

We estimated the background level due to K_L decays by using GEANT3-based Monte Carlo (MC) simulations. The inputs of the MC simulations were finely adjusted to properly reproduce the obtained data. The distribution of the four-photon invariance mass, shown in Fig. 15, has a special interest because it includes all contributions of the veto counters. The contamination of the $K_L \to \pi^0\pi^0\pi^0$ decays in the four-photon events is due to the escape of two photons among the six photons. Our MC simulation reproduced the data well, which means the MC estimated the photon detection inefficiency correctly and that the MC could estimate the background level due to the K_L decays.

We generated $K_L \to \pi^0\pi^0$ decays in MC with 11 times larger statistics than the data. After imposing all the cuts, the background level was estimated to be 0.11 events. For charged decay modes ($K_L \to \pi^+\pi^-\pi^0$, $K_L^0 \to \pi l\nu(l = e, \mu)$), we studied the rejection power of the kinematic cuts to these backgrounds. Multiplying the expected inefficiency of charged particle counters with their rejection factor, their contribution was estimated to be negligible.

Neutron Backgrounds

There are three different categories of neutron backgrounds: *CC02 BG*, *CV-*π^0 *BG*, and *CV-*η *BG* as shown in Fig. 16. The CC02 BG appeared when a halo neutron hits the CC02 and produced a π^0. The CV-π^0 BG and CV-η BG occurred when a halo neutron entered the CV and produced a π^0 or η. Ideally, their decay vertex, Z, should be reconstructed properly as the position of the detector producing the π^0, which is outside the signal box. However, the events can be backgrounds as a result of wrong reconstruction due to

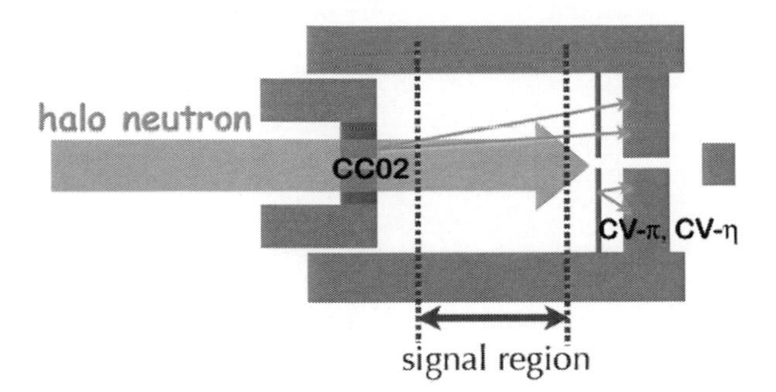

Fig. 16. Schematic view of the sources of the neutron backgrounds. Even though the generated positions of the π^0 and η were far from the signal region, the events could fall into the signal region due to misreconstruction.

mismeasurement of photon energy or when the detected two photons were not from the same π^0 decay.

- **CC02 Background**: Background from CC02 was mainly caused by energy mismeasurement due to shower leakage or photonuclear interactions in the calorimeter. When the energy of photon was mismeasured to be lower, the reconstructed vertex moved downstream (closer to the calorimeter) and entered the fiducial region. In order to estimate the CC02 background level, we have to understand a tail in the reconstructed vertex distribution, which was deduced from the data obtained in a special run using an Al plate in the beam (*Al plate run*). With a 0.5-cm-thick aluminum plate inserted into the beam at 6.5 cm downstream of the rear end of CC02, we accumulated π^0 events and obtained a distribution of reconstructed vertex from the fixed generated position. Then, it was convoluted with the Z distribution of π^0's production points within CC02 so as to match the peak position with that observed in the physics run, as shown in Fig. 17. The distribution was normalized to the number of events in $Z < 300$ cm. We estimated the number of CC02 BG events inside the signal box to be 0.16.
- **CV-π^0 Background**: Events in this category can be shifted upstream when either cluster in the calorimeter was overlapped by other associated particles and its apparent energy was enhanced or when the clusters were in fact not due to photons from the same π^0 in case of multi-π^0 generation. In order to evaluate the background level inside the signal box, we performed a bifurcation study with data [26, 27]. In the simulations studied earlier, the cuts against extra particles along with the shower-shaped cut were found to be efficient for background reduction; these cuts were chosen as two uncorrelated cut sets in the bifurcation study. The rejection power of one cut set was evaluated by inverting another cut set, and vice versa. Multiplying the obtained rejection factors, the number of CV background events inside the signal box was estimated to be 0.08.

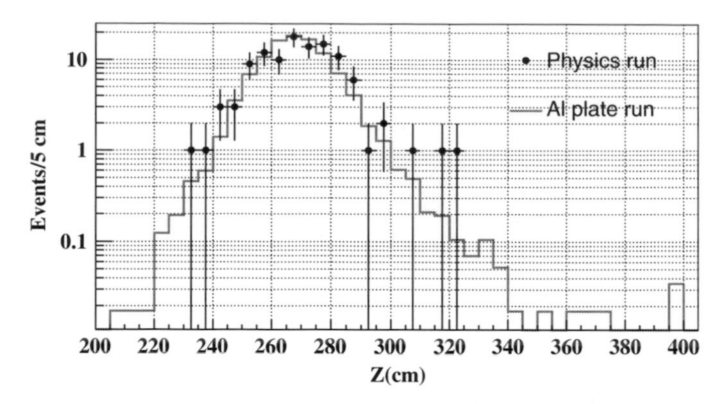

Fig. 17. Reconstructed Z vertex distribution of π^0s produced within CC02. The points show the data in the physics run in the upstream region ($Z < 340$ cm) and the histogram indicates the distribution from the Al plate run.

- **CV-η Background**: Even though we measured energies of the photons correctly, the reconstructed vertex shifted upstream when the two photons were from $\eta \to \gamma\gamma$ decay. Because the vertex position was calculated by assuming a π^0-mass, ηs were reconstructed about four times further away from the calorimeter and could fall into the signal box. To simulate the η production, we used a GEANT4-based simulation with the binary cascade hadron interaction model [28].[1] Figure 18 demonstrates the simulation which reproduced the invariant mass distribution of the events with two photons in the calorimeter from the *Al plate run*. We then simulated the η production at CV and estimated the number of CV-η background events inside the signal box to be 0.06.

Table 1 summarizes the estimated numbers of background events inside the signal box. We also examined the numbers of events observed in several regions around the signal box, and they were statistically consistent with the estimates.

2.5 Results and Prospects

The number of collected K_L^0 decays was estimated by using the $K_L \to \pi^0\pi^0$ decay, based on 1,495 reconstructed events, and was cross-checked by measuring $K_L \to \pi^0\pi^0\pi^0$ and $K_L \to \gamma\gamma$ decays. The 5% discrepancy observed between these modes was taken to be an additional systematic uncertainty. The single event sensitivity for $Br(K_L \to \pi^0\nu\bar{\nu})$ is given by

[1] The detector simulation in E391a was based on GEANT 3.21, which did not provide a hadronic interaction package for the η production. In order to simulate the η production in the detector, we used the GEANT 4.8.3 with the QBBC physics list. The η events obtained by GEANT4 were transferred to the E391a Monte Carlo code.

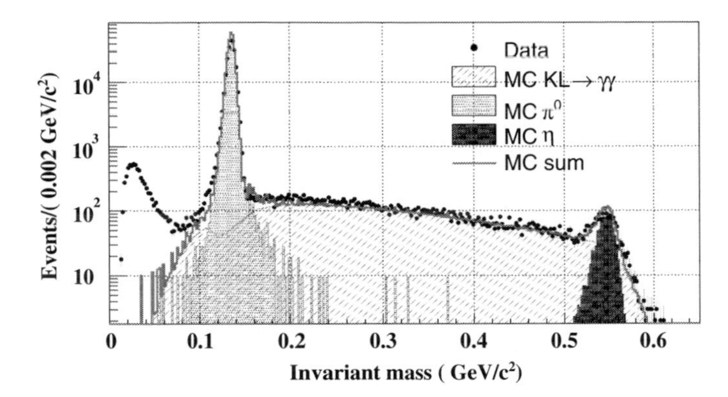

Fig. 18. Reconstructed invariant mass distribution of the two photon events in the Al plate run. Points with error bars show the data. Histograms indicate the contributions from π^0 and η produced in the Al plate, $K_L \to \gamma\gamma$ decays, and their sum, respectively, from the simulation. Events in the low-mass region were considered to be due to neutron interactions accompanying neither π^0s nor ηs, which were not recorded in the simulation.

Table 1. Estimated numbers of background events (BG) inside the signal box.

Background source	Estimated number of BG
$K_L \to \pi^0\pi^0$	0.11 ± 0.09
CC02	0.16 ± 0.05
CV-π^0	0.08 ± 0.04
CV-η	0.06 ± 0.02
Total	0.41 ± 0.11

$$S.E.S.(K_L \to \pi^0 \nu \bar{\nu}) = \frac{1}{\text{Acceptance} \cdot N(K_L^0 \text{decays})},$$

where the acceptance includes the geometrical acceptance, the analysis efficiency, and the acceptance loss due to accidental hits. By using the total acceptance of 0.67% and the number of K_L^0 decays of 5.1×10^9, the single event sensitivity was $(2.9 \pm 0.3) \times 10^{-8}$, where the error includes both statistical and systematic uncertainties.

After determining all the selection criteria and estimating the background levels, we examined the events in the signal box; we found no candidate events, as shown in Fig. 19. We set an upper limit as

$$Br(K_L \to \pi^0 \nu \bar{\nu}) < 6.7 \times 10^{-8} \text{ (90\% CL)},$$

based on the Poisson statistics.

After Run II data analysis, we proceeded to the Run III data analysis. The goal of the Run III analysis is to obtain a more concrete understanding of the backgrounds and the role of offline cuts developed in the Run II analysis.

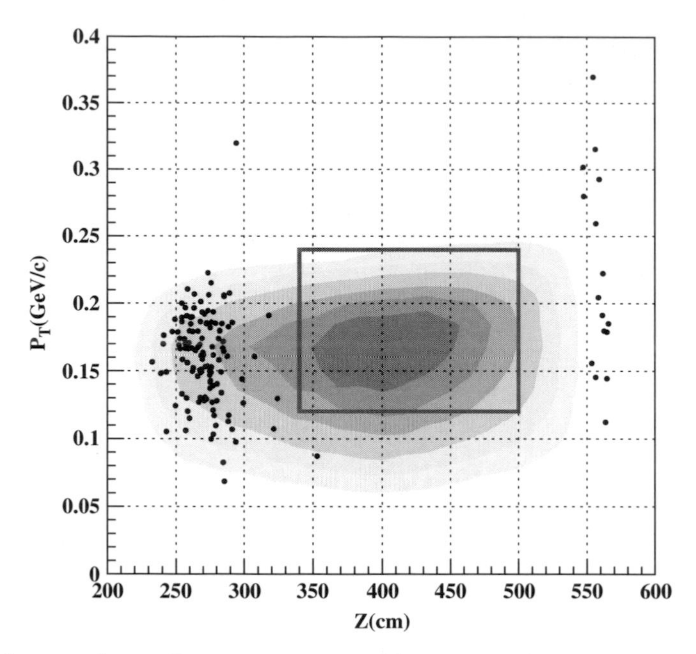

Fig. 19. Scatter plot of P_T vs. reconstructed Z position after imposing all the cuts. The *box* in the figure indicates the signal region.

The Run II results encouraged us that the experimental method with a *pencil beam* and *hermetic veto system* is a promising means for measuring $Br(K_L \to \pi^0 \nu\bar{\nu})$ precisely. We will continue our effort to measure this important physics process, the $K_L \to \pi^0 \nu\bar{\nu}$ decay at the J-PARC.

3 J-PARC E14

The E14 experiment aims at a precise measurement of $Br(K_L \to \pi^0 \nu\bar{\nu})$ [29]. The measurement will be performed as a two-step process: the first observation by using the upgraded E391a detector at J-PARC as Step 1 and the branching ratio measurement with a precision better than 10% as Step 2.

Because the E391a result indicates that the halo neutron interaction is a crucial background source, E14 (Step 1) is designed to suppress the background to less than that of the K_L decays. The halo neutron background is due to mismeasurement of energy and/or wrong pairing; an improvement of the photon energy measurement with an upgraded calorimeter is required, in addition to further suppression of flux of halo neutrons.

In order to measure the highly suppressed decay, we need a lot of K_Ls which will be provided by the newly constructed high-intensity accelerator at J-PARC. Accordingly, the readout system needs to be upgraded to fit such a high-intensity environment.

3.1 Beamline

Figure 20 shows a layout of the experimental area (hadron hall) at J-PARC and the K_L beamline for Step 1. The primary 30-GeV protons are extracted to the hadron hall during a 0.7-s spill with 3.3-s repetition and hit a secondary production target (T1) made of six Ni discs. Three secondary beamlines (K1.8 or K1.8BR, K1.1 or K1.1BR, and KL) will share the secondaries produced at the T1, simultaneously. The extraction angle of the K_L beam was determined to be 16° due to geometrical constraints for radiation shielding in the hadron hall, and it could not be optimized for E14. With the large extraction angle, the K_L momentum distribution becomes soft compared to that of E391a even though the primary proton has higher energy. As shown in Fig. 21, the K_L momentum has a peak at 1.3 GeV/c and the average momentum is 2.1 GeV/c. The large extraction angle has an advantage that high-momentum neutrons are significantly suppressed as shown in Fig. 21, because only high-energy neutrons are able to make background events.

The KL beamline is being designed and will be constructed by the fall of 2009. Its design principle is similar to that of the K0 beamline; however, the KL beamline has two different conditions. One is a finite target image. Because the T1 target has a sizable length (65 mm) along the primary beamline, it is seen as a 19-mm-long rod from the KL beamline. We expect a beam having $\sigma = 1$ mm in size at the T1, and the generation points looking at the KL beamline would be very asymmetric, having a large size in the horizontal (19 mm) and small size in the vertical (2 mm) directions. The other condition is that there are a lot of materials between the target and the collimators. As

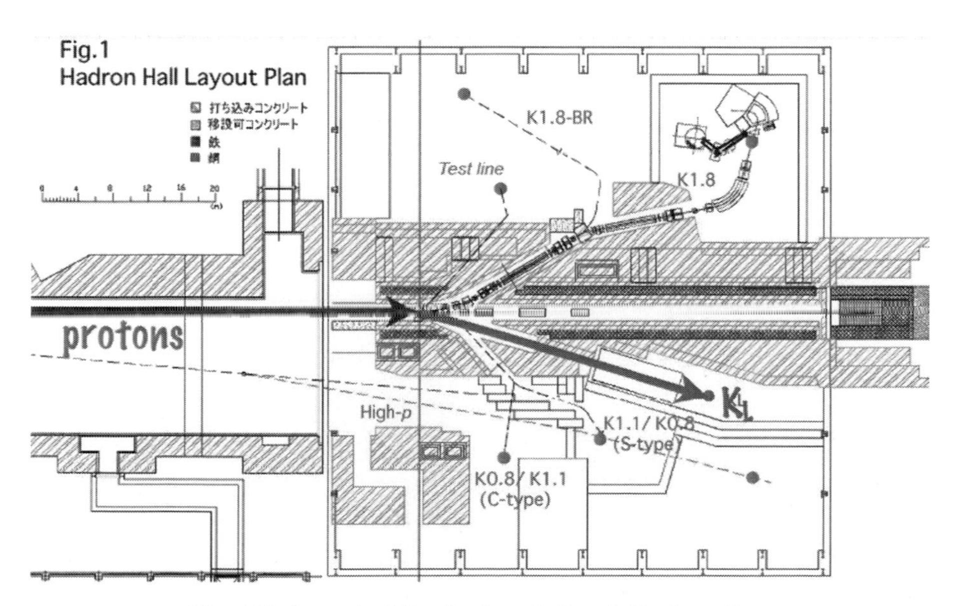

Fig. 20. Layout of the hadron hall and K_L beamline.

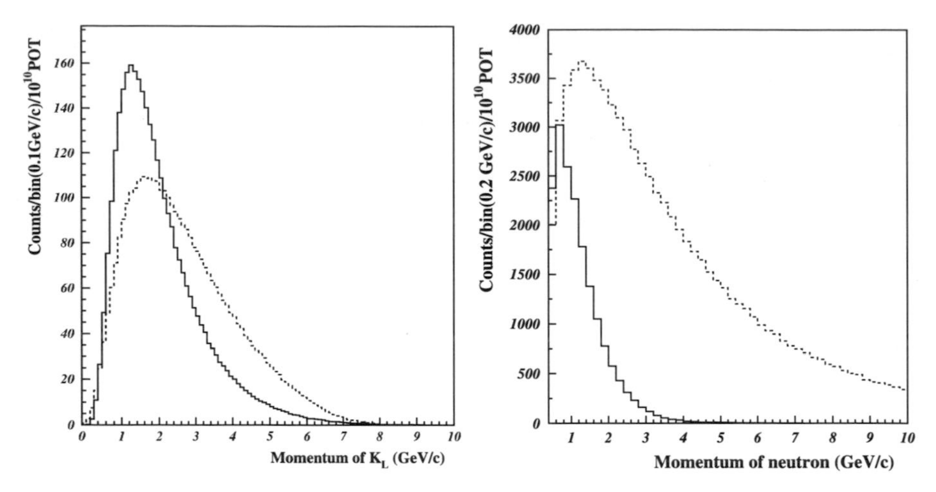

Fig. 21. K_L momentum distribution (*left*) and neutron momentum distribution (*Right*) in E391a (*dashed line*) and E14 (*solid line*).

shown in Fig. 22, there are many magnets for the K1.1 line. Thus, the collimator for the KL beamline should start far from the target (6.5 m downstream), and all materials between the T1 and the first collimator will be a scattering source to produce halo neutrons. One of the main subjects of beamline design is to reduce the effect.

Figure 23 shows a drawing for the design of the KL beamline. It consists of two collimators made of iron and a bending magnet to weep out charged particles. A 7-cm-long lead block reduces the photon flux and is located upstream in order to avoid the block from becoming a scattering source of halo neutrons. As mentioned above, the effective target image is significantly different in the

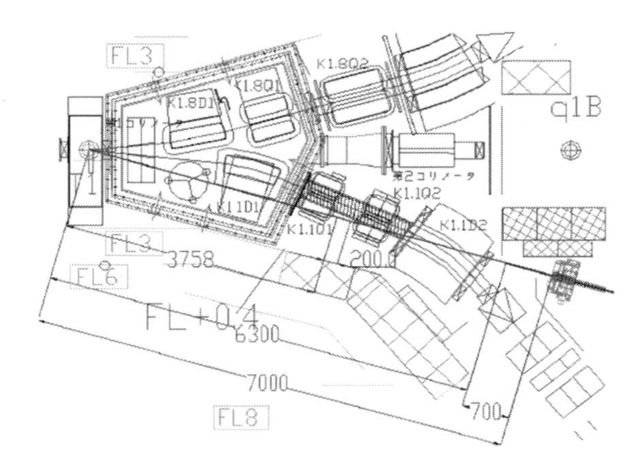

Fig. 22. Configuration of magnets for the K1.1 beamline. A straightline crossing the magnets represents the K_L beamline.

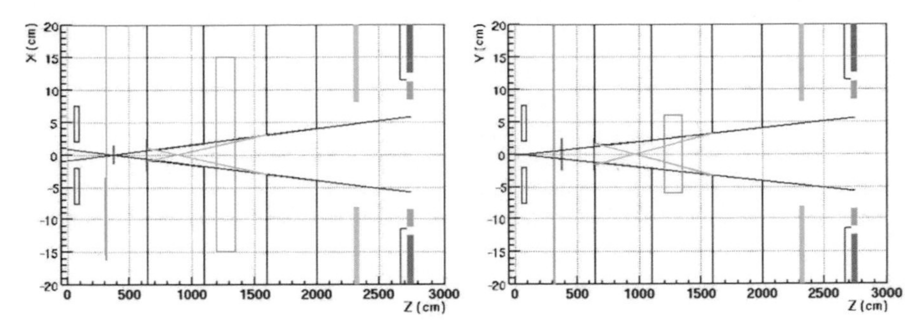

Fig. 23. A layout of the E14 beamline components for the horizontal direction (*left*) and the vertical direction (*right*).

horizontal and vertical directions, which suggested us to use a rectangular beam instead of a circular one. By adjusting collimation lines in the horizontal and vertical directions independently, we can suppress the halo neutrons. In order to remove the halo neutrons generated from the magnets for the K1.1 effectively, we adjust the later part of the downstream collimator.

3.2 Detector Upgrades

Calorimeter

The E14 will upgrade the electromagnetic calorimeter with the undoped CsI crystals, called "KTeV CsI" hereafter, used in the Fermilab KTeV experiment. By comparing with the crystals used in E391a, the KTeV CsI crystals are smaller in the cross section and longer in the beam direction (50 cm, 27 X_0). Figure 24 shows the layout of the new calorimeter, which consists of 2576 crystals. These crystals are of two sizes: $2.5 \times 2.5 \times 50$ cm^3 for the central region (2,240 blocks) and $5.0 \times 5.0 \times 50$ cm^3 for the outer region (336 blocks) of the calorimeter.

The reasons for replacing the crystals are as follows:

- *To reduce the probability of missing photons due to fusion clusters*: If two photons hit the calorimeter close to each other, the generated showers will overlap and be misidentified as a single photon. By using the KTeV CsI crystals, two photons as close as ~5 cm can be identified as separate clusters, compared to ~15 cm in E391a. Each cluster is required to have a transverse energy distribution consistent with an electromagnetic shower, which is easier to achieve with KTeV CsI crystals.
- *Eliminate the photon detection inefficiency due to punch-through*: In the case of 30-cm-long (16 X_0) crystals used in E391a, the probability that a photon passes through the crystal without interaction is comparable to the inefficiency due to photonuclear interactions. However, with the 50-cm-long (27 X_0) crystals, the probability is reduced to a negligible level.

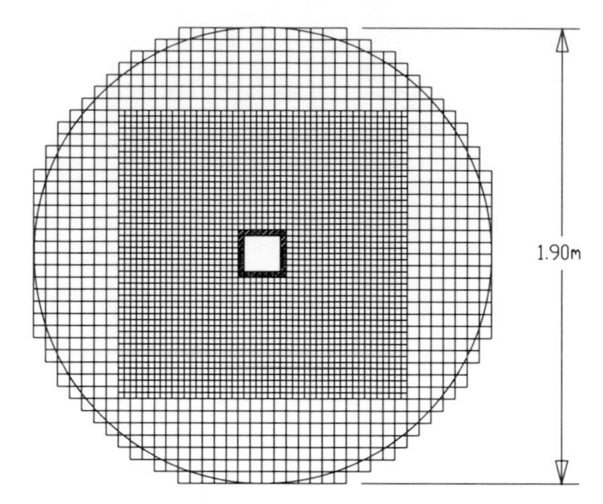

Fig. 24. Layout of the calorimeter for the E14 experiment with the KTeV CsI crystals. The $2.5 \times 2.5 \times 50$ cm^3 crystals are used for the inner region and $5.0 \times 5.0 \times 50$ cm^3 crystals are used for the outer region.

- *Reduce background due to shower leakage*: When a photon hits a calorimeter, some part of the shower leaks from the rear end, depending on the depth where the first e^+e^- pair in the shower is created. Because the decay vertex is reconstructed by assuming a π^0-mass, the energy leakage shifts the decay vertex downstream to be a background as explained in the previous section. As shown in Fig. 25 (left), the longer KTeV CsI crystals will

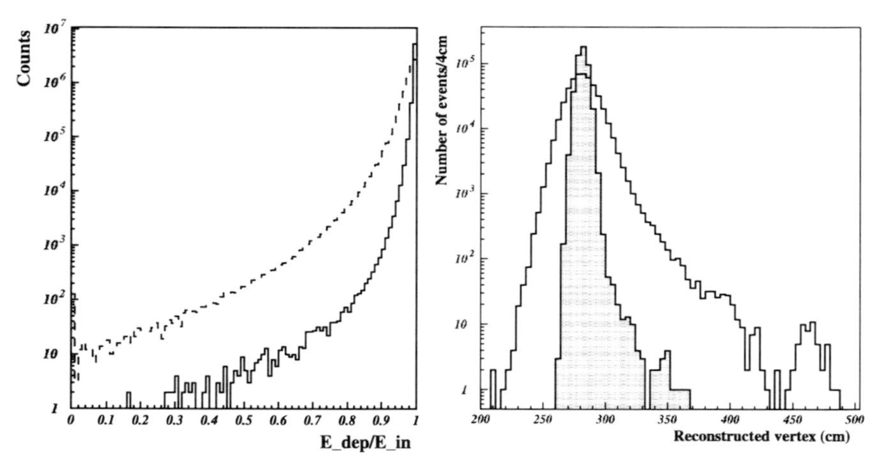

Fig. 25. Monte Carlo study of the shower leakage. *Left*: the ratio between the deposited energy in the CsI crystals and the true incident energy for the photons in the energies between 0.1 and 1 GeV. The *dotted line* is for E391a crystals (16 X_0) and the *solid line* is for KTeV crystals in Step 1 (27 X_0). *Right*: reconstructed vertex distribution for π^0 events produced by an aluminum plate located downstream of CC02 (true position = 280.5 cm in the plot) only during calibration runs for E391a (*open histogram*) and for Step 1 (*hatched histogram*).

have significantly small shower leakage, and thus suppress the background by reducing the tail of the vertex distribution (right).

Beam Hole Photon Veto

In E14, we have to use a new type of detector for the BA.[2] Because it is exposed to a high flux of neutrons, K_Ls, and beam photons, there would be a high probability of generating accidental hits, which inhibit a proper function of the detector. Thus, the detector should be insensitive to those unwanted particles, while keeping a high sensitivity to the products from K_L decays in the fiducial volume.

The beam hole photon veto (BHPV) is designed on a new concept of photon detection. One feature utilizes the Čerenkov radiation to detect electrons and positrons produced by an electromagnetic shower. This enables us to be blind to heavy (and thus slow) particles, which are expected to be the main products in neutron interactions. The other feature is to use directional information. In the background events which should be removed, photons are generated at the fiducial region 6 m or more upstream and enter the BHPV through a narrow beam hole. Thus, their electromagnetic showers are spread out in the forward direction. On the other hand, secondary particles from neutron interactions tend to spread isotropically. Thus, neutron signals can be reduced further, without losing photon detection efficiency, by requiring shower development along the beam direction.

Based on this concept, BHPV is designed to be an array of Pb–aerogel counters. Figure 26 shows the schematic view of the BHPV module. Each module is composed of 2-mm-thick lead as a photon converter, a stack of aerogel tiles as a Čerenkov radiator, a light collection system with a mirror and a Winston cone, and a 5-in. photomultiplier tube. In order to identify a genuine signal when it is smeared by accidental hits, 500-MHz waveform digitizers are used in the readout. The aerogel array has a cross section of 30 cm $\times$ 30 cm, a thickness of 5 cm, and the refractive index of 1.03. These modules are lined up in 25 layers along the beam direction, each of which is placed 35 cm apart as shown in Fig. 27. The total length of the BHPV is 8.75 m which corresponds to 8.9 X_0. This configuration enables us to select events that have their shower development in the forward direction, and thus reduce the sensitivity to neutrons.

Main Barrel (MB)

We will improve the main barrel in two ways:

- Pipeline readout of the output pulse shape with waveform digitizers
- Increasing the thickness of the main barrel by attaching additional layers

[2] We call the BA as beam hole photon veto for E14.

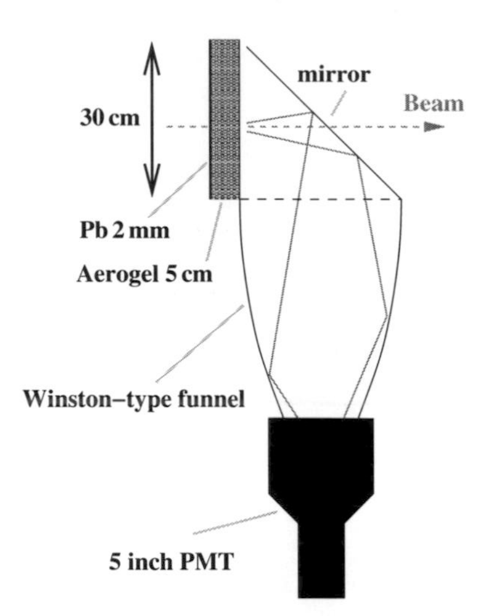

Fig. 26. Schematic view of the BHPV module.

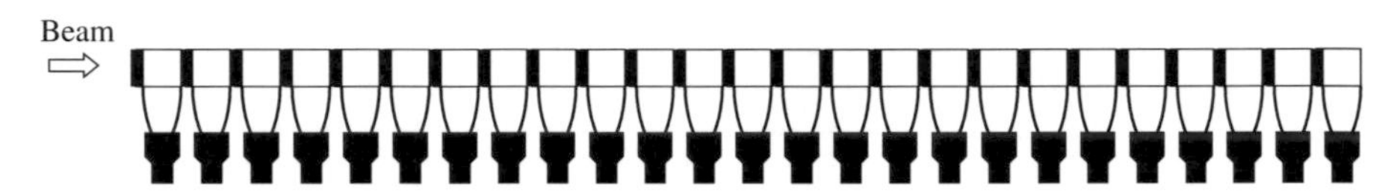

Fig. 27. Schematic side view of the BHPV arrangement.

Because the main barrel (MB) is a long detector, the timing information is important to prevent acceptance loss due to accidental hits. In E391a, we observed *back-splash* that arose from particles generated as a part of the shower at the calorimeter which enter back into the MB. This back-splash, which is a main source of acceptance loss due to the main barrel, is well identified when we use timing information. The correlation between the incident position and the time of the back-splash is opposite to that of real photons. However, the timing information obtained from TDC is frequently degraded in a high-rate environment due to pileup of signal, in particular for a large detector like the MB. It is important to get timing information for small signals without significant dead time caused by discrimination, which will be realized through a pipeline readout of the pulse shape with waveform digitizers.

To improve the efficiency for photon detection, more layers will need to be added because the thickness of the main barrel in E391a was only 13.5 X_0. We are studying two options to increase the thickness of MB. One is to add 5 X_0 behind the current main barrel, which will greatly reduce the photon detection inefficiency due to punch-through. Another option is to add a module having better visible coverage inside the current one in order to reduce the inefficiency due to sampling fluctuation. Because many low-energy photons can enter the main barrel, it is important to improve the detection efficiency against the low-energy photons.

Electronics and DAQ

E14 will use new readout system to perform the experiment under the high-rate conditions. A major change is to adapt the waveform digitization to distinguish pile-up signals from legitimate two-photon signals. To optimize the cost and performance, we will use 125-MHz flash analog-to-digital converter (FADC) with 8-pole Bessel filter [30] for most of the detectors including the (new) CsI calorimeter. By using the Bessel filter, we can convert the PMT signal of the detector into a Gaussian shape with a fixed width. The pile-up signals will be identified by fitting with pre-defined Gaussian function. In addition, the precise timing information obtained by the FADC will reduce losses arising from accidentals in the high counting rates of the detectors.

A multi-stage triggering scheme will be adopted to reduce the event rates so that the events of interests can be selected with high efficiency. A sum of energy deposit at the calorimeter and data from several veto counters will be used for the first-level trigger, whose rate is expected to be a few hundred kHz. A clustering scheme to count the number of photons entering the calorimeter will be made by a higher level trigger. With an online PC farm, a more complicated event reconstruction will be made and a few kHz of events will be recorded to the long-term storage for offline analysis. In order to achieve data acquisition without dead time, fully pipelined electronic modules will be used.

3.3 Expected Sensitivity

We expect that J-PARC will provide 2×10^{14} protons every 3.3 s and they will be slowly extracted during 0.7 s to the hadron hall. Assuming three Snowmass years (3×10^7 s) of data taking, we expect a total of 7.3×10^{13} K_Ls at the exit of the beamline.

For the K_Ls that exit the beamline at 20 m from the production target, 3.6% of them decay within the fiducial region ($300\,\text{cm} < Z < 500\,\text{cm}$ in the E14 coordinate system). The signal acceptance for the $K_L \to \pi^0 \nu \bar{\nu}$ decays is estimated to be 8% and the single event sensitivity is 4.7×10^{-12}. With a standard model prediction of $Br(K_L \to \pi^0 \nu \bar{\nu}) = 2.8 \times 10^{-11}$, we expect to observe 6.0 events.

When we reject background events by using veto counters with very low-energy threshold, we have to take into account the acceptance loss. Even though it surely depends on experimental conditions, we simply estimated the effects by scaling the E391a values. There are three major sources of acceptance loss:

- Accidental activities: Accidental activities in the detector originate from the high intensity of the beam and back-splash from the calorimeter. With a total accidental rate of 26 MHz and the 10-ns resolving time for veto system, we expect 26% loss in the acceptance.

- Shower shape cut: In order to remove the fusion events, we have to strictly require a proper distribution of electromagnetic shower for a given photon energy. From Monte Carlo calculations for the new calorimeter, we estimate the loss as 20% to be enough rejection of the fusion events.
- Collateral cluster: Sometimes a photon cluster in the calorimeter is associated with an isolated low-energy cluster nearby (*collateral cluster*). Because we have to reject events that have additional activities in the calorimeter beyond those for the reconstructed two photon clusters, the collateral cluster becomes a source of acceptance loss. We developed a highly efficient veto method by using activities in the calorimeter, which applies different veto thresholds according to the distance from the photon cluster. We expect the loss to be 10%.

As a result, we expect/assume a 56% acceptance loss; the SES is 1.0×10^{-11} and 2.7 standard model events are expected.

3.4 Expected Backgrounds

In E14, the neutron-related backgrounds are suppressed by a better K_L to neutron ratio (K_L/n) and a soft momentum distribution of neutrons. With the new calorimeter having finer segmentation and longer CsI crystals, we expect the background level related to the neutron interactions by extrapolating the E391a data, and it was estimated to be less than that from the K_L decay.

With increased experimental sensitivity and highly suppressed neutron backgrounds, we will be faced with backgrounds caused by the K_L decays. The $K_L \to \pi^0\pi^0$ ($K_{\pi 2}$) decay is considered as the main background source because of its relatively large branching ratio (8.83×10^{-3}) and its final state with four photons. Due to finite detection inefficiency, two photons are possibly undetected. As a result, the event will be a two-photon event and fake to signal. The $K_L \to \pi^+\pi^-\pi^0$ decay is another background source, because it is one of the main decay modes of K_L (12.54% of branching ratio) including single π^0. Also, the $K_L \to \pi^-e^+\nu_e$ decay has a potential to become a background source due to detection inefficiency for both π^+ and e^-. Contribution from other decay modes was estimated to be negligible.

Table 2 summarizes the estimated background levels. Brief explanations of how these levels are estimated for important processes are described in the subsequent sections.

$K_L \to \pi^0\pi^0$ ($K_{\pi 2}$) Decay

There are two kinds of $K_{\pi 2}$ backgrounds: *even pairing* and *odd pairing* according to which two photons are missed, and they have different characteristics in the event reconstruction. For the even-pairing events, two photons generated from a single π^0 decay enter the calorimeter, and the other two photons from another π^0 enter the veto system. As a result, the π^0s are correctly

Table 2. Estimated number of background events for E14 (Step 1). The single event sensitivity is 1.0×10^{-12}, with which 2.7 standard model events are expected including a 56% of acceptance loss.

Background source	Estimated number of BG
K_L decay backgrounds	
$K_L \to \pi^0\pi^0$	1.7 ± 0.1
$K_L \to \pi^+\pi^-\pi^0$	0.08 ± 0.04
$K_L \to \pi^- e^+ \nu_e$	0.02 ± 0.01
Neutron backgrounds	
CC02	0.01
CV-π^0	0.08
CV-η	0.3

reconstructed and have similar vertex and transverse momenta distribution to those of the signal. On the other hand, the odd-pairing events have two photons from different π^0 decays in the calorimeter.

Figures 28 and 29 show the difference between the two categories. The even-pairing events have at least one photon having relatively high energy (>50 MeV) entering veto counters, which provide a high efficiency for vetoing. The odd-pairing events tend to locate at low P_T and can be rejected by the high-P_T selection. In addition, the incident angles of photons deduced from the (wrong) vertex are incorrect and are located in an unphysical region for a given incident energy. Also, there are large imbalances between energies of the two photons in order to satisfy the high-P_T selection.

We have to take *fusion events* into account also. When three photons enter the calorimeter and two photons among them are very close, the two photons

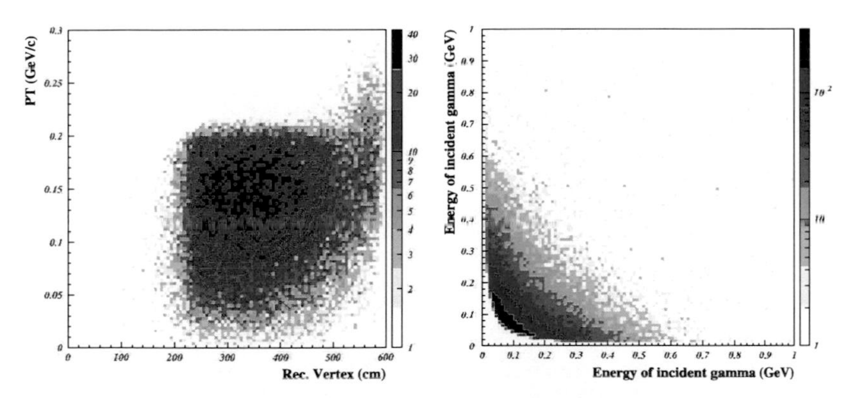

Fig. 28. *Left*: Reconstructed vertex and P_T distribution for even-pairing $K_L \to \pi^0\pi^0$ background. It has a distribution similar to that of the $K_L \to \pi^0\nu\bar{\nu}$ decay. *Right*: Energy distribution of gammas that enter the veto counters. At least one photon has sufficiently high energy to trigger the counter with a high detection efficiency.

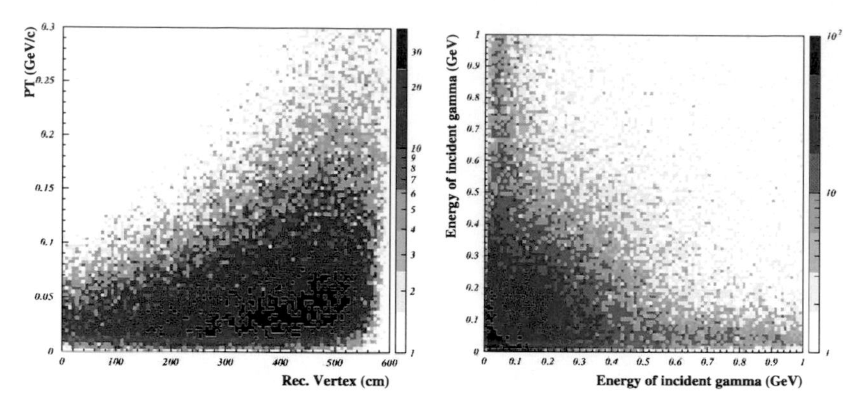

Fig. 29. *Left*: Reconstructed vertex and P_T distribution for odd-pairing $K_L \rightarrow \pi^0\pi^0$ background. The reconstructed vertex is not correct which makes the P_T lower than the signal box. *Right*: Energy distribution of photons that enter the veto counters. Even though many events have low energies for both the photons, the events are rejected through the high P_T selection. As a result, the photons needed to be rejected by the veto counters have distributions similar to those for even pairing.

are reconstructed as one electromagnetic shower. As a result, we miss one photon at the calorimeter and increase the background. In order to reject the background, we have studied the energy distribution among the CsI crystals that form a (correct single) photon cluster. A fused cluster can be identified if the distribution is far from the standard one (shower-shaped cut).

The $K_{\pi 2}$ background level is estimated by using Monte Carlo simulation. For each event, we assigned an event weight as a product of detection inefficiencies for all the photons that entered veto counters. The inefficiencies are a function of the incident energy and the angle of photon for a given detector. The inefficiency functions were obtained by combining Monte Carlo calculations (for electromagnetic processes) and obtained data (for photonuclear interaction). The background level is estimated from the sum of the weights for events that passed all the signal selection cuts.

$K_L \rightarrow \pi^+\pi^-\pi^0$ Decay

There are two effective tools to reject this background. One is the kinematic limit of this three-body decay. The transverse momentum of the π^0 (P_T) is relatively low, limited to 0.133 GeV/c, and thus the high P_T selection can greatly reduce this background. Second, the charged particle veto counters surrounding the decay region can detect additional charged pions and reject faked events.

We estimated the background level due to the $K_L \rightarrow \pi^+\pi^-\pi^0$ decay with the similar method of the $K_{\pi 2}$ decay. We used inefficiency functions of plastic scintillator for π^+ and π^- which are parametrized by obtained data from the experiments performed at the KEK and PSI [31, 32].

$K_L \to \pi^- e^+ \nu_e$ Decay

When π^- and e^+ are not detected at the CV due to annihilation processes, $\pi^- + p \to \pi^0 + n$ and $e^+ + e^- \to \gamma + \gamma$, the decay becomes a four-photon final state and has a potential to be a background. In order to estimate the background level due to the decay, we generated the annihilation processes according to the $K_L \to \pi^- e^+ \nu_e$ decay kinematics. The background level was estimated as a product of both annihilation processes and possibility to miss the two photons for the selected events.

4 Summary

A precise measurement of the branching ratio for the $K_L \to \pi^0\nu\bar{\nu}$ decay will provide the most crucial test of the standard model and various new physics scenarios beyond it. Owing to many theoretical efforts in the last decades, $Br(K_L \to \pi^0\nu\bar{\nu})$ becomes a physical quantity that can be calculated most accurately in the standard model.

In the experimental point of view, the measurement is quite challenging and we are taking a step-by-step approach to reach the final goal. The KEK-PS E391a experiment was a good starting point of this long journey, and J-PARC will enable us to complete it with an essential tool, high-intensity K_L beam. At J-PARC, we intend to test various models with improved experimental sensitivity down to the level of the standard model expectation (Step 1) and later to reach a final goal of performing a precise measurement of $Br(K_L \to$

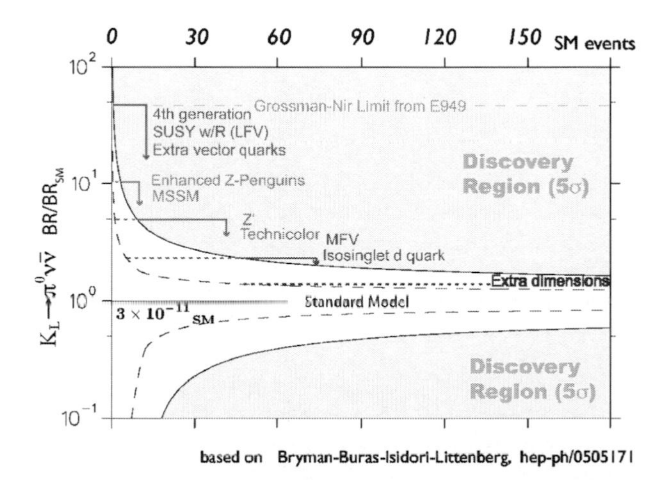

Fig. 30. A 5σ discovery region (*shaded area*) and 95% CL upper and lower exclusion limits vs. the number of $K_L^0 \to \pi^0\nu\bar{\nu}$ events assuming the SM [33]. The maximal enhancement of the branching ratio (BR) expected in various non-SM scenarios are also included.

$\pi^0\nu\bar{\nu}$) (Step 2). With improved sensitivity, we can discover or discriminate against several new physics scenarios as shown in Fig. 30.

References

1. The KTeV experiment http://kpasa.fnal.gov:8080/public/ktev.html
2. The NA48 experiment http://na48.web.cern.ch/NA48/Welcome.html
3. The Belle experiment http://belle.kek.jp/
4. The Babar experiment http://www-public.slac.stanford.edu/babar/
5. Kobayashi, M., Maskawa, K.: Prog. Theor. Phys. **49**, 652 (1973)
6. Littenberg, L.S.: Phys. Rev. **D39**, 3322 (1989)
7. Buras, A.J., Schwab, F., Uhlig, S.: *hep-ph/0405132*, and references therein
8. Isidori, G.: hep-ph/0307014, eConf **C0304052**, WG304 (2003), and references therein
9. Marciano, W.J., Parsa, Z.: Phys. Rev. **D53**, R1 (1996)
10. Buchalla, G., Buras, A.J.: Nucl. Phys. **B548**, 309 (1999)
11. Buchalla, G., Buras, A.J.: Nucl. Phys. **B412**, 106 (1994)
12. Buchalla, G., Isidori, G.: Phys. Lett. **B440**, 170 (1998)
13. Isidori, G., Mescia, F., Smith, C.: Nucl. Phys. **B718**, 319 (2005)
14. Isidori, G., Martinelli, G., Turchetti, P.: Phys. Lett. **B633**, 75 (2006)
15. Buras, A.J., Gorbahn, M., Haisch, U., Nierste, U.: hep-ph/0603079; Phys. Rev. Lett. **95**, 261805 (2005)
16. Inami, T., Lim, C.S.: Progr. Theor. Phys. **65**, 297 (1981)
17. Inami, T., Lim, C.S.: Progr. Theor. Phys. **65**, 1172 (1981) (Erratum)
18. Tarantino, C.: hep-ph/0706.3436v3 and references therein
19. Mescia, F.: CKM2006, Nagoya, Japan, 12–16 Dec 2009
20. D'Ambrosio, G., Giudice, G.F., Isidori, G., Strumia, A.: Nucl. Phys. **B645**, 155 (2002)
21. Watanabe, H., et al.: Nucl. Instr. Meth. **A545**, 542–553 (2005)
22. Doroshenko, M., et al.: Nucl. Instr. Meth. **A545**, 278–295 (2005)
23. Itaya, M., et al.: Nucl. Instr. Meth. **A522**, 477–486 (2004)
24. Ahn, J. K., et al.: Phys. Rev. D **74**, 051105(R) (2006)
25. Ahn, J. K., et al.: Phys. Rev. Lett **100**, 201802 (2008)
26. Adler, S., et al.: Phys. Rev. Lett. **79**, 2204 (1997)
27. Nix, J., et al.: Phys. Rev. D **76**, 011101 (2007)
28. Agostinelli, S., et al.: Nucl. Instr. Meth. Phys. Res. A **506** 250 (2003); "Geant4 8.3 Release Notes", CERN
29. P14: Proposal for $K_L \to \pi^0\nu\bar{\nu}$ Experiment at J-PARC, available from http://j-parc.jp/NuclPart/Proposal_e.html
30. Anderson, K., et al.: Nucl. Instr. Meth. **A551**, 469–471 (2005)
31. Inagaki, T., et al.: Nucl. Instr. Meth. **A359**, 478–484 (1995)
32. KOPIO Conceptual Design Report (2005)
33. Bryman, D., Buras, A.J., Isidori, G., Littenberg, L.S.: hep-ph/0505171, Int. J. Mod. Phys. **A21**, 487 (2006), and references therein

Search for T Violation in $K^+ \rightarrow \pi^0\mu^+\nu$ Decays

J. Imazato

Institute of Particle and Nuclear Physics, High Energy Accelerator Research Organization (KEK), Oho 1-1, Tsukuba, Ibaraki, 305-0801 Japan
jun.imazato@kek.jp

1 Introduction

Time reversal (T) symmetry has long been a subject of interest from pre-modern physics times, since it implies the reversibility of motion – for instance, an identical trajectory of an object when time runs back in classical mechanics. In modern quantum field theories it has received renewed attention as a discrete symmetry of space/time along with charge conjugation (C) and parity reflection (P) [1]. Although C and P are each maximally violated in weak interactions, T (and CP) is almost exact symmetry in all the interactions including the weak interaction. The violation of T would have a great impact [2] since it would mean that the physics laws in the time-reversed world are different from ours. T violation has also an important meaning in particle physics since it is equivalent to CP violation according to the CPT theorem. We can study the sources of CP violation which are necessary to explain the baryon asymmetry in the universe.

The transverse muon polarization (P_T) in $K \rightarrow \pi\mu\nu$ ($K_{\mu3}$) decays with T-odd correlation was suggested by Sakurai [3] about 50 years ago to be a clear signature of T violation. Unlike other T-odd channels in, e.g., nuclear beta decays, P_T in $K_{\mu3}$ has the advantage that the final state interactions (FSI), which may mimic T violation by inducing a spurious T-odd effect, are very small [4, 5]. This argument applies most particularly to $K^+_{\mu3}$ decay with only one charged particle in the final state where the FSI contribution is only from higher order loop levels and is calculable. Thus, it is not surprising that over the last two decades, dedicated experiments have been carried out in search of non-zero P_T in $K_{\mu3}$ decays [6]. An important feature of a P_T study is the fact that the contribution to P_T from the standard model (SM) is nearly zero ($\sim 10^{-7}$). Therefore, in a P_T search we are investigating new physics beyond the SM.

The most recent research of P_T has been performed at KEK as the E246 experiment. This experiment was carried out by an international collaboration whose core members continue the current J-PARC experiment. The E246

Imazato, J.: *Search for T Violation in* $K^+ \rightarrow \pi^0\mu^+\nu$ *Decays*.
Lect. Notes Phys. **781**, 75–104 (2009)
DOI 10.1007/978-3-642-00961-7_4

result was consistent with no T violation but provided the world best limit of $P_T = -0.0017 \pm 0.0023\ (stat) \pm 0.0017\ (syst)$ [7] and constrained the parameter spaces of several contender models. It was, however, statistics limited, mainly due to insufficient accelerator beam intensity in spite of smaller systematic errors.

Now we intend to continue the P_T experiment further at J-PARC where higher accelerator beam intensity will be available and a higher experimental sensitivity is promised, in order to search for new physics beyond SM. We aim for a sensitivity of $\delta P_T \sim 10^{-4}$ [8].

2 $K^+ \to \pi^0 \mu^+ \nu$ Decay and Muon Transverse Polarization

2.1 $K^+ \to \pi^0 \mu^+ \nu$ Decay

The decay matrix element of the $K_{\mu 3}$ decay based on the V-A theory can be written as [9–12]

$$M = \frac{G_F}{2} \sin\theta_c \left[f_+ \left(q^2\right) \left(p_K^\lambda + p_\pi^\lambda\right) + f_- \left(q^2\right) \left(p_K^\lambda - p_\pi^\lambda\right) \right] \cdot \left[\overline{u}_\nu \gamma_\lambda \left(1 - \gamma_5\right) v_\mu \right], \tag{1}$$

with two form factors $f_+(q^2)$ and $f_-(q^2)$ of the momentum transfer squared to the lepton pair, $q^2 = (p_K - p_\pi)^2$. Here, G_F is the Fermi constant, θ_c the Cabibbo angle, p_K, p_π, p_μ, and p_ν are the four momenta of the kaon, pion, muon, and anti-neutrino, respectively. Using $p_K = p_\pi + p_\mu + p_\nu$, this amplitude can be rewritten as follows:

$$M = \frac{G_F}{2} \sin\theta_c f_+ \left(q^2\right) \left[2p_K^\lambda \cdot \overline{u}_\nu \gamma_\lambda \left(1 - \gamma_5\right) v_\mu + \left(\xi \left(q^2\right) - 1\right) m_\mu \overline{u}_\nu \left(1 - \gamma_5\right) v_\mu \right], \tag{2}$$

where the parameter $\xi(q^2)$ is defined as $\xi(q^2) = f_-(q^2)/f_+(q^2)$. The first term of Eq. (2) corresponds to the vector (and axial vector) amplitude and the second term corresponds to the scalar (and pseudo-scalar) amplitude. The parameters f_- and f_+ depend on q^2 as $f_\pm(q^2) = f_\pm(0)[1 + \lambda_\pm(q^2/m_\pi^2)]$. Both f_+ and f_- can, in general, be complex. If time reversal (T) is a good symmetry, the parameter ξ is real. Any non-zero value of Imξ would imply T violation. As we show below, an experimentally observed T-violating muon polarization P_T is directly proportional to Imξ. The currently adopted values are $\lambda_+ = 0.0284 \pm 0.0027$, $\xi(0) = -0.14 \pm 0.05$, and $\lambda_- = 0$. The Dalitz distribution for $K_{\mu 3}$ decay is given by

$$\rho \left(E_\pi, E_\mu\right) \propto f_+^2 \left(q^2\right) \left[A + B\xi \left(q^2\right) + C\xi^2 \left(q^2\right) \right], \tag{3}$$

with

$$A = m_K \left(2E_\mu E_\nu - m_K E'_\pi\right) + m_\mu^2 \left(\frac{1}{4}E'_\pi - E_\nu\right), \tag{4a}$$

$$B = m_\mu^2 \left(E_\nu - \frac{1}{2}E'_\pi\right), \tag{4b}$$

$$C = \frac{1}{4}m_\mu^2 E'_\pi, \tag{4c}$$

$$E'_\pi = \left(m_K^2 + m_\pi^2 - m_\mu^2\right)/(2m_K) - E_\pi, \tag{4d}$$

where E_π, E_μ, and E_ν are the energies of the pion, muon, and neutrino in the kaon center-of-mass frame, and M_K, m_π, and m_μ the masses of the kaon, pion, and muon, respectively.

2.2 Transverse Polarization P_T

In three-body decays such as $K_{\mu 3}$, one defines three orthogonal components of the muon polarization vector: the longitudinal (P_L), normal (P_N), and transverse (P_T) as the components parallel to the muon momentum $\boldsymbol{p_\mu}$, normal to P_L in the decay plane, and normal to the decay plane, respectively. They are scalar products of the polarization vector ($\boldsymbol{\sigma}_\mu$) with three corresponding combinations of the unit momentum vector as follows:

$$P_L = \frac{\boldsymbol{\sigma}_\mu \cdot \boldsymbol{p}_\mu}{|\boldsymbol{p}_\mu|}, \tag{5a}$$

$$P_N = \frac{\boldsymbol{\sigma}_\mu \cdot \left(\boldsymbol{p}_\mu \times (\boldsymbol{p}_\pi \times \boldsymbol{p}_\mu)\right)}{|\boldsymbol{p}_\mu \times (\boldsymbol{p}_\pi \times \boldsymbol{p}_\mu)\,|}, \tag{5b}$$

$$P_T = \frac{\boldsymbol{\sigma}_\mu \cdot (\boldsymbol{p}_\pi \times \boldsymbol{p}_\mu)}{|\boldsymbol{p}_\pi \times \boldsymbol{p}_\mu|}. \tag{5c}$$

As can be seen, the P_T changes sign under the time reversal operation, thus making it a T odd observable. Using the decay probability (Eq. (3)) one can write the muon polarization in the kaon rest frame as [9–12] $\boldsymbol{\sigma}_\mu = \boldsymbol{P}/|\boldsymbol{P}|$, where $\boldsymbol{P}$ is determined as follows:

$$\boldsymbol{P} = \left\{a_1(\xi) - a_2(\xi)\left[(m_K - E_\pi) + (E_\mu - m_\mu)\left(\boldsymbol{p}_\pi \cdot \boldsymbol{p}_\mu\right)/|\boldsymbol{p}_\mu|^2\right]\right\}\boldsymbol{p}_\mu$$
$$- a_2(\xi)\, m_\mu \boldsymbol{p}_\pi + m_K m_\mu \mathrm{Im}(\xi)\left(\boldsymbol{p}_\pi \times \boldsymbol{p}_\mu\right), \tag{6}$$

with

$$a_1(\xi) = 2m_K^2\left[E_\nu + \mathrm{Re}\left(b\left(q^2\right)\right)\left(E_\pi^* - E_\pi\right)\right], \tag{7a}$$

$$a_2(\xi) = m_K^2 + 2\mathrm{Re}\left(b\left(q^2\right)\right) m_K E_\mu + |b\left(q^2\right)|^2 m_\mu^2, \tag{7b}$$

$$b\left(q^2\right) = \frac{1}{2}[\xi\left(q^2\right) - 1], \quad \text{and} \tag{7c}$$

$$E_\pi^* = \left(m_K^2 + m_\pi^2 - m_\mu^2\right)/\left(2m_K\right). \tag{7d}$$

One has to look for P_T in the presence of predominant in-plane components of the polarization P_L and P_N. P_T (Eq. (5)) can be further rewritten in terms of Imξ and a kinematical factor as follows:

$$P_T = \mathrm{Im}\xi \cdot \frac{m_\mu}{m_K} \frac{|\boldsymbol{p}_\mu|}{[E_\mu + |\boldsymbol{p}_\mu| \boldsymbol{n}_\mu \cdot \boldsymbol{n}_\nu - m_\mu^2/m_K]}. \tag{8}$$

The quantity Imξ, sensitive to the T violation, can thus be determined from a P_T measurement. The kinematic factor as a function of π^0 energy and μ^+ energy has an average value of ~0.3 yielding a full detector acceptance relation of $< P_T > \sim 0.3\,\mathrm{Im}\xi$.

It is of interest to establish the connection between the Imξ and effective parameters of new physics appearing in the coefficients of generic exotic interactions. To this end, an effective four-fermion Lagrangian can be used:

$$\begin{aligned} L = &- \frac{G_F}{\sqrt{2}} \sin\theta_C \, \bar{s}\gamma_\alpha (1-\gamma_5) u \, \bar{\nu}\gamma^\alpha (1-\gamma_5) \mu \\ &+ G_S \, \bar{s}u \, \bar{\nu} (1+\gamma_5) \mu + G_P \, \bar{s}\gamma_5 u \, \bar{\nu} (1+\gamma_5) \mu \\ &+ G_V \, \bar{s}\gamma_\alpha u \, \bar{\nu}\gamma^\alpha (1-\gamma_5) \mu + G_A \, \bar{s}\gamma_\alpha\gamma_5 u \, \bar{\nu}\gamma^\alpha (1-\gamma_5) \mu + \mathrm{h.c.}, \end{aligned} \tag{9}$$

where G_S and G_P are the scalar and pseudo-scalar coupling constants and G_V and G_A are the exotic vector and axial vector coupling constants, respectively. Tensor interactions are neglected. Imξ is found to be caused only by the interference between the SM term and the scalar term, namely by the complex phase of G_S [13, 14], which can be written as

$$\mathrm{Im}\xi = \frac{(m_K^2 - m_\pi^2)\, \mathrm{Im}G_S^*}{\sqrt{2}\,(m_s - m_u)\, m_\mu G_F \sin\theta_C}, \tag{10}$$

where m_s and m_u are the masses of the s-quark and u-quark, respectively. Thus, P_T can constrain the exotic scalar interactions. The situation is different in the similar transverse muon polarization P_T in the radiative kaon decay $K^+ \to \mu^+\nu\gamma$, which is caused by pseudo-scalar interactions G_P [14].

2.3 P_T and CP Violation Beyond the Standard Model

P_T in the Standard Model

In order to formulate physics motivation for this experiment, we look first at what is predicted in the SM for P_T. A T-violating (or CP-violating) amplitude arises from the relative phases between diagrams or complex coupling constants in a diagram. Since only a single element of the CKM matrix V_{us} is involved for the W-exchanging semi-leptonic $K_{\mu 3}$ decay in the SM, no CP violation appears in the first order. As is discussed in [15] this is a general feature for vector (and axial vector)-type interactions. The SM contribution comes from only higher order effects. The possible size of its contribution was

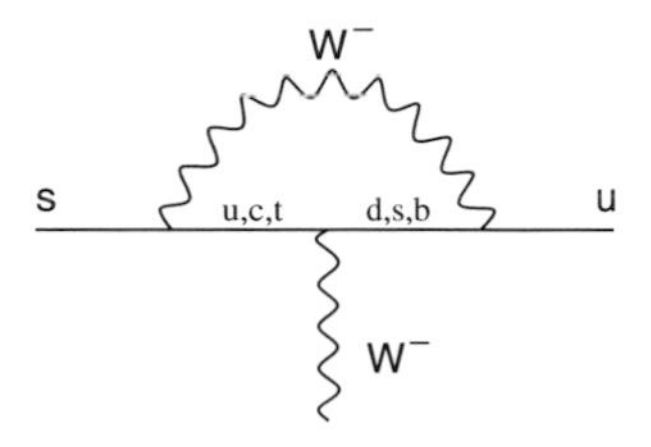

Fig. 1. Radiative corrections in the $K_{\mu3}$ decay which provide a standard model contribution to P_T [1].

once suggested qualitatively in [16] to be $P_T < 10^{-6}$. An actual value based on the lowest order vertex radiative corrections to the $\bar{u}\gamma_\mu(1-\gamma_5)sW^\mu$ vertex (Fig. 1) was presented in the textbook of Bigi and Sanda [1]. This has been estimated to be less than 10^{-7}. This fact constitutes the main motivation of the physics background for P_T experiment as a search for new physics. Since the effect arising from FSI is known to be of the order of 10^{-5} [4, 5] and it is calculable, an observation of a non-zero P_T implies unambiguously the existence of CP violation mechanisms beyond the SM, namely new physics. Assuming 10^{-6} for the ability of the FSI estimation, there is a large window to explore from the current limit of $P_T \sim 10^{-3}$, while several new physics models allow the appearance of P_T in the ranges of $10^{-4} - 10^{-3}$ level at any time (Fig. 2). This situation is very similar to the study of the neutron electric dipole moment (n-EDM) which is also a T violation quantity, in which the current experimental limit of $d_n = 3.1 \times 10^{-26} e$ cm [17] is slowly approaching the SM prediction of $d_n \sim 10^{-31} e$ cm.

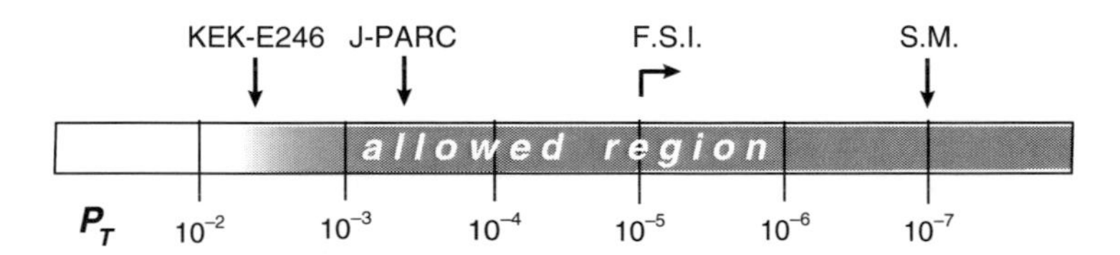

Fig. 2. Experimental status of $K_{\mu3}$ P_T physics relative to the SM prediction is illustrated. The experimental limits are in 90% confidence limits. "Allowed region" means that some of the non-SM CP violation models allow $P_T < 10^{-2}$–10^{-3} without conflicting with other experimental limits.

2.4 CP Violation and Physics of P_T

Today there are intensive studies underway at accelerator laboratories searching for CP violation beyond the SM. In addition to the direct search for new particles at high-energy colliders, there are also many precision measurements to search for small deviations from the SM predictions. In B meson decays, e.g., the observation of a slight difference between the CP asymmetry of the $B \to J/\psi K_S^0$ process and that of the penguin diagram process [18] might be a hint for new physics. In the kaon sector, the CP-violating $K_L \to \pi^0 \nu \bar{\nu}$ rare decay study has been performed at KEK [19] and this will also be proposed at J-PARC. It is attempted to measure possible difference of the unitarity triangle from that of the B physics, but it can only be achieved with sufficient event statistics. Considering the current situation of our understanding of CP violation, the importance of a P_T search is increasing. There are several important characteristics of P_T physics. They are briefly summarized as follows:

- If P_T is found at the level of 10^{-4} which cannot be explained as FSI, it will correspond to "direct CP violation" in contrast to "indirect CP violation" due to the K_1^0–K_2^0 state mixing in the case of the K_0 system. The amount of direct CP violation in neutral kaons is found as the ratio $\varepsilon'/\varepsilon(\sim 10^{-3})$ which is consistent with SM predictions [20]. However, one should note that the agreement between the experiment and the theory, while good, still leaves large theoretical uncertainties. Therefore, the observation of CP violation in the charged kaon system is very much desired.
- Since the $K_{\mu 3}$ is a semi-leptonic process with W exchange in the SM resulting in a significant branching ratio, even a small amplitude of a new physics diagram with large mass scale can contribute to P_T through interference with this W exchange process with a consequence of a relatively large effect. (This is in strong contrast to new physics contribution to flavor changing neutral current rare decay processes in which the effect appears from loop diagrams.) In terms of the effective Lagrangian, $P_T \sim 1/\Lambda^2$ with the mass scale Λ, while the direct detection of rare decays, such as $K_L^0 \to \pi^0 \nu \bar{\nu}$, should scale as $1/\Lambda^4$.
- Since the three Higgs doublets may allow a sizable value [21, 22], P_T can probe CP violation through the Higgs sector, of which, however, very little is known with regard to the structure and dynamics [1]. There is no constraint even on the number of Higgs doublets so far inferred theoretically. Models with an extended Higgs sector with more than one SM Higgs doublet allow many new sources of CP violation.
- If LHC finds the charged Higgs boson, the P_T measurement will become even more important to look for associated CP-violating couplings. It is expected that an MSSM Higgs boson should be discovered with 5σ significance after an integrated luminosity of 30 fb^{-1} if the condition of $\tan\beta/m_{H^+} \geq 0.06\ (\text{GeV})^{-1}$ is satisfied. If this limit is applied to the lightest charged Higgs boson in the context of the multi-Higgs model, the P_T corresponds to $|P_T| < 3 \times 10^{-4}$. This roughly corresponds to the aim of the J-PARC experiment of δP_T (one σ limit) $\sim 10^{-4}$.

2.5 Theoretical Models for P_T

In this section we briefly describe a few models which might lead to a sizable P_T value. We also give a more general discussion based on the effective field theory to clarify the difference of P_T physics from other T- and CP-violating observables.

Multi-Higgs Doublet Model

As the minimum and natural extension of the SM with one Higgs doublet, multi-Higgs doublet models have been considered, and a number of papers [21–29] have applied this model to P_T as one of the promising candidate theories. In the class of models without tree-level flavor changing neutral current, new CP-violating phases are introduced in the charged Higgs mass matrix if the number of doublets is more than 2. The coupling of quarks and leptons to the Higgs boson is expressed in terms of the Lagrangian [21, 22]:

$$L = \left(2\sqrt{2}G_F\right)^{\frac{1}{2}} \sum_{i=1}^{2} \{\alpha_i \bar{u_L} V M_D d_R H_i^+ + \beta_i \bar{u_R} M_U V d_L H_i^+ + \gamma_i \bar{\nu_L} M_E e_R H_i^+\} + \text{h.c.}, \tag{11}$$

where M_D, M_U, M_E are diagonal mass matrices, V is the CKM matrix, and α_i, β_i, and γ_i are the new complex coupling constants associated with the charged Higgs interactions. For the three-doublet case a natural flavor conservation can be arranged. The coefficients α_i, β_i, and γ_i can have complex phases, and P_T is calculated as follows:

$$\text{Im}\xi = \frac{m_K^2}{m_H^2} \text{Im}\left(\gamma_1 \alpha_1^*\right), \tag{12}$$

where α_1 and γ_1 are the quark and lepton couplings to the lightest charged Higgs boson (Fig. 3). The E246 result [8] yielded $|\text{Im}(\gamma_1 \alpha_1^*)| < 0.066(m_H/\text{GeV})^2$ as the most stringent limit for this parameter. $(\gamma_1 \alpha_1^*)$ is also constrained by the semi-leptonic decay of the B meson $B \to \tau\nu X$ [30, 31] as the deviation from the SM value, but the constraint on $|\text{Im}(\gamma_1 \alpha_1^*)|$ is less stringent than the P_T constraint. The recent observation of $B \to \tau\nu$ [32] constrains the model in a similar manner [33] but less stringent at the moment. Other constraints to this model come from the neutron EDM (d_n) [17], $b \to s\gamma$ [30, 31], and $b \to s l \bar{l}$ [34] complementing the P_T result in a different manner, since these channels limit $\text{Im}(\alpha_1 \beta_1^*)$. These two parameters are related as $\text{Im}(\alpha_1 \beta_1^*) = -(v_3/v_2)^2 \text{Im}(\gamma_1 \alpha_1^*)$ through the ratio of the Higgs field vacuum expectation values v_2 and v_3. An interesting scenario assumed in [21] is $v_2/v_3 \sim m_t/m_\tau \sim 95$, thus making P_T the most sensitive test of the three Higgs doublet model over d_n and $b \to s\gamma$ (Fig. 4).

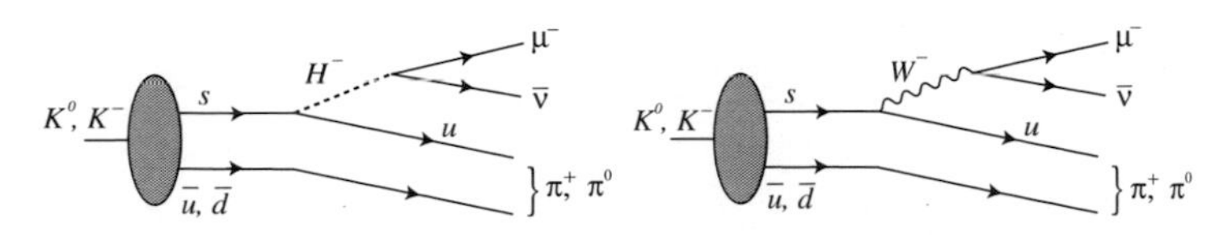

Fig. 3. P_T appears as the interference of the charged Higgs boson exchange with the standard model W boson exchange.

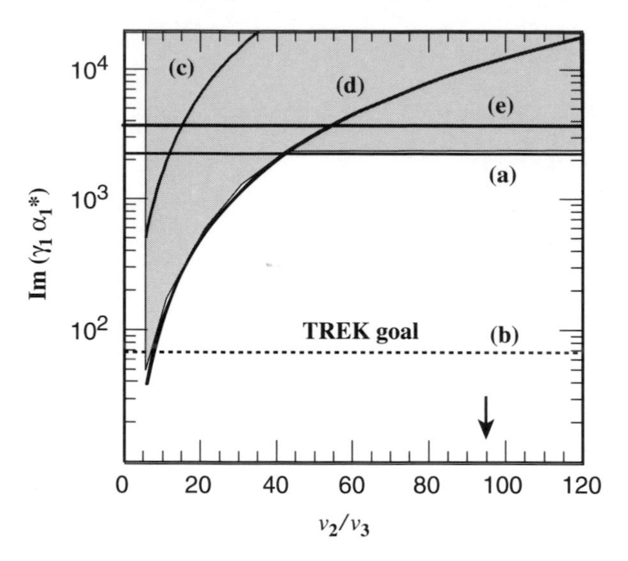

Fig. 4. Constraint to the three Higgs doublet model parameters of $|\mathrm{Im}(\gamma_1\alpha_1^*)|$ and v_2/v_3, the ratio of Higgs field vacuum expectation values, with the assumption of $m_{H^+} \cong 2m_Z$: (**a**) the P_T limit from the E246 experiment, (**b**) P_T expectation in TREK, (**c**) neutron electric dipole moment (EDM) only with the d-quark contribution, (**d**) $b \to s\gamma$, and (e) $b \to X\tau\nu$. The *gray region* is the excluded region. The *arrow* shows the most probable point of v_2/v_3 to be $m_t/m_\tau = 95$.

SUSY Models

A number of other models also allow P_T at observable level without conflicting with other experimental constraints. In other words, non-observation of P_T can constrain those models. Some models of minimal supersymmetric standard models (MSSM) allow sizable values. One interesting case is the model discussed by Wu and Ng [35]. In this model the complex coupling constant between the charged Higgs boson and strange- and up-quarks is induced through squark and gluino loops. Then, the P_T value when the muon and neutrino momenta are at right angles is given as

$$P_T^{H^+} \approx 3.5 \times 10^{-3} I_{H^+} \frac{p_\mu}{E_\mu} \frac{(\mu + A_t \cot\beta)}{m_g} \times \frac{(100\ \mathrm{GeV})^2}{M_H^2} \frac{\mathrm{Im}[V_{33}^{H^+} V_{32}^{D_L^*} V_{31}^{U_R^*}]}{\sin\theta_c} \tag{13}$$

for $\tan\beta \approx 50$. (For the meanings of various symbols, see Ref. [35] except to note that we assumed the top quark mass to be 180 GeV.) If we allow large flavor mixing coupling in the squark–quark vertices, there is an allowed parameter region for large P_T. The E246 P_T upper bound corresponds to $M_H > 140$ GeV. In view of several assumptions made, this bound should be considered as a qualitative estimate [36]. It is noteworthy that $P_T(K_{\mu 3})$ and $P_T(K_{\mu\nu\gamma})$ have opposite signs in this model.

Although MSSM predicts only unobservably small P_T [37] without tuning of relevant flavor parameters, the SUSY with R-parity violation [36] can give rise to a sizable value of P_T. If R-parity violation is allowed, a superpotential is defined [38] as $W = W_{MSSM} + W_{RPV}$ with

$$W_{RPV} = \lambda_{ijk} L_i L_j \bar{E}_k + \lambda'_{ijk} L_i Q_j \bar{D}_k + \lambda''_{ijk} \bar{U}_i \bar{D}_j \bar{D}_k, \tag{14}$$

where $i, j = 1, 2, 3$ are generation indices with a summation implied. $L_i(Q_i)$ are the lepton (quark) doublet superfields and $\bar{E}_j(\bar{D}_j, \bar{U}_j)$ are the electron (down- and up-quark) singlet superfields. λ, λ', and λ'' are the Yukawa couplings; the former two relevant to lepton number violation and the latter relevant to baryon number violation. There are altogether 48 independent λ_{ijk}s which should be determined experimentally.

Our P_T can be expressed in this formalism as a sum of two components of slepton exchange and down-type squark exchange as $\mathrm{Im}\xi = \mathrm{Im}\xi^{\tilde{l}} + \mathrm{Im}\xi^{\tilde{d}}$ with

$$\mathrm{Im}\xi^{\tilde{l}} = \sum_i \frac{\mathrm{Im}[\lambda_{2i2} (\lambda'_{i12})^*]}{4\sqrt{2} G_F \sin\theta_c \left(m_{\tilde{l}_i}\right)^2} \cdot \frac{m_K^2}{m_\mu m_s}, \tag{15a}$$

$$\mathrm{Im}\xi^{\tilde{d}} = \sum_i \frac{\mathrm{Im}[\lambda'_{21k} (\lambda'_{22k})^*]}{4\sqrt{2} G_F \sin\theta_c \left(m_{\tilde{d}_k}\right)^2} \cdot \frac{m_K^2}{m_\mu m_s}. \tag{15b}$$

Thus, we constrain $\mathrm{Im}[\lambda_{2i2}(\lambda'_{i12})^*]/m^2_{\tilde{l}_i}$ and $\mathrm{Im}[\lambda'_{21k}(\lambda'_{22k})^*]/m^2_{\tilde{d}_k}$ (summation is implied). The E246 limit corresponds to 1.8×10^{-8} for these quantities, which will be improved by a factor of 20 at J-PARC. There are a number of experimental and theoretical efforts to analyze λ, λ', and λ'' in other channels. In a recent paper [39] and review papers [40, 41] currently updated limits are compiled also for the products $\lambda\lambda'$, $\lambda'\lambda'$, etc. There are six relevant parameters for P_T as shown in Eq. (15) which are summarized in Table 1.

Table 1. The R-parity violating SUSY parameters relevant to P_T and the constraints from other experiments.

	Parameter	Upper bound	Experiment
	$\lvert\lambda^*_{232}\lambda'_{312}\rvert$	$3.8 \times 10^{-6}\mathrm{m}^2$	$K_L \to \mu^+\mu^-$ [41]
$\mathrm{Im}\xi^{\tilde{l}}$	$\lvert\lambda^*_{212}\lambda'_{112}\rvert$	No constraint	
	$\lvert\lambda^*_{222}\lambda'_{212}\rvert$	No constraint	
	$\lvert\lambda'^*_{211}\lambda'_{221}\rvert$	$2.8 \times 10^{-5}\mathrm{m}^2$	$K^+ \to \pi^+\nu\bar{\nu}$ [42]
$\mathrm{Im}\xi^{\tilde{d}}$	$\lvert\lambda'^*_{212}\lambda'_{222}\rvert$	$2.8 \times 10^{-5}\mathrm{m}^2$	$K^+ \to \pi^+\nu\bar{\nu}$ [42]
	$\lvert\lambda'^*_{213}\lambda'_{223}\rvert$	$2.8 \times 10^{-5}\mathrm{m}^2$	$K^+ \to \pi^+\nu\bar{\nu}$ [42]

3 Experimental Status of P_T

3.1 Early Experiments

The measurement of P_T in $K_{\mu 3}$ decays has a long history. Early measurements of P_T were carried out at the Bevatron [43, 44] and Argonne [45] in $K^0_{\mu 3}$ decays but they lacked statistical significance. More advanced experiments prior to our work were done at the 28-GeV AGS at the Brookhaven National Laboratory (BNL). Morse et al. [46] measured P_T of muons from in-flight decay of $K^0_{\mu 3}$. From a data sample of 12 million events, they deduced $\mathrm{Im}\xi = 0.009 \pm 0.030$. This result, while consistent with zero, has a central value compatible with a prediction of 0.008, from the T-conserving final state interactions. As mentioned above, the final state interactions in $K^0_{\mu 3}$ decays obscure the real value of P_T. At the same facility, Blatt et al. [6] measured P_T of $K^+_{\mu 3}$ for the first time by detecting neutral particles from the in-flight decay of an unseparated 4 GeV/c K^+ beam. From a data sample of 21 million events, they deduced $\mathrm{Im}\xi = -0.016 \pm 0.025$, consistent with T invariance. The most recent result was from the KEK-PS E246 experiment [7] on which the current J-PARC experiment is based. The details of this experiment are presented next. Table 2 presents the world data as of today.

Table 2. Early experiments and their Imξ results.

Laboratory	Decay	Year	Imξ	Ref.
Bevatron	$K^0_{\mu 3}$	1967	-0.02 ± 0.08	[43, 44]
Argonne	$K^0_{\mu 3}$	1973	-0.085 ± 0.064	[45]
BNL-AGS	$K^0_{\mu 3}$	1980	0.009 ± 0.030	[46]
BNL-AGS	$K^+_{\mu 3}$	1983	-0.016 ± 0.025	[6]
KEK-PS	$K^+_{\mu 3}$	2004	-0.005 ± 0.008	[7]

3.2 KEK E246 Experiment

The most recent and highest precision experiment was performed at the KEK proton synchrotron. The experiment used the stopped K^+ beam at the K5 low-momentum beam channel [47] with the superconducting toroidal spectrometer [48] setup. An elaborate detector (Fig. 5) consisting of a large-acceptance CsI(Tl) barrel, tracking chambers, an active target, and muon polarimeters was constructed and the data were taken between 1996 and 2000 for a total of 5,200 hours of beam time. A precise field mapping was made [49] before the measurements. Since K5 was equipped only with a single stage of electrostatic separator, the channel provided a beam with substantial π^+ contamination with a π/K ratio of about 8 for a 660 MeV/c beam.

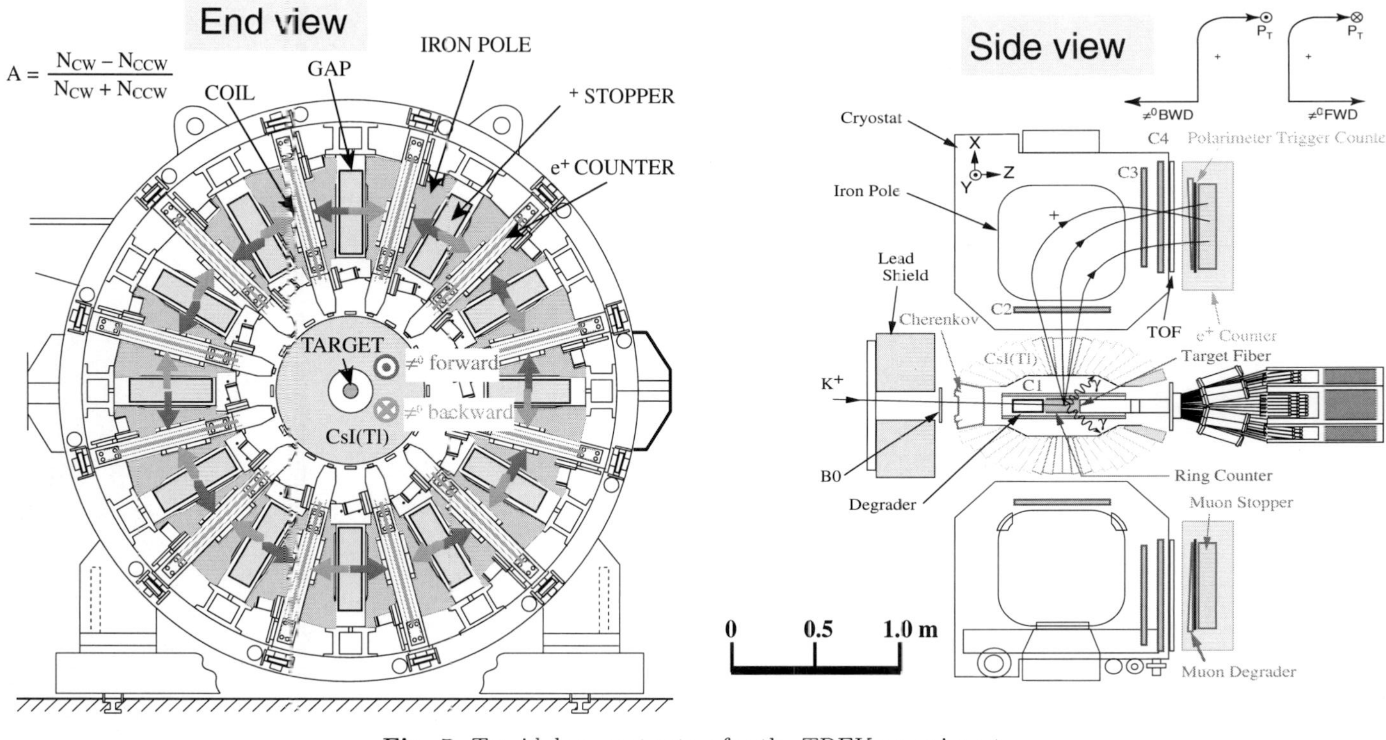

Fig. 5. Toroidal magnet setup for the TREK experiment.

Table 3. Experimental condition of E246.

Parameter	Value	Expectation at J-PARC
K^+ beamline	K5 with 660 MeV/c	K0.8 with 800 MeV/c
Proton intensity	1.0×10^{12}/s	0.54×10^{14}/s
K^+ beam intensity	1.0×10^5/s	2×10^6/s
π/K ratio	$\sim$8	$<$0.5
Beam duty factor	0.7 s/2.0 s	$>$0.7 s/3.5 s
Net runtime	$\sim$5200 hours (1.8×10^7s)	1.0×10^7s

(At J-PARC we expect $\pi/K < 0.5$.) The performance of the CsI(Tl) calorimeter was limited by this π^+ contamination with a halo resulting in accidental hits. The relatively low K^+ intensity of typically 10^5/s was the consequence of the maximum available proton beam intensity of 10^{12}/s from the slow extraction of the accelerator. The main parameters of the experiment are summarized in Table 3. Current estimates of these parameters at J-PARC are also shown.

The details of the detector were presented in [50]. Also, some individual elements have been documented in the literature, e.g., the CsI(Tl) calorimeter [51], its readout electronics [52], and the target ring counter system [53]. The features of the $K_{\mu 3}$ detection are (1) μ^+ (charged particle) detection by means of a tracking system with 3 MWPC and a fiber bundle target with the momentum analysis by the toroidal spectrometer and (2) π^0 detection as two photons or one photon with relatively large energy by the CsI(Tl) calorimeter. The muon polarization measurement relied on the sensitivity of the decay positron emission asymmetries in a longitudinal magnetic field with $< \mathbf{B} > \parallel \boldsymbol{P_T}$ using "passive polarimeters," where $< \mathbf{B} >$ is the average of muon magnetic field vector. Two independent teams separately carried out careful data analyses and the two results were combined at the end. Thanks to the stopped beam method which enabled a so-called forward (fwd) and backward (bwd) symmetric measurement with regard to the π^0 emission direction (see Fig. 5), and the high rotational symmetric structure of the toroidal spectrometer system, the systematic errors could be substantially suppressed. A full description of the experiment is given in [7]. The final result was

$$P_T = -0.0017 \pm 0.0023\,(stat) \pm 0.0011\,(syst) \tag{16}$$

$$\mathrm{Im}\xi = -0.0053 \pm 0.0071\,(stat) \pm 0.0036\,(syst) \tag{17}$$

corresponding to the upper limits of $|P_T| < 0.0050$ (90% CL) and $|\mathrm{Im}\xi| < 0.016$ (90% CL), respectively. This E246 result was statistics limited, i.e., the total systematic error was less than half of the statistical error. A remarkable point here is that most of the systematic error sources cancelled out after the 12-gap summation due to the rotational symmetry of the system and the fwd–bwd ratio due to its symmetry. The largest error was the effect of multiple scattering of muons through the Cu degrader of the polarimeter. The J-PARC experiment will be free from this error as we will employ an active polarimeter (see below). There were two items that were not cancelled out by any of the two cancellation mechanisms, the effect from the decay plane rotation, θ_z, and the rotation of the muon magnetic field, δ_z, which might remain as the most serious errors in the J-PARC experiment.

4 J-PARC TREK Experiment

4.1 Goal of the Experiment

Considering the current experimental situation of direct CP violation studies and searches for new physics as discussed in Sect. 2, we believe that it is essential to perform the P_T measurement in $K^+_{\mu 3}$ as the TREK[1] experiment [8] at J-PARC, which offers a far superior experimental environment when compared to KEK-PS. The 40-year history of P_T experiments shows a rather slow improvement in the upper limit. This is due to two reasons: the first point is that the statistical sensitivity of asymmetry measurements scales as $1/\sqrt{N}$ (N is the total number of events), while the single-event sensitivity in rare decay experiments scales as $1/N$. The second reason is the nature of this high-precision experiment which must be conducted and analyzed very carefully. The understanding and reduction of systematic errors can only be achieved step by step. We prefer to follow this approach for the J-PARC experiment and to proceed in a steady way to improve the sensitivity. We plan eventually a long-range strategy to attain the goal of SM+FSI signal region of 10^{-5} in a few steps.

The E246 result was essentially statistics-limited. (The largest systematic error in the error list was due to multiple scattering and was also statistical in nature.) This result was foreseen at the start of the E246 experiment. We propose to improve the E246 result by at least a factor of 20 ($\delta P_T < 1.2 \times 10^{-4}$), by improving both statistical error (by a factor of 20 at least) and systematics uncertainty (by a factor of 10 at least). This sensitivity puts the experiment well into the region where new physics effects can be discovered, and even a null result would set tight constraints on theoretical models. If warranted, further sensitivity improvement toward 10^{-5} will be proposed in the next stage after we have been convinced of the possibility to pursue this experiment to such a high-precision region. In that sense, the TREK experiment may be considered as a prelude to precision frontier experiments at J-PARC.

4.2 Stopped Beam Method

The salient feature of the E246 and TREK experiments is the use of a stopped beam and it was conceptually distinct from the previous BNL-AGS

[1] TREK is an acronym of time reversal experiment with kaons.

experiment [6] where in-flight K^+ decays were adopted for the K^+ decay. The advantages of using decays at rest are briefly summarized below:

- Isotropic decay of K^+ at rest involves all the kinematic conditions covering the full decay phase space. By using a symmetric detector like E246/TREK one can look for the T-odd asymmetry effect in its positive value Dalitz-plot region as well as in its negative region. A double ratio measurement, namely the comparison scheme between forward-going pion events and backward-going pion events was possible. Such a double ratio measurement is essential for high-precision experiments.
- The kinematical resolution is determined in the center-of-mass system in at-rest decays. The energy regions extend up to about 250 MeV and the energies can be easily measured with sufficiently high resolution. Decay particles are detected in the entire 4π sr solid angle region and the relative angles of the particles are measured with good accuracies, limited only by the detector resolutions.
- The isotropic decay at rest with large solid angle coverage ensures that the counting rates are distributed over many segmented detectors, thus keeping the counting rates low enough to minimize the pileup problems. This is an important feature, especially for the electromagnetic calorimeter.
- For a stopped beam experiment, we need not be concerned about the beam history nor the finite emittance of the K^+ beam. The latter is usually large in the case of a low-momentum beam and can cause severe problems in the case of in-flight decay experiments. The kaon stopping distribution is a consequence of the beam emittance in this experiment. The asymmetries in the stopping distributions, however, involve only three parameters of coordinates and they are easy to handle.
- Most parts of the detector are located outside the beam region. Hence, the beam-related pileup effects or background problems are less serious.

In particular, the use of the superconducting toroidal spectrometer setup in the stopped beam experiment offers many advantages.

- The relatively large bending power for charged particles in the field of 0.9 T provides sufficiently good resolution for charged particle analysis. An upgraded charged particle tracking system can make a full use of it.
- The bending of nearly 90° by the sector-type magnet creates a quasi-focal plane with large dispersion at the exit of the magnet. This arrangement effectively prevents most of the background channels such as $K_{\mu 2}$ and $K_{\pi 2}$ from entering the polarimeter.
- The presence of a quasi-focal plane enables a relatively small muon stopping volume when a wedge-shaped momentum degrader is inserted. This is a rather important condition since the relevant momentum range of $K_{\mu 3}$ is wide – 100–200 MeV/c.

The basic analysis in TREK will be done essentially in the following scheme as in the E246 experiment although an extended analysis using all the π^0 region will also be conceivable. The T-odd asymmetry is deduced as

$$A_T = (A_{fwd} - A_{bwd})/2, \tag{18}$$

where the π^0-fwd and bwd asymmetries were calculated using the "clockwise" and "counter-clockwise" positron emission rate N_{cw} and N_{ccw} (red arrows and green arrows in Fig. 5 (left), respectively) as

$$A_{fwd(bwd)} = \frac{N^{cw}_{fwd(bwd)} - N^{ccw}_{fwd(bwd)}}{N^{cw}_{fwd(bwd)} + N^{ccw}_{fwd(bwd)}}. \tag{19}$$

Then, P_T is deduced using the analyzing power α, which is determined by the structure of the polarimeter and its analysis, and the average kinematic attenuation factor $< \cos\theta_T >$, which is determined by the $K_{\mu 3}$ kinematics in the detector, to be

$$P_T = A_T/(\alpha < \cos\theta_T >). \tag{20}$$

Using a conversion coefficient Φ ($\sim$0.3 for a full detector acceptance) calculated in a detector Monte Carlo calculation, the result of Imξ is deduced as

$$\mathrm{Im}\xi = P_T/\Phi. \tag{21}$$

4.3 Upgraded E246 Detector

There are strong arguments which favor performing the experiment in the J-PARC phase 1 with the toroidal spectrometer system by upgrading the E246 detector system.

- The basic performance of the spectrometer is well known to us. Although some upgrades are required in the charged particle tracking, the motion of charged particles in the toroidal magnet is well understood and reproduced well in our simulation calculations. We have carried out, over a period of many years, intensive studies of the CsI(Tl) calorimeter response incorporating its complex geometry of the muon hole and the entrance/exit holes. This system is very well understood.
- The spectrometer magnet and the CsI(Tl) calorimeter are the components that were designed and manufactured with high precision. They were assembled very carefully in order to ensure their performance in a high-precision experiment and have been proven to work without any problem.
- We have full knowledge of the overall performance of the total system including the beam collimation, kaon stopping target, muon slowing-down, and the polarimeter system. For example, we can utilize the existing wedge-shaped muon degrader system which has been well tuned. We know the sources of systematic errors very well and we are able to design the upgrade detector based on the experimental data.
- Finally, the relatively low cost for starting the step 1 experiment as early as possible also favors an upgrade of the E246 detector rather than a new detector.

Also the fact that the beam intensities in the phase 1 of J-PARC will not be very high makes it unnecessary to develop a completely new calorimeter. If we increase the speed of the readout scheme by changing to a faster system based on recent technical developments, we will be able to handle a few times 10^6 per second kaon beam.

Our aim is to perform an experiment which, in comparison to E246, will have about 10 times more acceptance (using active polarimeter described below), about 20 times the integrated beam flux, and a few times higher analyzing power to achieve nearly a factor of at least 20 improvements in the statistical sensitivity, i.e., δP_T (one σ limit) $\sim 10^{-4}$. We feel confident that we can accomplish this task with the modifications described in the next section.

5 TREK Detector

In order to optimize the performance of the toroidal spectrometer system, several improvements must be undertaken. We will keep the principal concept of the experiment, namely the application of the muon magnetic field in the azimuthal direction parallel to the P_T component, and the object of the measurement in this experiment is primarily the *cw–ccw* positron asymmetry in the azimuthal direction, i.e., we keep the *fwd/bwd* scheme. As for the π^0 detection we cannot use 1γ events anymore since this gives rise to significant background contamination. It is expected that the statistics from the 2γ events only is sufficient to achieve the goal of our experiment in phase 1.

5.1 Active Polarimeter

The most important feature of the TREK experiment is the adoption of an active polarimeter in contrast to E246 where a passive polarimeter with a separate system of a muon stopper and positron counters was used. The advantage of this passive system was the simplicity in the analysis with the consequence of very small systematic errors associated with the analysis. The systematic cancellation scheme when the asymmetry was summed over the 12 sectors was also based on the use of positron counters as clockwise and counter-clockwise counter function at the same time. However, this was done at the cost of e^+ detection acceptance and polarization analyzing power. We now aim for higher detector acceptance and higher sensitivity by introducing the active polarimeter. The suppression of the systematic errors ensured in the E246 passive polarimeter will be guaranteed by a different method. The active polarimeter should have the following functions and advantages.

- An active polarimeter determines the muon stopping position for each event, which in turn renders the experiment free from the systematic error associated with the ambiguities in the muon stopping distribution. As the decay positron tracks are measured, the decay vertices will be determined event by event. In contrast to the situation in E246, where the positron signal time spectra were associated with non-negligible constant background events, the active polarimeter data will be relatively background free.

- Detection of decay positrons in all directions by a polarimeter with a large acceptance with nearly 4π solid angle. In E246 the positron counter solid angle was limited to about 10% on each side. The detector acceptance becomes 10 times larger, even though the sensitivity does not scale by this factor. The ability to measure positron emission angle provides the possibility to use not only the fwd/bwd pion scheme but also the $left/right$ pion scheme (Fig. 6) which was not possible in E246.
- Measurement of the positron emission angle and rough positron energy by means of the number of penetrating stopper plates. The asymmetry changes as the positron energy varies. A weighted analysis brings about a significant increase in the analyzing power resulting in the highest sensitivity. It is of interest to note that this superior performance is achieved with only moderate energy resolution.

Needless to say the stopper should have a large muon collection/stopping efficiency and the ability to preserve the polarization. In order (1) to ensure preservation of muon spin polarization, in particular the P_T component, and (2) to decouple stray fields such as the earth field, a magnetic field with a strength of at least 300 G is applied at the stopper. As in E246 the azimuthal field arrangement is adopted favoring the fwd/bwd scheme. The muon field magnet produces a uniform field distribution on the stopper.

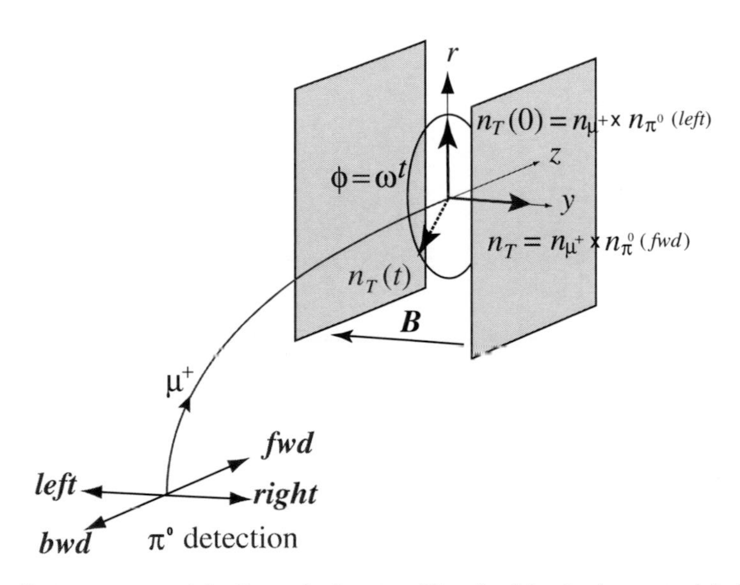

Fig. 6. P_T measurement in the polarimeter. The fwd–bwd scheme and $left$–$right$ scheme are shown.

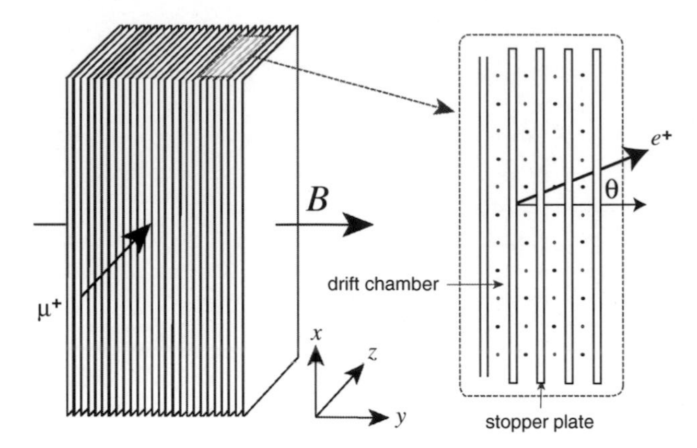

Fig. 7. Schematic view of the active polarimeter with stopper plates and drift chambers.

A parallel plate configuration will be adopted as shown in Fig. 7. The plates are made of light metal (alloy) such as Al or Mg providing the average density of $\sim 1/4\rho_{Al}$. They are arranged in a parallel orientation to the spectrometer gap, namely parallel to the incident muon momentum. Due to the multiple scattering when passing through the Cu degrader in front of the polarimeter, the divergence of the incoming muon beam (in the y direction also) is significant ($\Delta\theta \sim 0.1$) and there is no problem with the muon stopping. A stopping efficiency higher than 0.85 will be obtained. A drift chamber is constructed with these plates as ground potential forming cells with a size of the plate gap. Clearly, this is the best arrangement for the *cw*/*ccw* asymmetry measurement, as a single cell acts both as *cw* cell and a *ccw* cell for the two neighboring plates. The channel inefficiency cancellation works perfectly.

5.2 Muon Field Magnet

A uniform muon field with a large enough strength is essential in the TREK experiment, whereas a passive field was used by guiding and trimming the main field of the superconducting magnet in E246. The unavoidable consequence was that there was a non-uniform strength distribution and a curved flux distribution at the stopper. A uniform field parallel to the P_T component provides the maximum analyzing power. A new muon field magnet will be made to provide such a field (Fig. 8). To accommodate the polarimeter the parallel gap of the dipole magnet must be about 30 cm. The area is determined to produce a uniform field distribution in the stopper region.

From the point of view of (1) spin relaxation suppression and (2) stray field decoupling, a stronger field is preferable. However, the field is limited by the interference with the toroidal magnet, in particular with its SC coils. Point (2) is regarded as the determining factor; assuming 0.3 G of an unwanted

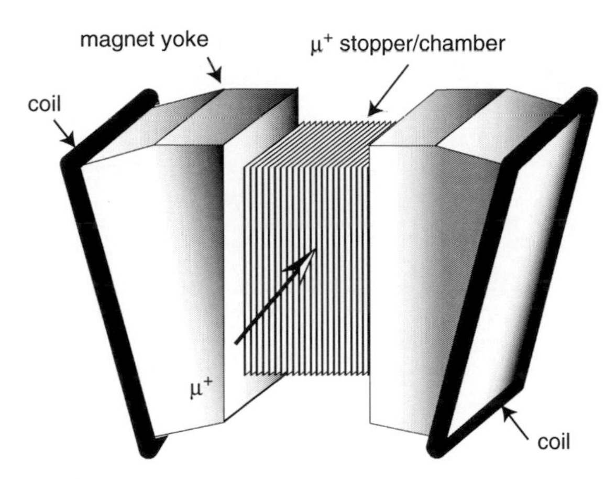

Fig. 8. Schematic view of the muon field magnets sandwiching the active polarimeter.

component in the shielded magnet gap, a field strength of at least 300 G is necessary to obtain a field alignment of 10^{-3}. Further alignment calibration is done by using experimental data.

The field symmetry across the median plane is important but a non-uniformity of a few times of 10^{-2} in strength as well as in the vector distribution is tolerable in the positron energy analysis. A parallel filed arrangement to the SC magnet field rather than the anti-parallel configuration will be selected in order to achieve flatter field uniformity.

5.3 Tracking System

In the TREK experiment two sources of systematic errors will dominate. While one source is the misalignment of the detector elements, in particular of the muon polarimeter, the other source will be given by the background contamination of muons from the decay-in-flight of $K_{\pi 2}$ pions ($K^+_{\pi 2}$-*dif* events). With the upgrade of the tracking system, the error from this background will be improved to meet the requirement of $< 10^{-4}$ for the total systematic error in P_T. These performance goals will be achieved both by reducing the material budget along the track and by rearranging existing and adding new tracking elements in replacement of the previous C1 chamber. The momentum uncertainty of 3.6 MeV/c in E246 can be reduced by at least a factor of 10 (1) by employing a 6-cm- instead of 9.3-cm-wide target with a segmentation of $3.0\times3.0\,\text{mm}^2$ fibers instead of $5\times5\,\text{mm}^2$, (2) by replacing the air volume in the magnet between C2 and C3 and before the C2 chambers with helium bags, and (3) by increasing the distance between the C3 and C4 elements to 30 cm from 15 cm. For sufficient identification and suppression of $K^+_{\pi 2}$-*dif* events we need to build a cylindrical tracking chamber ("C0") with a radius of 10 cm and a spatial resolution of <0.1 mm. The new C0 chamber will replace the previous cylindrical C1 chamber of the E246 setup. In order to increase tracking redundancy we propose to add a new planar tracking element (again named "C1") with <0.1 mm resolution to cover each of the 12 gaps at the outer surface of the CsI(Tl) calorimeter. By adding these additional elements to the track fitting procedure, the resulting χ^2 per degree of freedom will be much more effective to distinguish tracks from $K^+_{\pi 2}$-*dif* from regular tracks which do not have a kink along their path. In combination with the higher segmentation of the fiber target this will be sufficient to suppress the $K^+_{\pi 2}$-*dif* $/K^+_{\mu 3}$ ratio below 10^{-3}, rendering a spurious $P_T < 5 \times 10^{-5}$.

The planned modifications are in summary:

(1) Thinner target with higher segmentation.
(2) Helium gas bags in the magnet between C2 and C3, and before C2.
(3) Increase in the distance between C3 and C4 to 30 cm from 15 cm.
(4) Addition of new tracking elements: "C0" and "C1" chambers based on GEM technology.

Figure 9 and Table 4 show comparisons of the tracking system in E246 and the TREK experiment. The GEM technology on which both C0 and C1 will be based presents a new generation of position-sensitive counters that are reasonably cheap, radiation hard, and well suited to be operated in high-rate environments.

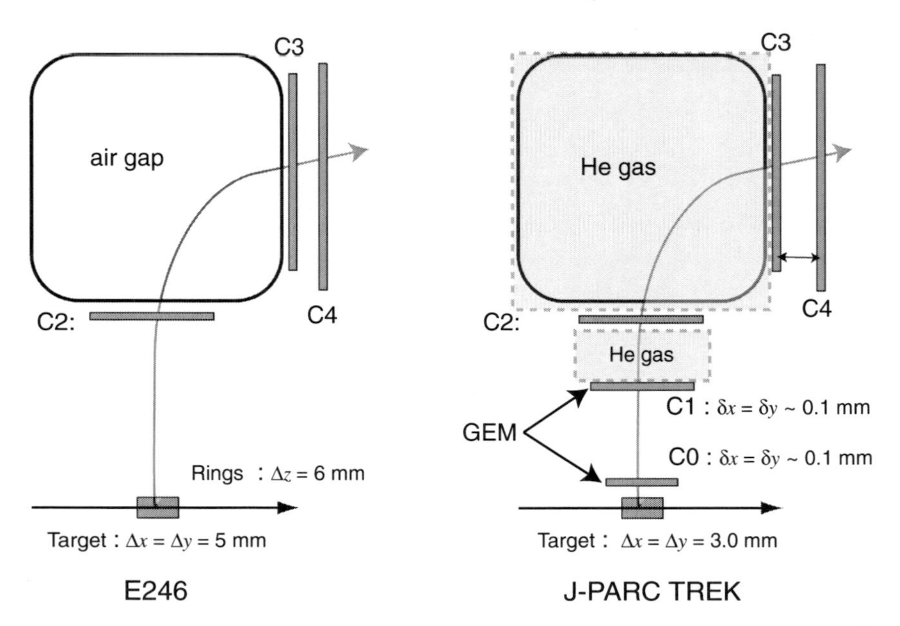

Fig. 9. Schematics of the tracking system in the TREK setup (*right*) compared with that of the E246 experiment (*left*).

Table 4. Main parameters of the charged particle tracking.

Item	Value	E246
High-resolution elements	C0, C1, C2, C3, and C4	C2, C3, and C4
Target fiber	3×3 mm	5×5 mm + rings
C0 chamber	Cylindrical GEM	MWDC
C1 chamber	Planar GEM chamber	–
C2 chamber	MWPC (not changed)	MWPC
C3 chamber	MWPC (not changed)	MWPC
C4 chamber	MWPC (not changed)	MWPC
C3–C4 distance	30 cm	15 cm
Magnet gap	He gas bag	Air
Total material thickness	$\sim 7 \times 10^{-3} X_0$	$6.6 \times 10^{-3} X_0$

5.4 CsI(Tl) Calorimeter

The photon calorimeter is a barrel of 768 CsI(Tl) crystals surrounding the target region (Fig. 10). There are 12 so-called muon holes to let the charged particle enter the spectrometer. The solid angle coverage is therefore not 4π sr but about 3π sr including also the losses due to the beam entrance and exit holes. The size of the muon hole was optimized for the $K^+_{\mu3}$ acceptance to be maximum [51]. The barrel structure is symmetric for the upstream (backward) and the downstream (forward) side, which is essential in the present experiment. The barrel was assembled very carefully ensuring a local as well as a global precision of better than 1 mm. The main parameters of the barrel are summarized in Table 5.

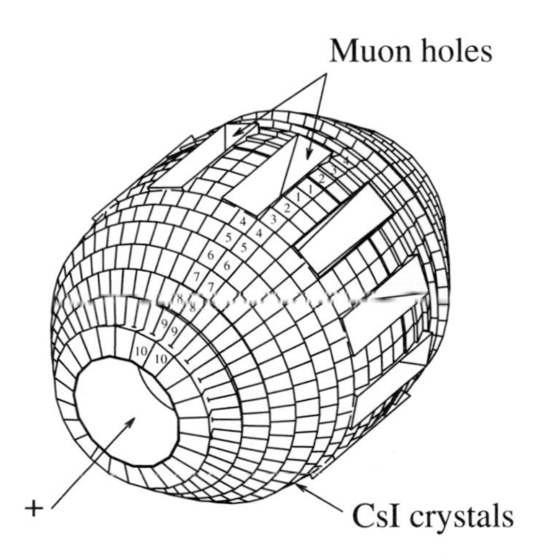

Fig. 10. CsI(Tl) barrel with 768 crystal modules. There are 12 muon holes.

Table 5. Main parameters of the CsI(Tl) calorimeter.

Parameter	Value
Number of CsI(Tl) crystals	768
Segmentation	$\Delta\theta = \Delta\phi = 7.5^\circ$
Inner/outer diameter	41/90 cm
Detector length	141 cm
Solid angle coverage	~75%
Crystal length	25 cm (13.5 X_0)
Typical size of crystals	$3 \times 3 - 6 \times 6$ cm^2
Wavelength at peak	560 nm
Light decay time	~900 ns

Considering the limitations in the PIN + preamplifier scheme for high-rate operation, we plan to adopt another CsI(Tl) readout method. One possibility is to use magnetic field-resistant photo-multiplier tubes now available with fairly large multiplication. However, the space limitation will still be a problem. There is also a cost question. Thus, we are led to consider avalanche photo-diodes (APD) of reverse bias type. Such APDs with a multiplication factor of about 100 with reasonably large sensitive areas are commercially available [54]. Several applications to calorimeters in high-energy physics experiments are being prepared. The wavelength matching to CsI(Tl) with the peak wavelength of 560 nm is better than PWO[55] and an average quantum efficiency greater than 80% can be expected. The readout of CsI(Tl) by an APD has already been studied in Ref. [56] by using Hamamatsu S8664 for a small crystal and it was found to work well.

A several times larger electron yield than the PIN (E246) with 5×10^6/100 MeV allows us to use a current preamplifier with high gain in place of the costly charge preamplifier which would also encounter the rate limitation due to the output voltage dynamic range. A high-rate performance test of APD readout has been performed using a positron beam for one crystal module with a prototype amplifier for S8664. About 500 kHz signal rate in one module will be tolerable and analyzable using an FADC (Table 6).

5.5 Alignments

Since the misalignments of detector elements may cause systematic errors, precise alignment is essential in the TREK experiment. The alignment of the tracking system and the CsI(Tl) calorimeter system relative to the reference system of the spectrometer will be performed using a set of calibration collimators for the former and $K_{\pi 2}$ events for the latter. Although careful designs are required for both, we regard the calibration procedure to be rather straightforward; the performance of the calibration can be easily checked with simulations. They are all *fwd/bwd* cancelling and thus controllable. On the

Table 6. Readout scheme using APDs compared with the E246 PIN readout.

Parameter	E246-PIN	APD readout
Diode	S3204-03	S8148 (or equiv.)
Total area	$18\times18\,\mathrm{mm}^2$	$5\times5\,\mathrm{mm}^2$
Quantum efficiency	$\sim$0.70	> 0.80
Photoelectron/GeV	1.1×10^7	1.0×10^6
Diode gain	1	50
Electron yield@ 100 MeV	1.1×10^6	5.0×10^6
Preamplifier	Charge sensitive	Current amplifier
Maximum rate	34 kHz	$\sim$500 kHz

contrary, the effect of polarimeter misalignments, which are direct systematics affecting the positron asymmetry, $A_{fwd(bwd)}$, is complicated with the entanglement of several factors including the muon field. Moreover, one of the misalignments, the rotation of the muon field around the z-axis δ_z, is the systematics which cannot be canceled out in the normal π^0 fwd/bwd subtraction scheme. In the following we present the alignment method of the polarimeter, which we regard as the most important in this experiment.

The misalignment of the polarimeter is characterized by four parameters: global rotation of the active stopper (1) around the r-axis, ϵ_r-and (2) around the z-axis, ϵ_z, and global rotation of the muon field distribution (3) around the r-axis, δ_r, and (4) around the z-axis, δ_z. They are only responsible for spurious $A_{fwd(bwd)}$; parallel displacements should not play a role as long as the active stopper covers the whole muon stopping region because of the parallel-shift symmetric structure. The rotation about the y-axis should not have any effect since it brings about only a rotation around the azimuthal axis.

In order to treat this problem an innovative method [57] will be performed using $K_{\mu3}$ data so as to remove any misalignment effects from the P_T analysis. The time integrated e^+ left/right asymmetry $\bar{A}$ due to the misalignments can be described for a certain initial muon spin phase in the median planes θ_0 measured from the direction of z beam axis as follows:

$$\bar{A}\,(\theta_0) = \alpha_0[\delta_r \cos\theta_0 - \delta_z \sin\theta_0 + \eta\,(\theta_0)], \tag{22}$$

where α_0 is the analyzing power for the polarization determination from the e^+ asymmetry. The oscillation terms with ωt are drastically reduced by the time integration and the remaining imperfect cancellation is expressed as $\eta(\theta_0)$. It is noted that the misalignments of the active stopper were integrated out into $\eta(\theta_0)$ and hence they are not relevant in the further discussion.

To extract the misalignment parameters δ_r and δ_z in the presence of a real P_T signal, we calculate two asymmetries A_{sum} and A_{sub} as a function of θ_0, i.e., the sum and difference of A_{fwd} and A_{bwd} – the asymmetries at the forward and backward pions, respectively. This leads to

$$A_{sum}(\theta_0) = \left[\bar{A}_{fwd}(\theta_0) + \bar{A}_{bwd}(\theta_0)\right]/2 \cong \alpha_0 \left[\delta_r \cos\theta_0 - \delta_z \sin\theta_0\right], \quad (23a)$$

$$A_{sub}(\theta_0) = \left[\bar{A}_{fwd}(\theta_0) - \bar{A}_{bwd}(\theta_0)\right]/2 = F(P_T, \theta_0), \quad (23b)$$

where $F(P_T, \theta_0)$ is the A_T asymmetry function only from a P_T origin and it does not involve any misalignment effects. Thus, we can extract P_T from $F(P_T, \theta_0)$ unaffected by the misalignments.

The validity of this method must be demonstrated with an MC calculation with high statistics. This MC study should show a unique determination of the misalignments when several misalignments are existing simultaneously and to check for any bias in the analysis including a possible bias in the tracking code. Assuming the existence of both δ_z and δ_r but no P_T, the A_{sub} analysis was performed using 25 billion events of $K_{\mu 3}$ decays, and P_{sub}^{av} was obtained from the averaged A_{sub} to be $P_{sub}^{av} = \alpha_0 A_{sub}^{av} = (0.3 \pm 0.6) \times 10^{-4}$. The result, which is consistent with zero within the statistical error, indicates the appropriateness of this analysis including the absence of any bias in the analysis code.

5.6 Beamline

In order to perform the TREK experiment using stopped K^+s, a separate low-momentum K^+ beamline with a good K/π ratio is required. In the phase 1 experimental hall of J-PARC, however, there will be only one primary proton line, A-line, and only one target station T1. All the secondary lines are going to be installed at this target. The construction of the beam facility is now proceeding according to the policy to accommodate the nuclear physics experiments which were nominated as “day-1 experiments.” Thus, the K1.8 and K1.1 lines are most likely to be built first, and the structure of the target station and the front end of the channels are now almost fixed. Considering the high-current operation of the facility, those designs with a very limited channel acceptance seem to be unique, and it will be difficult to modify the structure after starting a full-intensity operation. Even under such a situation, we pursue the possibility of a low-momentum beamline K0.8 with a beam momentum of 0.8 GeV/c which is optimum for the planned experiment. We have found a good solution, which is a branch line of the K1.1 line, and it can provide a high-quality low-momentum beam with a sufficiently good K/π ratio. The beam optics of this branch have been designed.

The quality of the 0.8 GeV/c separated kaon beam at this branch line, with a single stage of separation with a K^+/π^+ ratio larger than 2, is very good for our purposes. This is achieved by the existence of a vertical focus and also by the horizontal focus additionally designed for the K0.8 branch. The acceptance of K0.8 is, however, determined primarily by the upstream acceptance of K1.1, namely by the distance of the first focusing element from the target, which cannot be shorter. The calculated acceptance of ~4.5 msr(%$\Delta p/p$) is smaller by a factor of 10 than the C4 (LESB3) line at BNL-AGS of the nearly same length for 0.8 GeV/c operation. Although there might be some ambiguity due

Table 7. Main parameters of the K0.8 beam.

Parameter	Value
Momentum	800 MeV/c
Momentum bite	±2.5%
Channel length	19 m
Channel acceptance	4.5 msr%($\Delta p/p$)
π^+/K^+ ratio	< 0.5
K^+ intensity	3×10^6 @ 9 μA proton beam
Beam spot	$H = \pm 0.3$ (FWHM) cm, $V = \pm 0.4$ (FWHM) cm
	$H = \pm 1.6$ (total) cm, $V = \pm 1.6$ (total) cm
Dispersion at final focus	Achromatic

to the detailed target structure, we can roughly estimate the K^+ intensity to be 2×10^6/s at $I_p = 5.4 \times 10^{13}$/s (9 μA) and $E_p = 30$ GeV by scaling the known LESB3 intensity at $E_p = 24$ GeV. The main parameters of K0.8 are summarized in Table 7.

6 Sensitivity

6.1 Statistical Sensitivity

For a conservative estimate we assume the sensitivity coefficient for the integral analysis taking the *fwd* and *bwd* regions of the π^0 events. Several comments are presented for the other parameters.

1. The average beam intensity at the K0.8 beamline is assumed to be 2×10^6/s. Although it is stated that some beam commissioning period with a low accelerator beam intensity is necessary, our total beam request is 1.4×10^7 s of beam time with this kaon beam intensity. We estimate the sensitivity based on the total number of kaons of 3×10^{13}.
2. The fraction of *fwd* and *bwd* regions is 30% of the total good $K_{\mu 3}$ events including the *left* and *right* regions. It is somewhat difficult to estimate the analysis efficiency as it strongly depends on many details. However, at least for now, we can assume that it would be better than what we attained in E246. We use a conservative estimate of 0.67, the E246 efficiency.

The deduction of the statistical error is summarized in Table 8. In the standard *fwd*/*bwd* π^0 analysis a statistical error of $\delta P_T = 1.2 \times 10^{-4}$ will be obtained from a 1-year (1.4×10^7 s) run. An analysis including the *left* and *right* regions will provide a smaller error of $\delta P_T = 1.0 \times 10^{-4}$. A more ambitious weighted analysis event by event should attain the highest sensitivity of $\delta P_T = 0.8 \times 10^{-4}$ although the systematic errors have yet to be investigated carefully in this case.

Table 8. Deduction of statistical error.

Parameter	Value
Net runtime	1.4×10^7s
Proton beam intensity	9 μA on T1 target
K^+ beam intensity	2×10^6/s
Total number of good $K_{\mu 3}$	2.4×10^9
Total number of *fwd* and *bwd* (N)	7.2×10^8
Sensitivity coefficient	$3.34/\sqrt{N}$
δP_T for *fwd* and *bwd*	1.2×10^{-4}
δP_T for all π^0	1.0×10^{-4}
δP_T in weighted analysis	0.8×10^{-4}

6.2 Systematic Errors

As mentioned before the main sources of systematic errors in E246 must be suppressed substantially at least by 1 order of magnitude. The following summary observations are in order:

- The effects of polarimeter misalignments, in particular the field rotation δ_z, are not more relevant to the P_T determination. If necessary, they can be calibrated using data. Monte Carlo simulation studies assuming considerably large misalignments for the rotation parameters ϵ_r, ϵ_z, δ_r, and δ_z showed the associated systematic error to be smaller than 10^{-4} in the discrepancy between the fit value and the input value.
- The influence of decay phase space distortion parameterized by the decay plane angular parameters θ_r and θ_z should be corrected. The error associated with these corrections is essentially a statistical one and is estimated to be far less than 10^{-4} for both θ_r and θ_z. The validity of the correction method can be checked by introducing an artificial asymmetry in, for example, the kaon stopping distribution in the target to produce significant θ_r and θ_z.
- The error due to $K^+_{\pi 2}$-*dif* background contamination can be suppressed by means of the new upgraded tracking system down to less than 5×10^{-5}.
- There is a new potential source of error which was not present in E246, namely the error coming from the active polarimeter analysis. Although the detailed design of the polarimeter has yet to be done, the effects of E_{e^+} and θ_{e^+} ambiguities have to be suppressed to the level smaller than 10^{-4}.
- The largest systematic error in E246, which was the ambiguity of muon stopping point due to scattering, does not exist in the TREK polarimeter anymore.

Other potential sources such as the misalignments of the tracking elements are regarded as rather harmless since the correction based on the alignment calibration can be done accurately enough. Although the necessary Monte

Table 9. Expectation of the systematic errors in E06.

Source	$\delta P_T^{syst}(10^{-4})$	Method
Polarimeter misalignment ($\delta_z, \delta_r, \epsilon_z$, and ϵ_r)	<1	Confirmed by MC simulation
$K_{\pi 2}$-*dif* background	$\ll 0.5$	Confirmed by MC simulation
Decay plane rotations (θ_z and θ_r)	$\ll 1$	Correction and data symmetrization
Positron analysis (E_{e+} and θ_{e+})	<1	*Fwd/bwd* cancellation, *fwd/bwd* symmetrization, etc.
Total	$\delta P_T^{syst} \preceq 10^{-4}$	Quadratic sum

Carlo studies will be continued we believe that each correction is applied with an uncertainty of less than 10% of the correction values and that the total systematic error can be made much smaller than 10^{-4}. These expectations are summarized in Table 9.

7 Summary

In summary, we plan to perform a high-precision measurement (TREK experiment) of the transverse polarization of muons in the $K_{\mu 3}^+$ decays, which constitutes a T-odd observable. This observable is one of the few tests of T-invariance corresponding to direct CP violation in non-neutral meson system. We aim to improve the precision of this measurement at least by a factor of 20 compared to the best result from our own KEK-PS-E246 and reach a limit of $\delta P_T{\sim}10^{-4}$ (Fig. 11). The FSI contribution in the SM descriptions is significantly smaller than the sensitivity of this type of measurement; however, several exotic models inspired by multi-Higgs doublet, etc., allow P_T values within the sensitivity attainable to us. Thus, this experiment is likely to find new sources of CP violation, if any of these models are viable. It will certainly constrain the parameter space of the candidate models. The sensitivity of this experiment is comparable or complementary to that of the proposed new neutron EDM experiments and other rare decay processes, since P_T as a semi-leptonic process spans the parameter space which is not covered by other channels.

The experiment will be performed using the stopped K^+ beam method as the previous KEK-E246 experiment. This method is suited for a double ratio measurement in terms of π^0 emission direction, enabling the efficient suppression of systematic errors, which is essential for the high-precision experiment TREK. Although the basic toroidal setup of E246 is used, several upgrades of the detector element will be done in order to meet the requirements from higher rate performance, larger acceptance, better background rejection, and more efficient suppression of the systematic errors. The major upgrades are as follows:

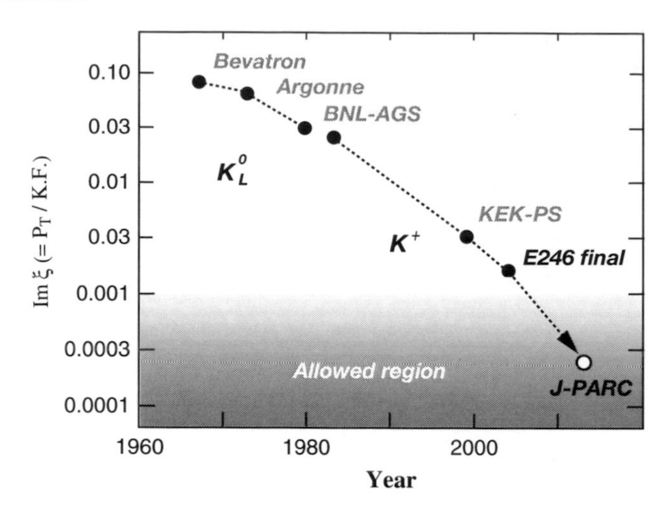

Fig. 11. Expected sensitivity (90% CL assuming a zero central value) of the TREK experiment.

- Adoption of active polarimeters to ensure the small systematic error, higher analyzing power using the energy and angle information of emitted positrons, and larger acceptance of nearly 4π. The decay positron measurement will be background free.
- Improvement of the charged particle tracking system by additional tracking elements using state-of-the-art GEM detectors for the innermost two detectors, C0 and C1. Together with a slimmer target with finer segmentation the new system can suppress the $K^+_{\pi 2}$-*dif* background drastically.
- Faster readout of the CsI(Tl) calorimeter by means of APD and FADC. The rate capability will be more than 10 times higher.
- Alignment of the detector elements with high precision. Especially, the global alignment of the polarimeter and the muon field which are the most serious systematics affecting the positron asymmetry will be done using real data as one of the processes to deduce P_T.

We are now planning to perform the TREK experiment in the early stage of the phase 1 period of J-PARC. The high-quality beam from the K0.8 beamline and the upgraded high-precision detector will enable us to attain the sensitivity of $\delta P_T \sim 10^{-4}$ in a 1-year run of the measurement.

Acknowledgments

The author would like to thank all the members of the TREK collaboration for valuable discussions. He would also like to acknowledge the discussions with the J-PARC hadron facility group on the beamline.

References

1. Bigi, I.I., Sanda, A.I.: CP Violation. Cambridge University Press (2000)
2. See, for example, Prologue of [1]
3. Sakurai, J.J.: Phys. Rev. **109**, 980 (1957)
4. Zhitnitkii, A.R.: Yad. Fiz. **31**, 1014 (1980)
5. Efrosinin, V.P., et al.: Phys. Lett. **B493**, 293 (2000)
6. Blatt, S.R., et al.: Phys. Rev. **D27**, 1056 (1983)
7. Abe, M., et al.: Phys. Rev. **D73**, 072005 (2006)
8. J-PARC proposal P06; Measurement of T-violating Muon Polarization in $\boldsymbol{K^+ \to \pi^0\,\mu^+\nu}$ Decays. http://j-parc.jp/NuclPart/Proposal e.html (2006)
9. Cabibbo, N., Maksymowicz, A.: Phys. Lett. **9**, 352 (1964)
10. Cabibbo, N., Maksymowicz, A.: Phys. Lett. **11**, 360(E) (1964)
11. Cabibbo, N., Maksymowicz, A.: Phys. Lett. **14**, 72(E) (1966)
12. Eidelman, S. et al.: [Particle Data Group], p.618 in "Review of Particle Physics". Phys. Lett. **B592**, 1 (2004)
13. Wu, G.-H., Kiers, K., Ng, J.N.: Phys. Rev. **D56**, 5413 (1997)
14. Kobayashi, M., Lin, T.-T., Okada, Y.: Prog. Theor. Phys. **95**, 361 (1995)
15. Leurer, M.: Phys. Rev. Lett. **62**, 1967 (1989)
16. Golowich, E., Valencia, G.: Phys. Rev. **D40**, 112 (1989)
17. Harris, P. [EDM collaboration]: Preliminary result presented at SUSY-2005, Durham, June 2005
18. Graham, M.: Talk given at "Flavor Physics and CP Violation", Vancouver, April 9–12, 2006
19. Ahn, J.K. et al.: Phys. Rev. Lett. **100**, 201802 (2008)
20. Particle Data Group: Review of particle physics, J. Phys. G: Nucl. Part. Phys. **33**,1, 688 (2006)
21. Garisto, R., Kane, G.: Phys. Rev. **D44**, 2038 (1991)
22. Bélanger, G., Geng, C.Q.: Phys. Rev. **D44**, 2789 (1991)
23. Sanda, A.I., Phys. Rev. **D23**, 2647 (1981)
24. Deshpande, N.G.: Phys. Rev. **D23**, 2654 (1981)
25. Cheng, H.-Y.: Phys. Rev. **D26**, 143 (1982)
26. Cheng, H.-Y.: Phys. Rev. **D34**, 1397 (1986)
27. Bigi, I.I., Sanda, A.I.: Phys. Rev. Lett. **58**, 1604 (1987)
28. Leurer, M.: Phys. Rev. Lett. **62**, 1967 (1989)
29. H.-Cheng, Y.: Phys. Rev. **D42**, 2329 (1990)
30. Grossman, Y.: Int. Mod, J. Phys. **A19**, 907 (2004)
31. Hurth, T.: Rev. Mod. Phys. **75**, 1159 (2003)
32. Ikado, K., et al. (Belle collaboration): Phys. Rev. Lett. **97**, 251802 (2006)
33. Grossman, Y.: Nucl. Phys. **B426**, 355 (1994)
34. Okada, Y., Shimizu, Y., Tanaka, M.: arXiv:hep-pf/9704223
35. Wu, G.-H., Ng, J.N.: Phys. Lett. **B392**, 93 (1997)
36. Fabbrichesi, M., Vissani, F.: Phys. Rev. **D55**, 5334 (1997)
37. Christova, E., Fabbrichesi, M.: Phys. Lett. **B315**, 113 (1993)
38. Dreiner H.: An introduction to explicit R-parity violation. In: Kane, G.L. (ed.) Perspectives on Supersymmetry (1998)
39. Chen, S.-L. et al.: JHEP **09**, 044 (2007); http://jhep.sissa.it/archive/papers/jhep092007044/jhep092007044.pdf
40. Barbier R., et al.: Phys. Rept. **420**, 1 (2005); [hep-ph/0406039]

41. Chemtob, M.: Prog. Part. Nucl. Phys. **54**, 71 (2005); [arXiv:hep-ph/0406029 2 June 2008]
42. Deandrea, A., et al.: JHEP **10**, 038(2004); [arXiv:hep-ph/0407216v2 20 Oct 2004]
43. Young, K., et al.: Phys. Rev. Lett. **18**, 806 (1967)
44. Longo, J.J., Ypung, K.K., Helland, J.A.: Phys. Rev. **181**, 1808 (1969)
45. Sandweiss, J. et al.: Phys. Rev. Lett. **30**, 1002 (1973)
46. Morse, W. M. et al.: Phys. Rev. **D21**, 1750 (1980)
47. Tanaka, K.H. et al.: Nucl. Instrum. Methods Phys, Res. Sect. **A363**, 114 (1995)
48. Imazato, J., et al.: Proc. 11th Int. Conf. on Magnet Technology, Tsukuba, Japan, August 1989
49. Ikeda, T., et al.: Nucl. Instr. Meth. Phys. Res. Sect. **A401**, 243 (1997)
50. Macdonald, J.A. et al.: Nucl. Instrum. Meth. Phys. Res. Sect. **A506**, 60 (2003)
51. Dementyev, D.V. et al.: Nucl. Instrum. Meth. Phys. Res. Sect. **A440**, 15 (2000)
52. Yu. G., Kudenko, et al.: Nucl. Instr. Meth. Phys. Res. Sect. **A411**, 437 (1998)
53. Ivashkin, A.P., et al.: Nucl. Instrum. Meth. Phys. Res. Sect. **A394**, 321 (1997)
54. Hamamatsu Photonics K.K.: Technical Information SD-28, Oct, 2001; Cat. No. KAPD1012J02, April, 2004
55. Britvitch, I., et al.: Nucl. Instrum. Meth. Phys. Res. Sect. **A535**, 523 (2004)
56. Ikagawa, T. et al.: Nucl. Instrum. Meth. Phys. Res. Sect. **A538**, 640 (2005)
57. Shimizu, S.: Proceedings of the Kaon International Conference 2007, Frascatti (2007), http://pos.sissa.it//archive/conferences/046/025/KAON 025.pdf

Gamma-Ray Spectroscopy of Λ Hypernuclei

Hirokazu Tamura

Department of Physics, Tohoku University, Sendai 980-8578, Japan
tamura@lambda.phys.tohoku.ac.jp

1 Introduction

Precision γ spectroscopy using germanium (Ge) detectors is one of the most powerful means to study nuclear structure. However, structure of Λ hypernuclei was long investigated mostly by reaction spectroscopy methods using the $(K^-,\ \pi^-)$ and $(\pi^+,\ K^+)$ reactions, and γ-ray spectroscopy did not play a major role because of technical difficulties in using Ge detectors for hypernuclear experiments. In 1998 we developed a large-acceptance Ge detector array equipped with fast readout electronics (Hyperball) and succeeded in observing hypernuclear γ-rays for the first time with Ge detectors [1, 2]. Since then we have observed more than 20 γ transitions from several Λ hypernuclei, ${}^{7}_{\Lambda}$Li, ${}^{9}_{\Lambda}$Be, ${}^{11}_{\Lambda}$B, ${}^{12}_{\Lambda}$C, ${}^{15}_{\Lambda}$N, and ${}^{16}_{\Lambda}$O in KEK E419, E509, E518, E566, and BNL E930 experiments [1–11] with an energy resolution of a few keV (FWHM), which is drastically improved from 1 to 2 MeV (FWHM) resolution in reaction spectroscopy. Precision hypernuclear γ spectroscopy has been established as a new frontier in strangeness nuclear physics.

Figure 1 shows all the hypernuclear γ-ray transitions observed and level schemes identified since 1998. The experiments have been carried out at KEK-PS via the (π^+, K^+) reaction using 1.05 GeV/c π^+ beams and at BNL-AGS via the $(K^-,\ \pi^-)$ reaction using 0.9 GeV/c K^- beams. The ${}^{13}_{\Lambda}$C hypernucleus was studied with an NaI detector array at BNL-AGS (E929) [12], while all the others were investigated with Hyperball (and Hyperball2 for KEK E566). For example, Fig. 2 shows γ-ray spectra for ${}^{9}_{\Lambda}$Be and ${}^{16}_{\Lambda}$O taken at BNL using Hyperball, revealing "hypernuclear fine structure" (see Sect. 4.2), the level structure typical in hypernuclei. See Ref. [13] for details of the Hyperball project and physics results.

At the 50-GeV PS in the J-PARC facility where intense K^- beams are available, we will be able to investigate various bound-state levels of a wide range of hypernuclei from ${}^{4}_{\Lambda}$He to ${}^{208}_{\Lambda}$Pb. Accumulated data on precise structure of hypernuclei will be used to extract quantitative information on detailed properties of the hyperon–nucleon (YN) interactions. Transition probabilities

Tamura, H.: *Gamma-Ray Spectroscopy of Λ Hypernuclei.*
Lect. Notes Phys. **781**, 105–138 (2009)
DOI 10.1007/978-3-642-00961-7_5

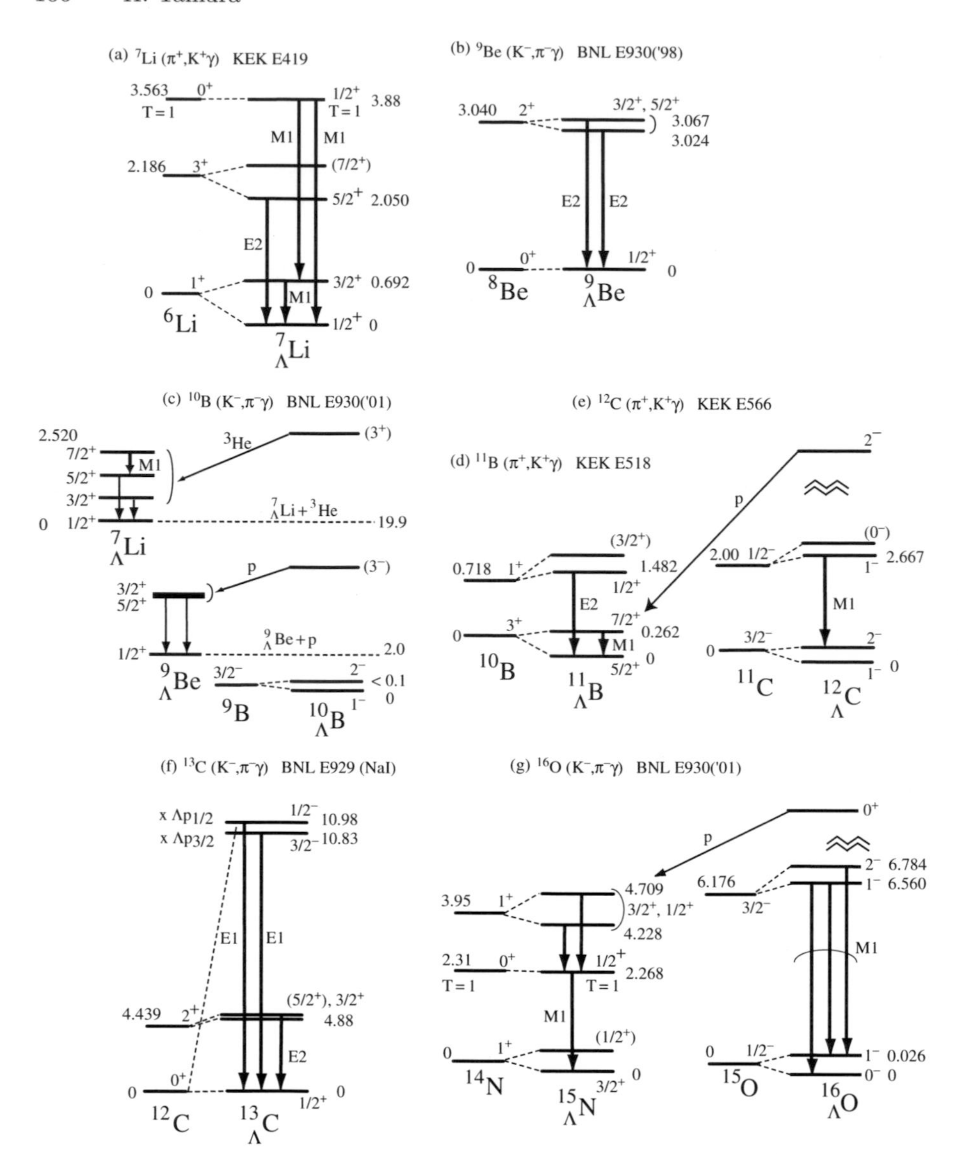

Fig. 1. γ-Ray transitions and level schemes of p-shell Λ hypernuclei that have been investigated by γ-ray spectroscopy since 1998. The Ge detector array, Hyperball (and Hyperball2 for KEK E566), was used to investigate all these hypernuclei [1–11], except for ${}^{13}_{\Lambda}$C which was studied with an NaI detector array [12].

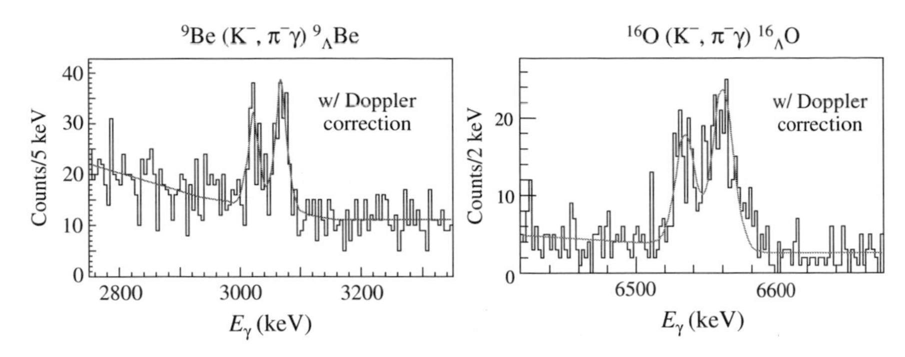

Fig. 2. γ-Ray spectra for ${}^{9}_{\Lambda}$Be [3, 7] and ${}^{16}_{\Lambda}$O [6], after Doppler shift correction, measured with Hyperball via the (K^-, π^-) reaction at BNL-AGS. They exhibit "hypernuclear fine structure" described in Fig. 3. *Solid lines* show results of the fitting using calculated peak shapes considering Doppler shift correction.

of $B(E2)$ and $B(M1)$ will also be measured for various hypernuclear transitions. $B(E2)$ values provide information on the size and deformation of hypernuclei and allow us to discuss nuclear structure change induced by a Λ hyperon as an impurity, while $B(M1)$ values provide a g-factor of Λ in nuclei, which may be modified from that in free space.

2 Physics Subjects and Experimental Plans

The physics subjects that can be pursued by precision hypernuclear γ spectroscopy are classified into

1. YN interactions
2. impurity effect in nuclear structure
3. medium effect in baryon properties

In the following, we describe details of the three physics subjects including present status of their studies.

2.1 YN Interaction

One of the most important motivations for hypernuclear physics is to investigate hyperon–nucleon (YN) and hyperon–hyperon (YY) interactions. Their direct studies using hyperon–nucleon scattering experiments are extremely difficult due to short lifetimes of hyperons. On the other hand, hypernuclear structure provides us with quantitative information on YN and YY interactions. In particular, combined with theoretical studies for nuclear structure using shell model calculations together with G-matrix method and few-body calculations with variational method, accumulated data on precise structure of Λ hypernuclei provided by γ-ray spectroscopy are used not only to understand spin dependence (spin–spin, spin–orbit, and tensor forces) in the ΛN interaction but also to investigate Λ–Σ coupling force and charge symmetry breaking (CSB) in the interaction.

These properties of the ΛN interaction play essential roles to discriminate and improve baryon–baryon interaction models, not only those based on meson exchange picture but also those including quark–gluon degree of freedom, toward unified understanding of the baryon–baryon interactions. For example, the very small Λ-spin-dependent spin–orbit force established by recent γ spectroscopy experiments of ${}^{9}_{\Lambda}$Be [3] and ${}^{13}_{\Lambda}$C [12] has aroused the discussion about the role of quark–gluon degrees of freedom [14], while the ΛN tensor force strength determined from our ${}^{16}_{\Lambda}$O γ spectroscopy data [9] is explained well in the meson exchange framework. In addition, good knowledge on the YN and YY interactions is essential to describe high-density nuclear matter containing hyperons which is expected to be realized in the center of neutron stars.

Spin-Dependent Interactions

Structure of a hypernucleus is understood in weak coupling scheme between a core nucleus and a Λ. As shown in Fig. 3, each level of the core nucleus with a spin J_C is split into a doublet with spins $J_C + 1/2$ and $J_C - 1/2$, of which spacing is governed by the spin-dependent components of the ΛN interaction. Due to small strengths of the spin-dependent ΛN interaction, the spacing is usually of the order of 10–100 keV, being generally smaller than level spacings of low-lying states of the core nucleus, and thus such a doublet is called "hypernuclear fine structure." In order to separate such a

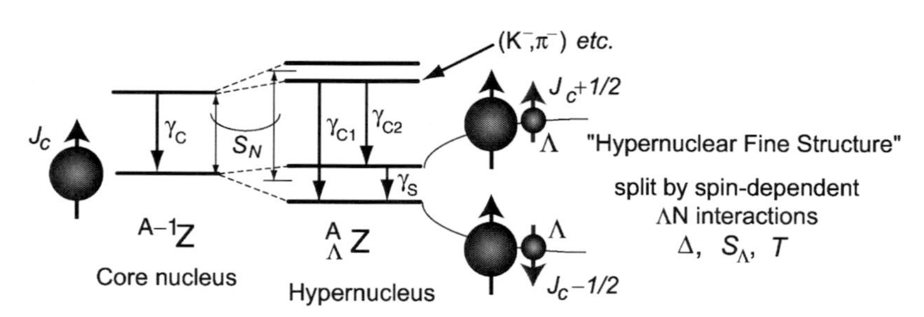

Fig. 3. Schematic scheme of low-lying levels of a Λ hypernucleus and their γ transitions. When a Λ is coupled to the core nucleus with spin J_c, the level is split into a doublet($J_c + 1/2$, $J_c - 1/2$). The excellent resolution of Ge detectors is essential to resolve each member of the doublet which is split by the ΛN spin-dependent interactions by a small spacing (typically <0.1 MeV) (called "hypernuclear fine structure"). γ_{C1} and γ_{C2} are called "core transitions," being essentially the same as γ_C transition in the core nucleus, while γ_S is called "Λ spin-flip $M1$ transition."

fine structure, precision γ spectroscopy using Ge detectors with a few keV resolution is indispensable.

The ΛN interaction may be expressed as follows:

$$V_{\Lambda N}(r) = V_0(r) + V_\sigma(r)\mathbf{s}_N\mathbf{s}_\Lambda + V_\Lambda(r)\mathbf{l}_{N\Lambda}\mathbf{s}_\Lambda + V_N(r)\mathbf{l}_{N\Lambda}\mathbf{s}_N + V_T(r)[3(\boldsymbol{\sigma}_N\hat{\mathbf{r}})(\boldsymbol{\sigma}_\Lambda\hat{\mathbf{r}}) - \boldsymbol{\sigma}_N\boldsymbol{\sigma}_\Lambda]. \qquad (1)$$

From level structure of p-shell Λ hypernuclei, we can investigate spin dependence of the effective interaction between a Λ in $0s$ orbit and a nucleon in $0p$ orbit. The effective $s_\Lambda p_N$ (and $p_\Lambda s_N$) interactions have five radial integrals corresponding to each of the five terms in Eq. (1). They are denoted by $\bar{V}$, Δ, S_Λ, S_N, and T, respectively [15, 16], which represent the strengths of the spin-averaged central force, the spin–spin force, the Λ-spin-dependent spin–orbit force, the nucleon-spin-dependent spin–orbit force, and the tensor force, respectively. These integrals (effective interaction parameters) can be determined phenomenologically from low-lying level structure of p-shell hypernuclei. Before we started the hypernuclear γ spectroscopy project with Hyperball, the spin-dependent strengths (Δ, S_Λ, S_N, and T) were not well known; the available hypernuclear data were not sufficient in quality and quantity to unambiguously determine them. The study of these ΛN spin-dependent interactions has been the main motivation of the previous Hyperball experiments.

By comparing the level data taken in KEK E419 and BNL E930 experiments with results of the shell model calculations by Millener [17], it is found that the level spacings of doublets (hypernuclear fine structure), ${}^{7}_{\Lambda}$Li($3/2^+$,$1/2^+$)[1], ${}^{7}_{\Lambda}$Li($7/2^+$,$5/2^+$)[9], ${}^{9}_{\Lambda}$Be($3/2^+$,$5/2^+$)[3, 7], ${}^{11}_{\Lambda}$B($7/2^+$,$5/2^+$) [10], ${}^{16}_{\Lambda}$O(1^-,0^-), [6] and ${}^{16}_{\Lambda}$O(2^-,1^-) [11] (see Fig. 1), can be well reproduced with a common set of the effective interaction parameters (Δ, S_Λ, T), with a slight modification in Δ depending on the mass number. In addition, most of the core-excited level energies in the p-shell hypernuclei are also explained from a common value of the S_N parameter. The determined values are

$$\Delta = 0.33 \text{ or } 0.43, \quad S_\Lambda = -0.01, \quad S_N = \quad 0.4, \quad T = 0.02 \text{ (MeV)}. \qquad (2)$$

Here, the spin–spin strength, which was determined to be $\Delta = 0.43$ MeV for $A = 7$, needs to be slightly changed to $\Delta = 0.33$ MeV for $A = 11, 16$ [9, 17].

This parameter set cannot reproduce some experimental level energies in $A = 10, 11, 12$ hypernuclei; the ${}^{10}_{\Lambda}$B data of $E(2^-) - E(1^-) < 100$ keV corresponds to $\Delta < 0.3$ MeV [7], the ${}^{11}_{\Lambda}$B data of $E(5/2^-) - E(1/2^-) = 1.48$ MeV to $S_N = -0.9$ MeV [8], and the ${}^{12}_{\Lambda}$C data of $E(1^-_2) - E(2^-) = 2.67$ MeV to $S_N = -0.7$ MeV [10]. To solve this inconsistency, the effect of the three-body force due to ΛN–ΣN coupling described below should be taken into account for each hypernuclear level. In addition, inaccuracy of the core-nuclear wave function might be manifested by the presence of a Λ hyperon.

ΛN-ΣN Coupling and Three-Body Force

It has been pointed out that the level energies of hypernuclei may not be able to be understood well only from the two-body effective interaction because of

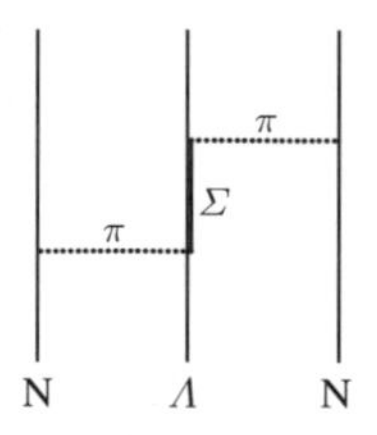

Fig. 4. Diagram for the ΛNN three-body force due to intermediate Σ states introduced by ΛN–ΣN couplings.

the ΛNN three-body force due to ΛN–ΣN coupling as shown in Fig. 4. It has been understood that the three-body force gives rise to a large effect to the binding energies of the 1^+ and 0^+ states of $A = 4$ hypernuclei [18, 19].

The ΛNN three-body force is expected to be stronger than the $3N$ force via the intermediate Δ states, since the Σ hyperon in the intermediate state is only 80 MeV heavier than the Λ hyperon compared to the 300 MeV difference in the case of nucleon and Δ. In addition, since the ΛNN force is mediated by two pion exchanges while one pion exchange between a Λ and a nucleon is forbidden by isospin conservation, importance of the ΛNN force relative to the ΛN force is expected to be larger than the case of the $3N$ force. Therefore, knowledge on the ΛNN three-body force as well as the two-body force is indispensable to understand the structure of hypernuclei.

The ΛNN force, which is not renormalizable into two-body effective ΛN interaction, changes hypernuclear level energies so that they cannot be expressed only in terms of the five ΛN two-body effective interaction parameters in Eq. (1). Therefore, when we determine the parameter set redundantly, the effect of the ΛNN force becomes visible from consistency of the parameters obtained from various hypernuclear data. Actually, our data of ${}^{10}_{\Lambda}\mathrm{B}$ taken in E930 cannot be explained by the parameter set in Eq. (1). To obtain more information on the ΛNN force, plenty of hypernuclear data, particularly those of non-zero isospin states, are required, because the effect of the three-body force is expected to be large for non-zero isospin states due to the isospin of the intermediate Σ. The effect of the ΛNN force is calculated for $A = 3, 4, 5$ hypernuclei by taking the ΛN–ΣN coupling explicitly in variational calculations [20, 21]. Such calculations for p-shell hypernuclei are also necessary.

Charge Symmetry Breaking (CSB)

Another important subject in the study of ΛN interaction is charge symmetry breaking (CSB). Since a Λ has no isospin and no charge, the Λp and Λn interactions are identical if the charge symmetry holds exactly. As shown in Fig. 5, however, the Λ binding energies of the lightest mirror pair of hypernuclei, ${}^{4}_{\Lambda}\mathrm{H}$ and ${}^{4}_{\Lambda}\mathrm{He}$, are reported to have a large difference, suggesting that the Λp interaction is more attractive than the Λn interaction. The CSB effect in ${}^{4}_{\Lambda}\mathrm{H}$ and ${}^{4}_{\Lambda}\mathrm{He}$ is related not only to the CSB in the YN interaction but also

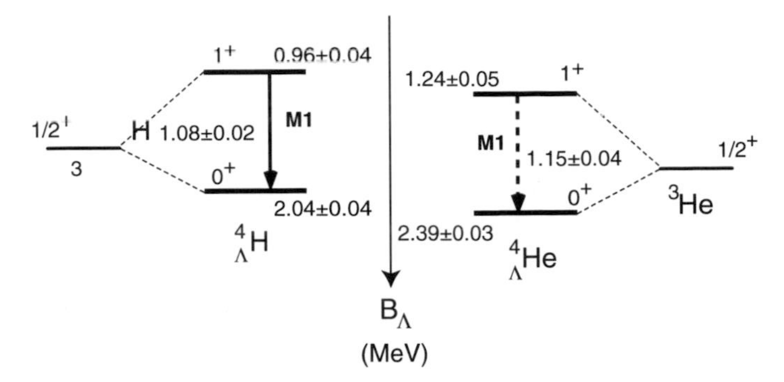

Fig. 5. Level schemes of the mirror hypernuclei, ${}^{4}_{\Lambda}$H and ${}^{4}_{\Lambda}$He. Λ binding energies (B_Λ) for the ground (0^+) and the excited (1^+) states are shown in MeV. The data suggest a large CSB in the ΛN interaction.

to the Coulomb effect from the core ^{3}H/^{3}He size and their distortion by a Λ. Therefore, it has attracted interest of theorists on few-body nuclear problems for a long time. However, the reported B_Λ difference in Fig. 5 has not been explained yet.

The origin of the reported CSB is not understood. The Λ–Σ^0 isospin mixing, which gives rise to one pion exchange force, leads to a spin-dependent CSB effect. The mass difference for $\Sigma^+/\Sigma^0/\Sigma^-$, which is about 10% of the Σ–Λ mass difference, also induces a CSB effect through the ΛN–ΣN coupling. The ΛNN three-body force via the ΛN–ΣN coupling may contribute to CSB effects differently between the 0^+ and 1^+ states. The $A = 4$ hypernuclear data, however, suggest an almost spin-independent CSB potential [22]. An accurate few-body calculation on ${}^{4}_{\Lambda}$H and ${}^{4}_{\Lambda}$He from various Nijmegen YN interaction models, including the Λ–Σ^0 mixing and the mass difference of Σs in ΛN–ΣN coupling, failed to reproduce the reported energies [23].

Systematic study of various pairs of mirror hypernuclei, as well as confirmation of the ${}^{4}_{\Lambda}$H/${}^{4}_{\Lambda}$He data, is indispensable to investigate the spin dependence in CSB interaction and to clarify the characteristics and thus the origin of the CSB interaction.

2.2 Impurity Effect in Nuclear Structure

Since a hyperon is free from Pauli effect in a nucleus and feels a hyperon–nucleon force different from the nucleon–nucleon force, only one (or two) hyperon(s) introduced in a nucleus may give rise to various changes of the nuclear structure, such as changes of the size and the shape, change of the cluster structure, emergence of new symmetries, and change of collective motions. Level schemes and $B(E2)$ values of Λ hypernuclei studied by γ spectroscopy will reveal such phenomena and a new field "impurity nuclear physics" will be exploited.

In KEK E419, we measured the lifetime of the ${}^{7}_{\Lambda}$Li($5/2^+$) state by analyzing the $E2(5/2^+\to 1/2^+)$ γ-ray peak shape which is partly broadened by Doppler shift according to the lifetime of ${}^{7}_{\Lambda}$Li($5/2^+$) and the stopping time of the recoiling hypernucleus in a target material. The high resolution of Ge detectors enabled us to apply this method called Doppler shift attenuation method (DSAM) for the first time to hypernuclei. From the measured lifetime we derived a $B(E2;\, 5/2^+\to 1/2^+)$ value of $3.6 \pm 0.5(\text{stat})^{+0.5}_{-0.4}(\text{syst})$ e^2fm^4 [2]. It is three times smaller than the $B(E2)$ value of the core nucleus, $B(E2;\, 3^+\to 1^+) = 10.9 \pm 0.9\ \text{e}^2\text{fm}^4$, indicating a significant shrinkage of ${}^{7}_{\Lambda}$Li compared to ^{6}Li. According to a cluster model calculation by Hiyama et al. [24], the reduction of $B(E2)$ is interpreted as a contraction of inter-cluster (α-d) distance by 19 $\pm$ 4%. Such a shrinking effect induced by the presence of a Λ in the $0s$ orbit was first predicted by Motoba et al. [25].

Such a nuclear structure change induced by a Λ may appear in various hypernuclei [26] as described in Sects. 5.2 and 5.4. In particular, drastic changes of nuclear structure are expected when a Λ is implanted into neutron-rich hypernuclei having neutron skin or halos (see Sect. 5.4).

2.3 Medium Effect in Baryon Properties

One of the recent topics in nuclear and hadron physics is the investigation of partial restoration of chiral symmetry from the modification of hadron properties in a nuclear medium. Changes of vector meson masses in nuclear matter have been theoretically predicted [27, 28] and experimentally observed [29, 30]. Such hadron modifications in medium are not easily observed for baryons because of the Pauli effect. A Λ hyperon in a hypernucleus can be a unique probe to investigate such medium effects for baryons. In particular, the magnetic moment of Λ may be modified in a nucleus from the free-space value.

Magnetic moments of baryons can be well described by the picture of constituent quark models in which each constituent quark has a magnetic moment of a Dirac particle having a constituent quark mass. Thus it is naively expected that the reduction of constituent quark masses in nuclei due to partial restoration of chiral symmetry results in enhancement of the magnetic moment of the baryon. However, since the origin of the baryon spin is not understood in terms of quarks and gluons, how the baryon magnetic moment behaves under the partial restoration of chiral symmetry is an open question.

A possible change of the magnetic moment of a Λ in a nucleus has attracted the attention of nuclear physicists. It was first pointed out [31, 32] that Pauli effect in the quark level, if it exists, may modify a hyperon in a hypernucleus and change its magnetic moment. Then a calculation with the quark cluster model [33] showed that the "quark exchange current" between two baryons at a short distance changes the magnetic moment, of which the effect depends on the confinement size of the hyperon in the nucleus. However, possible effects of the partial chiral symmetry restoration to the magnetic moments have not been theoretically studied yet.

Direct measurement of hypernuclear magnetic moments is extremely difficult because of its short lifetime for spin precession. Here we propose to derive a g-factor of Λ (g_Λ) in the nucleus from a probability ($B(M1)$ value) of a spin-flip $M1$ transition between hypernuclear spin-doublet states (γ_S in Fig. 3). In the weak coupling limit between a Λ and a core nucleus, the $B(M1)$ value can be expressed as [15] follows:

$$\begin{aligned} B(M1) &= (2J_{up}+1)^{-1} \mid < \phi_{low} \| \boldsymbol{\mu} \| \phi_{up} > \mid^2 \\ &= (2J_{up}+1)^{-1} \mid < \phi_{low} \| \; g_c \boldsymbol{J}_c + g_\Lambda \boldsymbol{J}_\Lambda \; \| \phi_{up} > \mid^2 \\ &= \frac{3}{8\pi} \frac{2J_{low}+1}{2J_c+1} (g_c - g_\Lambda)^2 \; \mu_N^2 \,, \end{aligned} \tag{3}$$

where g_c and g_Λ denote effective g-factors of the core nucleus and the Λ hyperon ($g_\Lambda^{free} = -1.226$), $\boldsymbol{J}_c$ and $\boldsymbol{J}_\Lambda$ denote their spins, and $\boldsymbol{J} = \boldsymbol{J}_c + \boldsymbol{J}_\Lambda$ is the spin of the hypernucleus. Here the spatial components of the wave functions for the lower state and the upper state of the doublet, ϕ_{low} and ϕ_{up} (with spins J_{low} and J_{up}), are assumed to be identical.

In this study, the effect of the meson exchange current between a Λ and a nucleon has also to be considered. The effect of meson exchange current is expected to be totally different from the NN case, because in the ΛN interaction one pion exchange is forbidden but ΛN–ΣN coupling makes another effect. $B(M1)$ values and magnetic moments of several s, p-shell hypernuclei were calculated in order to investigate such effects [34, 35]. These effects were found to change the $B(M1)$ value by several %. For example, $B(M1)$ of ${}^{7}_{\Lambda}$Li($3/2^+ \rightarrow 1/2^+$) is reduced by 7%. Accurate measurement of $B(M1)$ for various hypernuclei will also reveal the effect of the meson exchange current.

The reduced transition probability $B(M1)$ can be derived from the lifetime of the excited state via $1/\tau = 3.1483 \times 16\pi/9 \; E_\gamma^3 \; B(M1)$, where τ is in ps, E_γ is in MeV, and $B(M1)$ is in the unit of μ_N^2. The lifetime τ is obtained by Doppler shift attenuation method (DSAM) or the γ-weak coincidence method. The DSAM, which was successfully used in the $B(E2)$ measurement of ${}^{7}_{\Lambda}$Li, can be applied when the stopping time of the recoiling excited hypernucleus in the target material is of the same order as the lifetime of the γ-emitting excited state. The γ-weak coincidence method (Sect. 5.5) is applicable if the γ transition rate is of the same order as the weak decay rate $\sim (100\,\mathrm{ps})^{-1}$.

In our previous γ-spectroscopy experiments, we have attempted $B(M1)$ measurement but meaningful results have not been obtained yet. When the ${}^{7}_{\Lambda}$Li($3/2^+ \rightarrow 1/2^+$, 692 keV) transition was first observed at KEK [1], the $3/2^+$-state lifetime was not able to be derived, because the γ-ray peak was fully broadened by Doppler shift due to too slow stopping time of ${}^{7}_{\Lambda}$Li in the target compared with the lifetime. Later, in the BNL E930('01) experiment, the ${}^{7}_{\Lambda}$Li($3/2^+ \rightarrow 1/2^+$) γ-ray peak was observed in the ^{10}B(K^-, π^-) reaction when a highly unbound region of ${}^{10}_{\Lambda}$B mass was selected. Here the spin-flip state ${}^{7}_{\Lambda}$Li($3/2^+$) was produced as a hyperfragment by the reaction ^{10}B(K^-, π^-)${}^{10}_{\Lambda}$B*, ${}^{10}_{\Lambda}$B* $\rightarrow$ ${}^{7}_{\Lambda}$Li + ^{3}He. The peak shape of the

${}^{7}_{\Lambda}$Li($3/2^+ \rightarrow 1/2^+$) transition was partly broadened and the lifetime of the $3/2^+$ state was derived to be $0.58^{+0.38}_{-0.20}$ ps (statistical error only), from which the $B(M1)$ was obtained as $0.30^{+0.12}_{-0.16}$ [μ_N^2] (preliminary). This value corresponds to $g_\Lambda = -(1.1^{+0.4}_{-0.6})$ [μ_N], which is compared with the free-space value of $g_\Lambda = -1.226$ [μ_N]. In order to measure the spin-flip $B(M1)$ with better accuracy, we performed the KEK E518 and E566 experiments. In E518, we observed six γ-rays in ${}^{11}_{\Lambda}$B [8] but the complete level assignment was not possible because γ–γ coincidence method was not able to be applied due to low statistics. In E566, we observed the spin-flip $M1$ transition of ${}^{11}_{\Lambda}$B($7/2^+ \rightarrow 5/2^+$), but unfortunately this transition energy is found to be much lower than expected and therefore the transition is too slow to apply DSAM.

2.4 Plans of Experiments at J-PARC

The plans of hypernuclear γ spectroscopy experiments at J-PARC are listed below. They are classified by physics motivation and experimental methods:

(1) Structure of Λ hypernuclei via the (K^-, π^-) reaction
  (1-a) for s-shell and p-shell hypernuclei
  (1-b) for sd-shell hypernuclei
  (1-c) for medium and heavy Λ hypernuclei
(2) Study of n-rich/mirror hypernuclei via the (K^-, π^0) reaction
(3) $B(M1)$ measurements
  (3-a) using DSAM
  (3-b) using γ-weak coincidence method

Experiments (1) and (2) are planned to investigate the structure of various Λ hypernuclei in a wide mass range, particularly for the purpose of studying the YN interactions and impurity effect. Item (1-a) has already been pursued at KEK-PS and BNL-AGS, but more measurements are necessary. The others (1-b), (1-c), and (2), which have not been attempted so far, are feasible only at J-PARC using intense kaon beams. Experiments (3) study the nuclear medium effect of baryons through a g-factor of Λ in various nuclei. This subject was attempted in previous experiments but meaningful results have not been obtained yet as described in Sect. 4.1. The first experiment at J-PARC, E13, which is approved as one of the day-1 experiments in the Hadron Hall, covers some parts in (1-a), (1-b), and (3-a).

3 Experimental Apparatus

In most of the planned experiments at J-PARC ((1) and (3) in Sect. 2.4), we use the (K^-, π^-) reaction to produce Λ hypernuclei and detect γ-rays in coincidence with the reaction. In those experiments, we will employ the K1.8 or K1.1 beamline together with the SKS spectrometer to analyze the mass of produced hypernuclei and measure γ-rays from the hypernuclei with the germanium detector array called Hyperball-J.

3.1 Beamline

In the (K^-, π^-) reaction, only one member of the hypernuclear spin doublet (see Fig. 3) is, in many cases, produced from the target nucleus by the non-spin-flip amplitude of the elementary reaction $(K^- n \rightarrow \Lambda\pi^-)$ and the other member of the doublet by the spin-flip amplitude. Therefore, by using both spin-flip and non-spin-flip amplitudes, we can discriminate between the spin-flip and non-spin-flip states in the doublet, which gives essential information for reconstruction of the level scheme from observed γ-rays.

The spin-flip amplitude is larger than the non-spin-flip amplitude in the region of 1.1–1.45 GeV/c, while the non-spin-flip amplitude is larger in the other momentum region [36]. Therefore, in order to populate both spin-flip and non-spin-flip states without changing the experimental setup, the best K^- beam momentum is 1.1 and 1.45 GeV/c. In both cases, we require high purity K^- beams with the π/K ratio less than 0.5 to minimize the radiation damage to the Ge detectors.

The K1.1 beamline, in which the kaon beam intensity is optimized at 1.1 GeV/c, is planned to be constructed in the Hadron Hall but not funded yet. Therefore, in the early stage of our studies at J-PARC, we will use the 1.5 GeV/c beam at the K1.8 beamline, which will be available for day-1 experiments. In addition, we need a good spectrometer with a momentum resolution better than 0.2% FWHM to select the bound-state region of hypernuclei. For this purpose, the SKS spectrometer which will be installed at the K1.8 beamline is best fitted.

3.2 Magnetic Spectrometers

Figure 6 shows the experimental setup at the K1.8 beamline. It is similar to the previous setup for the (π^+, K^+) reaction at the KEK-PS K6 beamline, but some detectors are modified or newly introduced.

The kaon beam momentum is analyzed by the K1.8 beamline spectrometer composed of the QQDQQ magnets, 1 mm pitch MWPCs (BC1, BC2), drift chambers with 1.5 mm drift length (BC3, BC4), and timing counters (BH1, BH2). The momentum resolution of the K1.8 beamline spectrometer is estimated to be 0.014% in rms (H. Noumi, 2005, private communication), which gives a negligible effect to the energy resolution of the hypernuclear mass spectrum for our thick targets (10–20 g/cm^2). Just upstream and downstream of the target, aerogel Čerenkov counters (BAC and SAC) are located to identify kaons in the beam and pions in the scattered particles.

The scattered pions are identified and momentum analyzed by the use of the SKS spectrometer. In the 1.5 GeV/c (K^-, π^-) reaction, the pion momentum is around 1.4 GeV/c, being twice as large as the K^+ momentum in the (π^+, K^+) reaction with 1.05 GeV/c pion beam in the previous condition at KEK-PS using SKS. So we excite the SKS magnet to the maximum field (2.7 T) and install drift chambers and TOF stop counters covering a

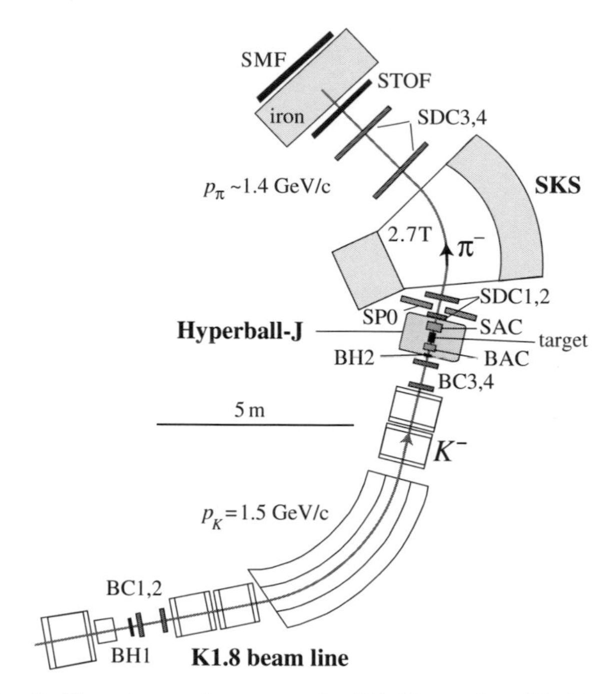

Fig. 6. Experimental setup at the K1.8 beamline (SksMinus).

much larger area ($\sim$ 2 m in horizontal). As shown in Fig. 6, the new SKS setup optimized for 1.4 GeV/c pions (called SksMinus) [37] includes drift chambers with 1.5 and 2.5 mm drift lengths (SDC1,2) located upstream of the SKS magnet, and drift chambers with 10 mm drift length (SDC3,4) together with time-of-flight hodoscopes (TOF) located downstream of the SKS magnet.

According to a simulation, the SksMinus setup has an acceptance of about 120 msr. Since the SKS covers a wide range of scattering angle ($|\theta| < 20°$), we can populate non-spin-flip states at $|\theta| < 10°$ and spin-flip states at $|\theta| > 10°$ simultaneously (see Fig. 9, for example). The bending angle for 1.4 GeV/c pions is typically 60°, significantly smaller than the angle for the usual SKS setting (100°), giving a mass resolution of 4.1 MeV (FWHM). Although it is much worse than those in usual SKS experiments at KEK for the (π^+, K^+) reaction ($\sim$2 MeV (FWHM)), our mass resolution is mainly determined by the energy loss effect in our thick (typically 20 g/cm^2) targets. By including the energy loss effect, the mass resolution is estimated to be 5.9 MeV (FWHM), which is sufficient to tag hypernuclear bound-state regions.

The beam kaons which decay in flight in the target region between BAC and SAC as $K^- \to \mu^- \nu$ and $\pi^- \pi^0$ cannot be separated from (K^-, π^-) reaction events because μ^- and π^- from K^- decay enter the same kinematical region (momentum and scattering angle) as the π^- associated with hypernuclear production. Those events cause a serious background, not only in the trigger level but even after full analysis of the hypernuclear mass. In order to reject muons, we install a set of muon filters (SMF in Fig. 6) made of an 80-cm-thick iron block and thin plastic counters behind it. According to a simulation, background muons of 1.25–1.5 GeV/c can be completely detected and removed by SMF, while all the π^-'s associated with the hypernuclear production do not hit SMF. As for the $K^- \rightarrow \pi^-\pi^0$ and other π^0 emitting decay modes, we detect π^0 by using several layers of lead–plastic-sandwiched counters installed between SDC1 and SDC2 and covering the forward angles except for the SKS entrance (SP0 in Fig. 6). A simulation showed that the π^0 counters reject 82% of the π^0-emitting K^- decays occurring at the target region.

At the K1.1 beamline available in the future, a spectrometer similar to SKS will also be used. For the (K^-, π^0) experiment (2), we need a π^0 spectrometer similar to the one (NMS) used for BNL E907/931 experiments [38]. It consists of an array of scintillation counters and a set of tracking devices to measure hit positions and energies of two energetic photons from π^0. It is required to have an acceptance of about 10 msr and an energy resolution of 2 MeV (FWHM). In the γ-weak coincidence experiment (3-b), we also need "decay arm" detectors which measure energies, tracks, and timings of protons and π^-s emitted from weak decays of Λ hypernuclei, as described in Sect. 5.5.

A New-Generation Ge Detector Array, Hyperball-J

The Ge detector array, Hyperball, was developed to overcome huge counting rates and energy deposit rates in hypernuclear experiments. It consisted of 14 sets of coaxial N-type Ge detectors equipped with special readout electronics. Each Ge crystal is surrounded by BGO scintillation counters to suppress background from Compton scattering as well as high-energy charged particles and γ-rays. The Hyperball array was recently upgraded to a larger array, Hyperball2, with a doubled efficiency but with similar components of detectors and electronics.

With the full proton beam intensity (50 GeV, 15 μA) at J-PARC, the K^- beam intensity will be $\sim 1 \times 10^7$ particles/s and the counting rate and the energy deposit rate will be five times higher than in the present condition at KEK-PS K6 beamline with the π^+ beam rate of $2–3 \times 10^6$ s^{-1}. In this condition, the present Hyperball2 array will have serious problems described in the following, but they will be solved in our new-generation Ge detector array, Hyperball-J, which is currently under construction [39].

Figure 7 illustrates a half part of Hyperball-J. Hyperball-J consists of about 30 sets of Ge detectors, each of which has a photo-peak efficiency of about 60% relative to $3''\phi \times 3''$ NaI detector. Each Ge detector is surrounded

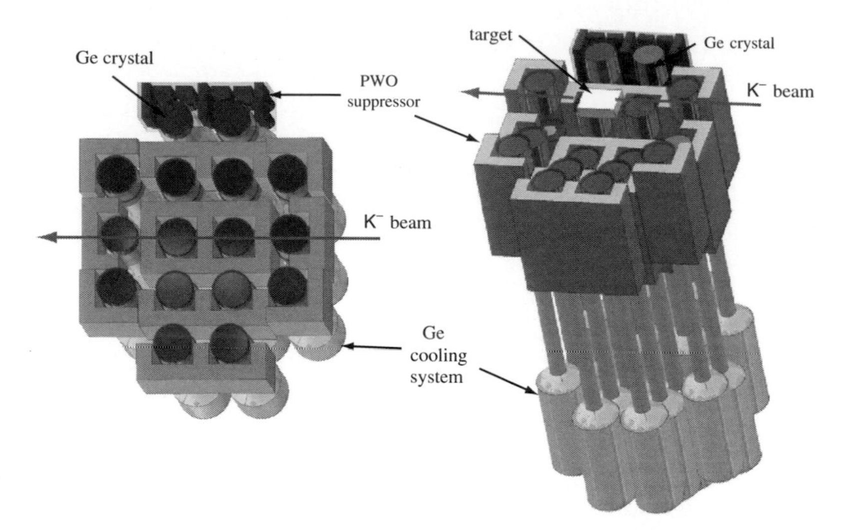

Fig. 7. Schematic view of Hyperball-J (lower half only) viewed from top (*left figure*) and from side (*right figure*).

by fast PWO counters for background suppression instead of the previous BGO counters. A half array (Fig. 7) is installed at the top of the target and another at the bottom, and the distance of each Ge detector from the target, typically 15 cm, can be adjusted according to the requirement of the experiment and beam conditions. According to a simulation, the photo-peak efficiency is ∼6% at 1 MeV for a point source.

The Hyperball-J array is characterized by the following new techniques:

1. *Mechanical cooling*: Severe radiation damage to the Ge detectors will deteriorate the resolution within 1 week beam time with the full-intensity beam at J-PARC and unrealistically frequent annealing of the Ge detectors will be necessary. The effect of radiation damage is much reduced when the Ge crystal temperature is kept lower than 85 K [40]. So we have developed a new mechanical cooling method instead of the ordinary liquid nitrogen method which limits the crystal temperature higher than 90 K. The compact mechanical cooler enables us to arrange the Ge detectors in a wall-like configuration as shown in Fig. 7, which significantly increases the photo-peak efficiency of the array.
2. *PWO counters*: The BGO scintillation counters in Hyperball2 reject ∼10% of good events due to accidental coincidence in the present condition at KEK, where the BGO counters are located at ∼13 cm from 20-g/cm^2-thick target with 2×10^6 beam particles per second. Since such an overkill rate is expected to be more than 50% at J-PARC, the BGO counters are not able to be used. We will replace BGO crystals by PWO crystals, which has a much shorter decay time (∼10 ns) than BGO (∼300 ns) and is tolerant against higher counting rates. A problem of a small detection efficiency

for low-energy γ-rays due to a small light yield of PWO has been solved by using doped PWO crystals at a temperature less than $-20\,^{\circ}$C.

3. *Waveform analysis*: In our previous KEK experiments using 2–3 $\times$ 10^6/s pion beam, our Ge detector readout electronics caused a dead time of 40–50%, and thus it cannot be used with the full beam intensity ($\sim 2 \times 10^7$ K^-/s) at J-PARC. We are developing a new readout method using high-resolution waveform digitizers, of which data will be analyzed off-line to decompose pile-up signals and to correct for baseline shifts after preamplifier reset. According to current R&D studies, the capability for counting rate will be improved by a factor of 5 or more. This new readout system will be installed to Hyperball-J before the full-intensity beam becomes available at J-PARC.

Details of Hyperball-J are described in Ref. [39]. Hyperball-J will also be used to measure X-rays from Ξ^- atoms in the energy range of 100–500 keV in order to investigate the ΞN interaction. The approved experiments for Ξ-atomic X-rays, E03 and E07 [41, 42], will run at the K1.8 beamline, using the KURAMA spectrometer to tag Ξ^- production by the (K^-, K^+) reaction.

4 The Day-1 Experiment: γ Spectroscopy of Light Λ Hypernuclei (E13)

For the first experiment of γ-ray spectroscopy at J-PARC, we proposed [43] to study structures of several light hypernuclei (${}^{4}_{\Lambda}$He, ${}^{7}_{\Lambda}$Li, ${}^{10}_{\Lambda}$B, ${}^{11}_{\Lambda}$B, and ${}^{19}_{\Lambda}$F), via the 1.5 GeV/c (K^-, π^-) reaction, employing the K1.8 beamline and the SKS spectrometer system together with the new Ge detector array, Hyperball-J. The experiment (E13) has been approved as one of the day-1 experiments.

One of the purposes of the experiment is to measure the reduced transition probability $B(M1)$ of the Λ spin-flip M1 transition ${}^{7}_{\Lambda}$Li($3/2^+ \rightarrow 1/2^+$) and extract a g-factor of Λ inside the nucleus. By comparing the result with the free-space g_Λ value, we expect to investigate a possible modification of baryon properties in a nucleus.

Another purpose is to investigate the ΛN interaction further than our previous studies. In order to establish the strengths of the ΛN spin-dependent (spin–spin, spin–orbit, and tensor) interactions and to understand the ΣN–ΛN coupling force, we take complete data for ${}^{10}_{\Lambda}$B and ${}^{11}_{\Lambda}$B in which inconsistency in the spin-dependent interaction strengths compared with the other p-shell hypernuclei have been found as described in Sect. 2.1. We will also run for a ^{19}F target and detect both ${}^{19}_{\Lambda}$F($1/2^- \rightarrow 3/2^+, 1/2^+$) transitions to determine the ground state doublet spacing. It gives the strength of the effective ΛN spin–spin interaction in sd-shell hypernuclei for the first time and provides information on the radial dependence of the interaction. In addition, we also plan to run with a ^{4}He target and determine the ground state doublet ($1^+, 0^+$) spacing precisely, in order to solve a long-standing puzzle of extremely large CSB in the ΛN interaction. The cross sections of the spin-flip ^{4}He(1^+) and non-spin-flip ^{4}He(0^+) states will also be measured for several K^- momenta to confirm the spin-flip/non-spin-flip property of hypernuclear production in the (K^-, π^-) reaction.

The proposed experiment will be performed at the K1.8 beamline using the SKS spectrometer and Hyperball-J. Details of the apparatus are described in Sect. 3.

4.1 $B(M1)$ Measurement in $^7_\Lambda$Li

As described in Sect. 2.3, measurement of $B(M1)$ values for Λ spin-flip $M1$ transition provides unique information on the g-factor of Λ in a nucleus. In the proposed experiment, we use the $^7_\Lambda$Li hypernucleus, because the bound-state level scheme is perfectly known and the feasibility of the $^7_\Lambda$Li $B(M1)$ measurement can be most unambiguously estimated. Figure 8 shows the level scheme of $^7_\Lambda$Li, where the excitation energies measured by our previous experiments are given. The calculated cross sections of the $^7_\Lambda$Li bound states by the 1.5 GeV/c (K^-, π^-) reaction are shown in Fig. 9. We plan to measure the $B(M1)$ of the spin-flip $M1(3/2^+ \rightarrow 1/2^+)$ transition at 692 keV. The spin-flip $3/2^+$ state is directly produced with a smaller cross section than those of the non-spin-flip $1/2^+$ and $5/2^+$ states. Fortunately, the $3/2^+$ state is also produced via the $1/2^+$; T=1 state at 3.88 MeV through the fast γ decay of $1/2^+$; T=1 $\rightarrow 3/2^+$. (In this chapter, isospin T is omitted only when T=0 or 1/2.) The $1/2^+$; T=1 state is populated by the non-spin-flip reaction with a large cross section. Thus the $3/2^+ \rightarrow 1/2^+$ yield is large, as experimentally known in our first Hyperball experiment [1].

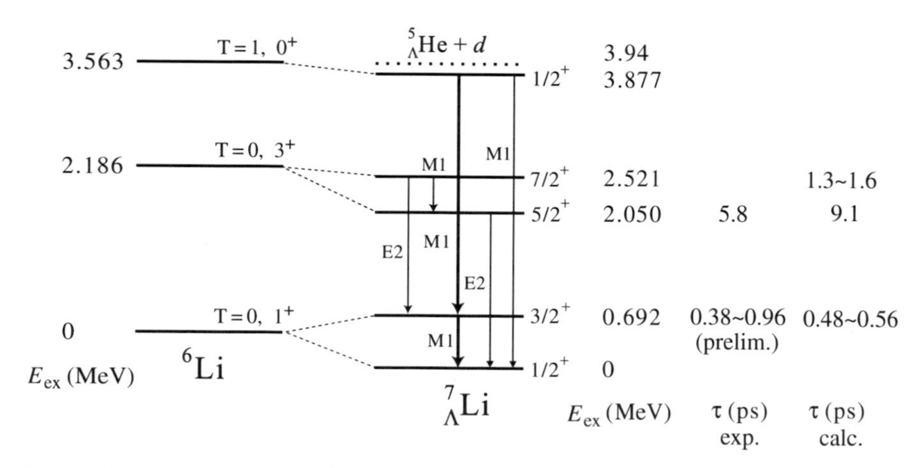

Fig. 8. Level scheme of the $^7_\Lambda$Li hypernucleus. Excitation energies of all the bound states are experimentally determined [1, 9].

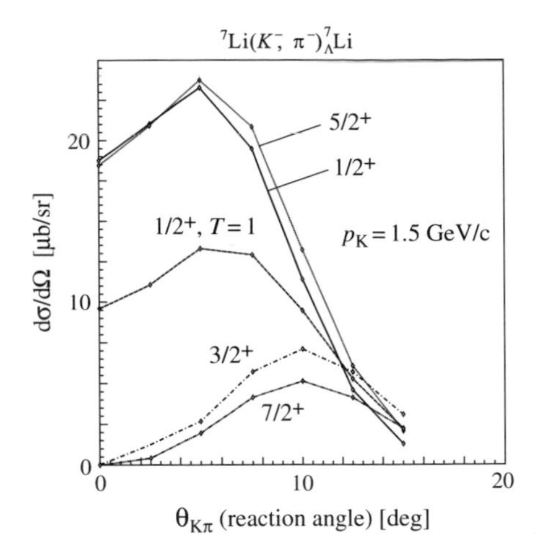

Fig. 9. Cross sections of all the bound states of the $^{7}_{\Lambda}$Li hypernucleus in the (K^-, π^-) reaction at p_{K^-} = 1.5 GeV/c calculated by Motoba (2006, private communication).

Without any anomalous effect in g_Λ, the $B(M1)$ value of this transition is estimated to be $B(M1) = 0.326$ [μ_N^2] from Eq. (3) and calculated to be $B(M1) = 0.322$ [μ_N^2] from a cluster model [24], which corresponds to 0.5 ps of the $3/2^+$ state lifetime. In order to measure the lifetime by DSAM, we use a Li_2O target with a density of 2.01 g/cm^3 instead of Li metal (density of 0.5 g/cm^3). The stopping time of the recoiling $^{7}_{\Lambda}$Li hypernuclei in Li_2O is calculated to be 2–3 ps. It is close to the expected lifetime of ∼0.5 ps and DSAM works well. Although we produce the $3/2^+$ state via the $1/2^+$; T=1→ $3/2^+$ transition, this transition is fast (theoretically estimated to be 0.1 fs) and does not affect the lifetime measurement of the $3/2^+$ state.

Figure 10 shows a simulated spectrum for the $^{7}_{\Lambda}$Li ($3/2^+ \rightarrow 1/2^+$) peak for 500 hours' run with 0.5×10^6 K^- per spill. Here, the lifetime of the $3/2^+$ state was assumed to be 0.5 ps, and the background was reliably estimated from our previous $^{7}_{\Lambda}$Li spectrum in KEK E419, with a contribution from ^{16}O included. By fitting the peak with calculated peak shapes for various lifetimes, the $3/2^+$ state lifetime was obtained to be 0.478 ± 0.026 ps in this simulation, demonstrating a good statistical accuracy of the proposed $B(M1)$ measurement. Thus, we expect to determine the lifetime of the $3/2^+$ state within 6% accuracy. It corresponds to a statistical error of 3% for $|g_\Lambda - g_c|$. According to our experience in the $B(E2)$ measurement of $^{7}_{\Lambda}$Li (KEK E419) [2], the systematic error of $B(M1)$ is estimated to be less than 5%; the systematic error here is mainly from inaccuracy of the stopping time calculation. The effect of possible alignment of hypernuclei, which was another main source of the systematic error in E419, does not exist here because of the $1/2^+$; T=1 spin of the populated $^{7}_{\Lambda}$Li level.

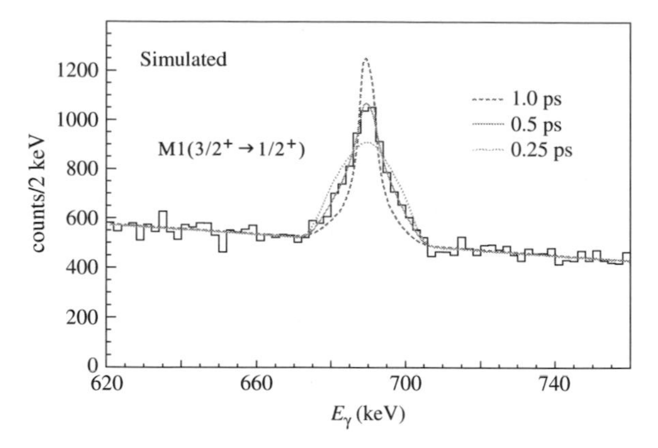

Fig. 10. Simulated γ-ray spectrum to be measured in the ^{7}Li $(K^-, \pi^-)^7_\Lambda$Li reaction with reaction angles from 2.5° to 5° in 500 hours beam time. The partly Doppler broadened peak is the spin-flip $M1(3/2^+ \to 1/2^+)$ transition in $^7_\Lambda$Li with a lifetime of the $3/2^+$ state assumed to be 0.5 ps. Calculated peak shapes for various lifetimes are also shown.

4.2 *YN* Interaction

$^{10}_\Lambda$B and $^{11}_\Lambda$B

As described in Sect. 2.1, the four parameters of the spin-dependent ΛN interaction strengths determined from $^7_\Lambda$Li, $^9_\Lambda$Be, and $^{16}_\Lambda$O data cannot consistently explain the $^{10}_\Lambda$B and $^{11}_\Lambda$B data. In order to solve this problem we need to accumulate more data for p-shell hypernuclei and clarify the properties of the ΛN–ΣN coupling force. A plenty of data will also provide clues to improve the wave functions of the core nuclei. For this purpose, $^{10}_\Lambda$B, $^{11}_\Lambda$B, and $^{12}_\Lambda$C data are particularly important, because experimental data for these hypernuclear bound states are still limited and their existing data seem to show inconsistency. Since the $^{12}_\Lambda$C γ-ray data have recently been taken at KEK (E566) and are under analysis, we propose to take high-quality data for $^{10}_\Lambda$B and $^{11}_\Lambda$B at J-PARC. These hypernuclei can be theoretically studied not only by shell model calculations but also by few-body calculation approach using cluster models.

The $M1$ transition between the $^{10}_\Lambda$B ground state doublet members ($2^- \to 1^-$) (see Fig. 1 (c)) was not observed in the previous experiments in spite of an expected large cross section of 2^- [7, 44]. This result suggests $E(2^-)-E(1^-) <$ 100 keV, where the γ-ray detection efficiency rapidly drops below 100 keV in these experiments. According to a shell model calculation [45], this spacing is expressed as

$$E(2^-) - E(1^-) = 0.579\Delta + 1.413S_\Lambda + 0.013S_N - 1.073T + \Lambda\Sigma, \quad (4)$$

where the ΛN–ΣN coupling effect is estimated to be $\Lambda\Sigma = -0.015$ MeV. The experimental result of $E(2^-) - E(1^-) < 100$ keV is inconsistent with the parameter values in Eq. (2) giving $E(2^-) - E(1^-) = 171$ keV; we need $\Delta < 0.30$ MeV to explain the data. Recently, Millener suggested that the value of Δ for $^{11}_{\Lambda}$B and $^{16}_{\Lambda}$O is smaller ($\Delta = 0.33$ MeV) than the value for $^{7}_{\Lambda}$Li ($\Delta = 0.43$ MeV). If $\Delta = 0.33$ MeV can also be applied to $^{10}_{\Lambda}$B, it gives $E(2^-) - E(1^-) = 121$ keV [17], which still seems inconsistent with the data.

Since the K^- beam profile at the K1.8 target position is vertically thin (7.5 mm FWHM), we can use a vertically thin target which reduces absorption of low-energy γ-rays in the target. With the thin target and the intense K^- beam, we will be able to detect γ-rays with energy down to ~50 keV. In addition, the previous experiments using the non-spin-flip reaction (0.8–0.93 GeV/c (K^-, π^-) reaction) populated only the 2^- state, but the proposed experiment using the 1.5 GeV/c (K^-, π^-) reaction can populate both spin-flip (1^-) and non-spin-flip (2^-) states. Therefore, the doublet spacing of $^{10}_{\Lambda}$B$(2^-, 1^-)$ can be determined regardless of the spin ordering in the doublet, unless the spacing is very small ($|E(2^-) - E(1^-)| < 50$ keV). The weak decay of the upper state may compete with the γ transition to the ground state. According to the expected $B(M1)$ value, the transition rate, which is proportional to E_γ^3 where E_γ is the transition energy, is (200 ps)$^{-1}$ for 67 keV and (400 ps)$^{-1}$ for 53 keV. Since the weak decay lifetime is around 200 ps [46], the γ transition is suppressed only by a factor of 3 for 53 keV.

The $^{11}_{\Lambda}$B hypernucleus is expected to have many bound states and is suitable for a test of the ΛN interaction parameters and the theoretical framework. In our previous experiment (KEK E518) by the (π^+, K^+) reaction we observed six transitions of $^{11}_{\Lambda}$B but most of them were not able to be assigned. Figure 11 is the expected level scheme of $^{11}_{\Lambda}$B. The γ-ray peak observed at 1483 keV was assigned to $E2(1/2^+ \to 5/2^+(gs))$ transition [8]. But the result of Millener's shell model calculation [45],

$$E(1/2^+) - E(5/2^+) = -0.243\Delta - 1.234S_\Lambda - 1.090S_N - 1.627T + E_{core}$$

with $E_{core} = 0.718$ MeV and the parameter values in Eq. (2) predict 1020 keV. To understand this inconsistency, we need more level energy data in $^{11}_{\Lambda}$B. In addition, the 264 keV γ ray, which was assigned to $^{11}_{\Lambda}$B$(7/2^+ \to 5/2^+)$ later in the ^{12}C(K^-, π^-) experiment (KEK E566) [10], suggests $\Delta = 0.33$ MeV [17], significantly smaller than the value ($\Delta = 0.43$ MeV) from $^{7}_{\Lambda}$Li. This smaller Δ value should be confirmed from other $^{11}_{\Lambda}$B levels. In the proposed experiment, both spin-flip and non-spin-flip populations are available, and a higher efficiency of Hyperball-J and larger cross sections of the (K^-, π^-) reaction allow γ–γ coincidence measurement. They enable us to reconstruct the level scheme.

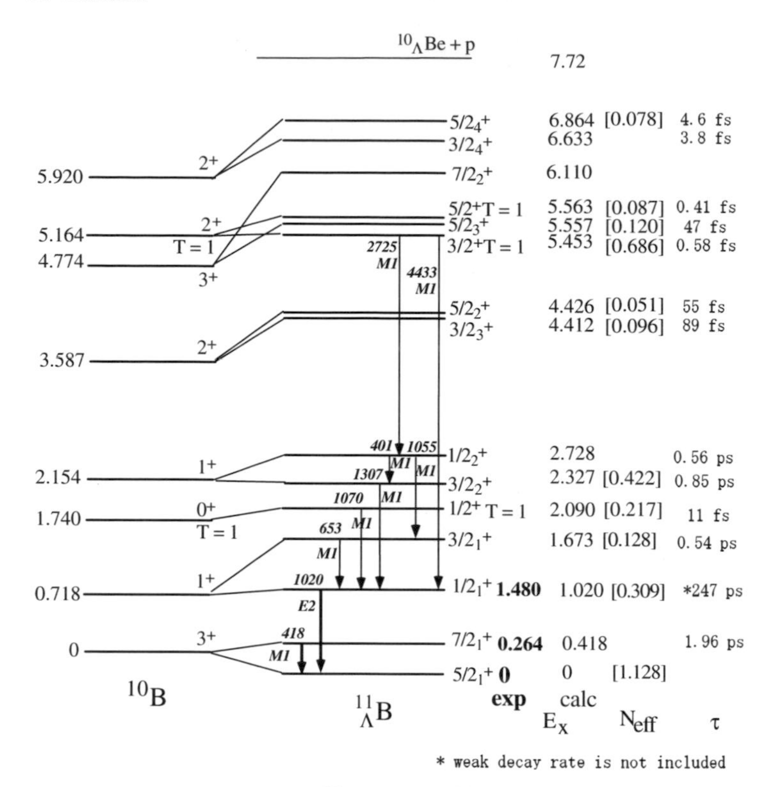

Fig. 11. Expected level scheme of ${}^{11}_{\Lambda}$B calculated by Millener [45]. Transitions and level energies measured by KEK E518 and E566 experiments are shown by *thick arrows* [8].

${}^{19}_{\Lambda}$F

We will also run with a ^{19}F target (using Teflon $(CF_2)_n$) to study *sd*-shell hypernuclei for the first time. Figure 12 is an expected level scheme and γ transitions in ${}^{19}_{\Lambda}$F. Since the cross section to populate the $1/2^-; T = 0$ state with $\Delta L = 1$ transition is expected to be large (of the order of 40 μb/sr at 2.5–10°) by the (K^-, π^-) reaction at 1.5 GeV/c, we can observe both $E1(1/2^- \to 3/2^+, 1/2^+)$ transitions in ${}^{19}_{\Lambda}$F and determine the ground state doublet $(3/2^+, 1/2^+)$ spacing, which gives the strength of the effective ΛN spin–spin interaction in the *sd*-shell hypernuclei. The other excited states, ${}^{19}_{\Lambda}\mathrm{F}(1/2^+; T = 1)$ and ${}^{19}_{\Lambda}\mathrm{F}(5/2^+)$, are also expected to be populated with $\Delta L = 0$ and 2 reactions, and the γ transitions of $M1(1/2^+; T = 1 \to 3/2^+, 1/2^+)$ and $E2(1/2^+ \to 3/2^+, 1/2^+)$ can be observed, respectively, with less intensities. Since the $1/2^+; T = 1$ and $1/2^-; T = 0$ states are populated by different ΔL reactions, the transitions from these two states can be clearly discriminated from each other using angular distribution of the (K^-, π^-) reaction. The

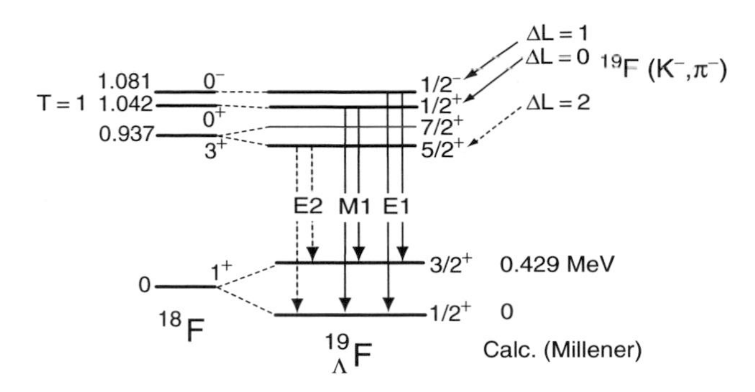

Fig. 12. Expected level scheme of $^{19}_{\Lambda}$F. Transitions expected to be observed in the proposed experiment are also shown.

$E2(1/2^+ \rightarrow 3/2^+, 1/2^+)$ transitions are also discriminated by their sharp peaks due to their slow transition rates.

The ground state doublet $(3/2^+, 1/2^+)$ spacing is roughly estimated to be 429 keV by Millener from the spin–spin interaction strength in the p-shell hypernuclei ($\Delta = 0.43$ MeV). But the distance between a Λ in the $0s$ orbit and nucleons in the $0d$ orbit in sd-shell hypernuclei is different from the distance between a Λ in the $0s$ orbit and nucleons in the $0p$ orbit in p-shell hypernuclei. Therefore, the interaction strength in $^{19}_{\Lambda}$F provides information on radial dependence of the ΛN interaction, namely, the shape of $V_\sigma(r)$ in Eq. (1).

$^{4}_{\Lambda}$He

In the proposed experiment, we measure the ground-state doublet spacing $(1^+, 0^+)$ of $^{4}_{\Lambda}$He using a liquid helium target. The $M1$ transition γ-ray of $^{4}_{\Lambda}$H $(1^+ \rightarrow 0^+)$ was measured several times with NaI detectors [47, 48] and the doublet $(1^+, 0^+)$ spacing was determined to be 1.08 ± 0.02 MeV (see Fig. 5). On the other hand, a measurement of $^{4}_{\Lambda}$He γ-rays was once reported in 1970s, claiming a $^{4}_{\Lambda}$He $(1^+ \rightarrow 0^+)$ energy of 1.15 ± 0.04 MeV [48]. Since the statistical quality of the $^{4}_{\Lambda}$He γ-ray spectrum is extremely poor, it has to be measured again with modern techniques. In the proposed experiment, we can precisely measure the energy of $^{4}_{\Lambda}$He$(1^+ \rightarrow 0^+)$. By comparing with the $^{4}_{\Lambda}$H$(0^+ \rightarrow 1^+)$ energy, valuable information on the spin dependence of CSB will be obtained as described in Sect. 2.1.

There is another important objective in the ^{4}He target run. The cross sections of Λ hypernuclear productions by the (K^-, π^-) reaction on a nucleus, particularly their spin-flip and non-spin-flip cross sections (or polarization properties), for K^- momentum higher than 1 GeV/c have not been measured yet, although they are important for further studies of hypernuclear physics at J-PARC. By using $^{4}_{\Lambda}$He, we will measure the cross sections of the spin-flip 1^+ and non-flip 0^+ states at several K^- momenta (e.g., 1.3, 1.5, 1.8 GeV/c) and confirm the theoretical calculations based on the elementary cross sections. Since the (K^-, π^-) reaction in nuclear matter is not completely understood, the data will play a significant role to establish the (K^-, π^-) reaction mechanism and to make more realistic estimates of hypernuclear production in the future. For this purpose, ${}^4_\Lambda$He is the best hypernucleus because the cross sections for the 0^+ and 1^+ states are particularly large, and the 1^+ state is produced purely by the spin-flip interaction. The absolute cross section summed up for the 1^+ and 0^+ states, of which peaks cannot be resolved, is measured from the peak counts in the (K^-, π^-) spectrum, and the $1^+ \rightarrow 0^+$ γ-ray yield provides the ratio of the 1^+ and 0^+ productions.

5 Future Experiments

After the E13 experiment, we will extend our study to a wide variety of hypernuclear species and to more physical information. Most of the following experiments will be best performed with 1.1 GeV/c K^- beam at the K1.1 beamline, but 1.5 GeV/c K^- beam at K1.8 is also usable.

5.1 Light (p-shell) Hypernuclei (1-a)

Most of the *p*-shell hypernuclei that can be produced by the (K^-, π^-) or (π^+, K^+) reactions from available target nuclei were already studied as shown in Fig. 1, but some of them have to be investigated again to determine more level energies, assign their spin parities, and measure $B(E2)$ values.

In the level assignment, γ–γ coincidence measurement and angular correlation of γ-rays, as well as angular distribution of pions in the (K^-, π^-) reaction, are essential. In the forward (K^-, π^-) reaction, the angular correlation between γ-rays and pions is calculated in Ref. [15]. In the case of hypernuclear productions of ΔL (angular momentum transfer) = 1, such as production of low-lying $(p_n^{-1} s_\Lambda)$ levels of *p*-shell hypernuclei, the correlation is expressed as $W(\theta_{\pi\gamma}) \propto 1 + A\cos^2\theta_{\pi\gamma}$. In the spin assignment, the π–γ angular correlation is more sensitive than the angular correlation between two γ-rays emitted in a cascade decay.

By assigning spin parities of various bound states in ${}^{12}_\Lambda$C, intershell coupling states [49] are expected to be identified. In ${}^9_\Lambda$Be and ${}^{13}_\Lambda$C, measurement of $B(E2)$ values allows systematic study of the shrinking effect of hypernuclei. The ground state doublet spacing of ${}^{14}_\Lambda$N or ${}^{15}_\Lambda$N provides valuable data to confirm the ΛN tensor interaction strength (T), which is determined only from the ${}^{16}_\Lambda$O ground state doublet spacing [6].

Example of ${}^{12}_\Lambda$C

In ${}^{12}_\Lambda$C, we will be able to observe several γ transitions and to reconstruct the level scheme by γ–γ coincidence method. In addition, measurements of angular correlation and polarization of γ-rays will enable us to assign spin parities of these states experimentally. The structure of ${}^{12}_{\Lambda}$C is particularly important to study the ΛN spin-dependent interactions and the ΛN–ΣN coupling force. In addition, intershell coupling states can be best investigated in ${}^{12}_{\Lambda}$C.

The ${}^{12}_{\Lambda}$C level observed at $E_{\rm ex} = 8.3\,$MeV, which is barely separated in the (π^+, K^+) spectrum (E369) (#4 in Fig. 13 top) [50], is interpreted as a 2^+ state having ${}^{11}{\rm C}(\frac{5}{2}^+) \otimes s_\Lambda$ $(d_N \otimes s_\Lambda)$ configuration with an admixture of ${}^{11}{\rm C}(J^-) \otimes p_\Lambda$ $(p_N^{-1} \otimes p_\Lambda)$ configurations [49], as illustrated in Fig. 13 (bottom). Such a coupling between two states with a Λ in different parity shell orbits is a unique phenomenon in hypernuclei. It is expected that a small admixture of $p_N^{-1} \otimes p_\Lambda$ component makes production cross section of this 2^+ state large. Due to such intershell coupling, a few positive-parity states around 7 MeV are expected to be populated strongly enough to be observed. Reference [49] predicts the first 2^+ and 0^+ states at 7.4 and 6.7 MeV, respectively. Since they are particle bound (threshold ~9.3 MeV), we can precisely determine their level energies from γ transitions. Their level energies contain information on the $p_\Lambda p_N$ effective interactions. The determination of spin parities by γ-ray spectroscopy is essential to identify these unique states.

Figure 13 shows level energies, cross sections, and γ-ray branching ratios assumed in the following simulation for the ${}^{12}_{\Lambda}$C experiment, and Fig. 14 shows simulated spectra for 20 days' run with 1.1 GeV/c K^- beam of 1.0×10^6/spill (the expected K^- intensity for the 30 GeV full-intensity proton beam of 9 μA), with a 20-g/cm^2-thick target, and using a spectrometer similar to SKS with a ~100 msr acceptance. If the 50 GeV 15 μA proton beam is available, we expect to have 4×10^6 K^- intensity at 1.1 GeV/c and the simulated spectra correspond to 5 days' run. We will be able to observe all the γ transitions from "a" to "m" in Fig. 13. Figure 15 shows some examples of γ–γ coincidence spectra. Such spectra enable us to reconstruct a level scheme as in Fig. 13. The spins of these levels can also be determined by angular correlation of γ-rays.

5.2 sd-Shell Hypernuclei (1-b)

One of the motivations to study *sd*-shell hypernuclei is to investigate ΛN effective interaction between a Λ in the $0s$ shell and a nucleon in the *sd* shell, which provides information on the radial dependence of the interaction, as described for the ${}^{19}_{\Lambda}$F experiment in E13. Another motivation is to study the impurity effect. Some *sd*-shell Λ hypernuclei are expected to exhibit interesting phenomena such as changes of cluster structure and deformation. For example, a result of a relativistic mean field calculation [54] implies a drastic change of nuclear deformation induced by a Λ in some of *sd*-shell hypernuclei. Such an effect can be studied by measuring excitation energies in the rotational band. If $B(E2)$ values are measured, the deformation is better determined.

In addition to ${}^{19}_{\Lambda}$F, we plan to study ${}^{20}_{\Lambda}$Ne, ${}^{28}_{\Lambda}$Si, ${}^{40}_{\Lambda}$Ca, and some more hypernuclei via the (K^-, π^-) reaction. In the case of ${}^{20}_{\Lambda}$Ne, a negative-parity

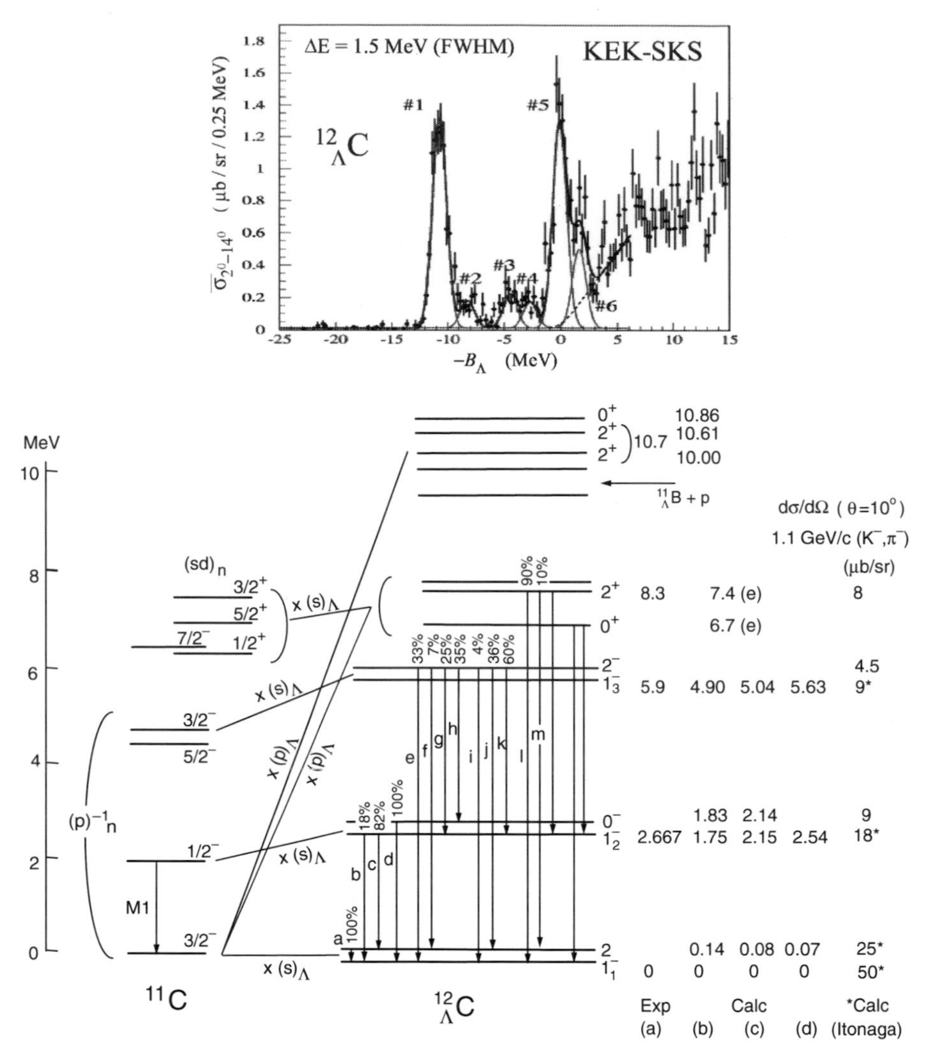

Fig. 13. *Top*: $^{12}_{\Lambda}$C spectrum measured by the (π^+, K^+) reaction (KEK E369) [50]. The peaks #1, #2, #3, and #4 are interpreted as 1_1^-, 1_2^-, 1_3^-, and 2_1^+ states, respectively. *Bottom*: Level scheme and expected γ transitions of $^{12}_{\Lambda}$C hypernucleus. (a) Experimental level energies (E369 and E566) and (b)–(e) calculated energies by Itonaga et al. [51], Fetisov et al. [52], Millener et al. [53], and Motoba (1995, 1996, 1999, private communication), respectively. The production cross sections except for the first 2^+ state were calculated by Itonaga [51]. The cross section of the first 2^+ state is estimated from the relative intensities of peak #4 to peak #5 in the *top figure*. The level energies are taken from the experimental values from E369 and E566, but the doublet spacing energies are estimated from the parameter set of the spin-dependent interactions by Millener [53]. The branching ratios are taken from Ref. [51] but corrected for the excitation energies. The branching ratios from the 6 MeV 2^- state and 2^+ state are calculated assuming weak coupling limit.

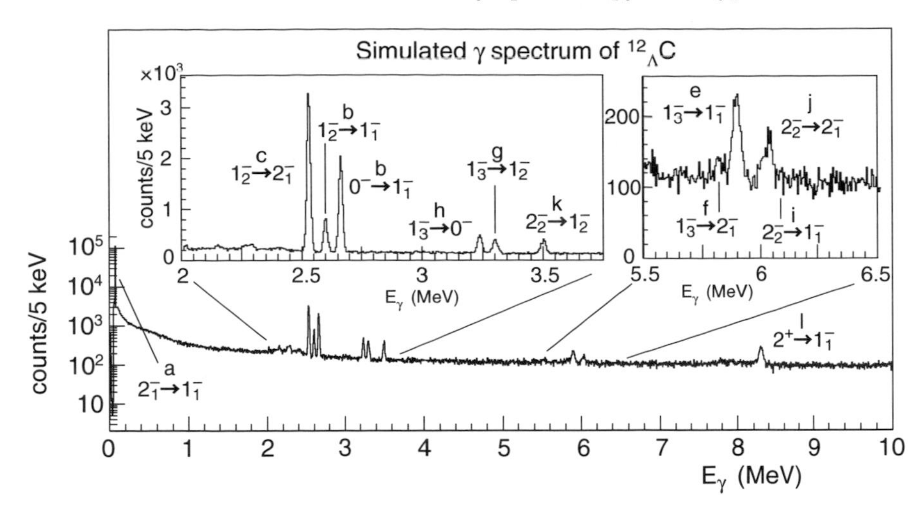

Fig. 14. Simulated spectrum of $^{12}_{\Lambda}$C γ-rays to be obtained with the (K^-, π^-) reaction at 1.1 GeV/c for 20 days' run under 30 GeV 9 μA operation. Compton/π^0 suppression and Doppler shift correction are applied.

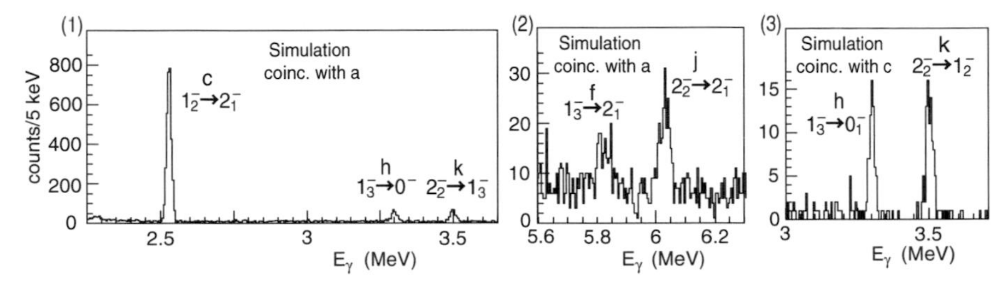

Fig. 15. Simulated γ–γ coincidence spectrum of $^{12}_{\Lambda}$C. (**1**) and (**2**) Coincidence with γ-ray "a" $(1_2^- \to 1_1^-)$ and (**3**) coincidence with γ-ray "c" $(1_2^- \to 2_1^-)$.

ground state of $^{20}_{\Lambda}$Ne is predicted [55], being contrary to the naively expected positive-parity ground state of $(sd)_n^{-1}(s)_\Lambda$ configuration. Figure 16 shows relevant low-lying levels and transitions of $^{20}_{\Lambda}$Ne. The left part of Fig. 16 is expected from a naive picture, where energies of $(0^+, 1^+)$ and 1^- states are taken from a shell model calculation [16]. The right part shows the result of a cluster model calculation [55]. Here the negative-parity states whose structure is $^{16}_{\Lambda}$O (p-hole state) + ^{4}He have lower energies than the $(0^+, 1^+)$ doublet states having the ^{16}O + $^4_{\Lambda}$He structure. It is because the negative-parity states in ^{19}Ne having a well-clusterized structure of ^{15}O (p-hole state of ^{16}O) + ^{4}He significantly shrinks when a Λ is added, while the positive-parity states having ^{16}O + ^{3}He structure is not well clusterized but rather spherical and thus shrinkage by a Λ is expected to be smaller. As shown in Fig. 16, expected γ transitions in each case are completely different. In the latter case, γ transitions stem mainly from the population of the 2^+ state via $\Delta L = 2$ which has

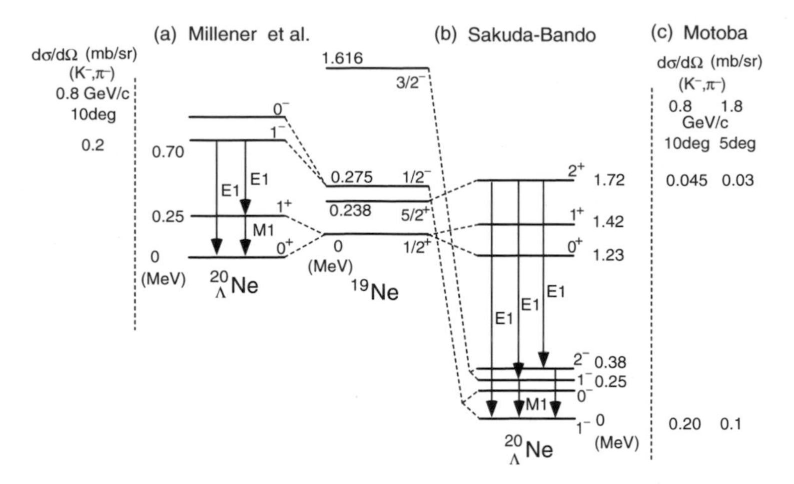

Fig. 16. Level scheme and transitions of ${}^{20}_{\Lambda}$Ne. The expected level scheme and the cross sections are calculated (**a**) with a shell model by Millener et al. [16] and (**b**) with a cluster model by Sakuda and Bando (**b**) [55]. (**c**) Estimated cross sections by Motoba (1995, 1996, 1999, private communication) from calculated ${}^{18}_{\Lambda}$O cross sections by Yamada et al. [56].

a sizable cross section in the $(K^-,\ \pi^-)$ reaction. By observing all the transitions in Fig. 16 we can reconstruct the level scheme and assign spin parities. The structure of ${}^{20}_{\Lambda}$Ne, particularly its ground state doublet, also provides us the effective ΛN spin-dependent interaction strengths in the *sd*-shell region.

5.3 Medium and Heavy Hypernuclei (1-c)

The p_Λ single-particle orbit is generally unbound in the p-shell hypernuclei except for ${}^{13}_{\Lambda}$C, but it is bound in medium and heavy hypernuclei because the $1\hbar\omega$ energy between p_Λ and s_Λ orbits becomes lower than the nucleon separation energy. The p_Λ states in a wide range of the mass number provide detailed information on the Λ single-particle potential, which is related to the ΛN interaction in nuclear matter. In addition, it presumably reflects possible modifications of a Λ hyperon lying deeply inside a nucleus. On the other hand, it is also interesting to measure the spin–orbit splitting of the p_Λ orbit, which is known to be extremely small in p-shell hypernuclei $(E(p_{1/2}) - E(p_{3/2}) =$ 0.15 MeV for ${}^{13}_{\Lambda}$C [12]), but should be examined in heavier hypernuclei.

We plan a γ spectroscopy study of several hypernuclei such as ${}^{89}_{\Lambda}$Y and ${}^{208}_{\Lambda}$Pb, where we will detect $E1(p_\Lambda \to s_\Lambda)$ intershell transitions and other transitions in coincidence with the intershell transitions. In order to give a large momentum transfer to populate deeply bound p_Λ states, we use 1.5–1.8 GeV/c $(K^-,\ \pi^-)$ reaction at large scattering angles at the K1.8 beamline. We require a beam time for about a week (with the full beam) per target.

Example of ${}^{208}_{\Lambda}$Pb

In ${}^{208}_{\Lambda}$Pb, the excitation energy of the p_{Λ} states was measured to be 4.5 ± 0.6 MeV [57], being smaller than the proton/neutron separation energy of 6.7 MeV. Figure 17 shows the expected low-lying level scheme and γ transitions of ${}^{208}_{\Lambda}$Pb. Here we assume one-hole configurations for ^{207}Pb core and weak coupling between a Λ and the core. γ transitions are characterized by several 4–5 MeV $E1(p_{\Lambda} \rightarrow s_{\Lambda})$ transitions, some of which are followed by $M1$ or $E2$ core transitions.

We estimate γ-ray yields for ${}^{208}_{\Lambda}$Pb. We use the reaction ^{208}Pb(K^-, π^-) ${}^{208}_{\Lambda}$Pb at 1.8 GeV/c to efficiently populate p_{Λ} states coupled to the $i_{13/2}$ hole state (1,633 keV isomer of ^{207}Pb). There are three possible non-spin-flip $n \rightarrow \Lambda$ transitions, namely $i_{13/2} \rightarrow p_{1/2}$ with $\Delta L = 7$, $i_{13/2} \rightarrow p_{3/2}$ with $\Delta L = 7$, and $i_{13/2} \rightarrow p_{3/2}$ with $\Delta L = 5$ (see Fig. 17). Their calculated cross sections (D. J. Millener, 1999, private communication) have peaks at 8° with 10 and 18 μb/sr and at 6° with 50 μb/sr, respectively. Those populated states undergo 4 MeV $E1(0p_{\Lambda} \rightarrow 0s_{\Lambda})$ transitions to the doublet $(7^+, 6^+)$. By assuming the full 1.8 GeV/c K^- intensity at the K1.8 line (1.6×10^6/spill for 30 GeV 9 μA), a 10 g/cm^2 target, and a pion acceptance of 100 msr, the yields for those transitions are estimated to be 200, 120, and 500 events, respectively, for a beam time of 10 days.

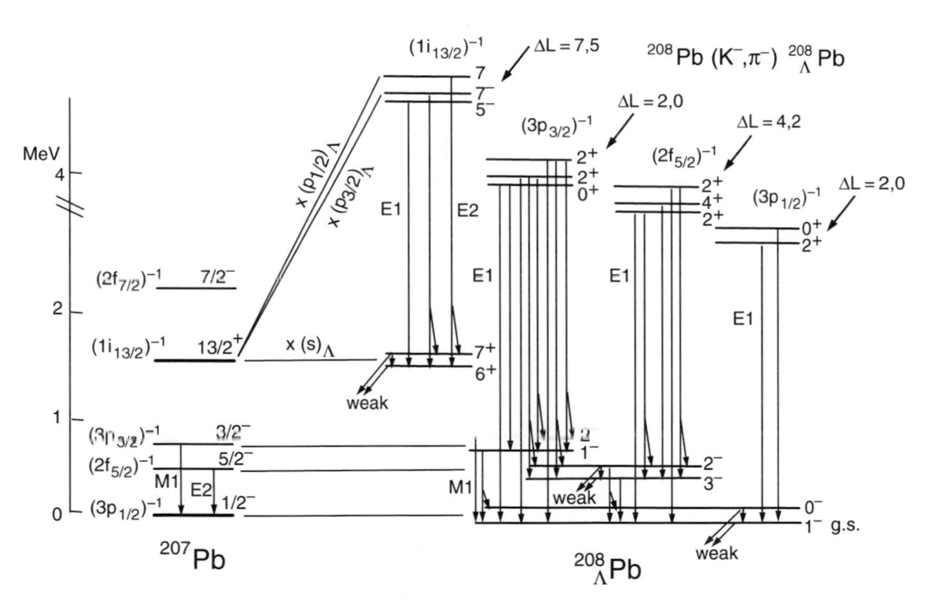

Fig. 17. Expected level scheme and γ transitions of ${}^{208}_{\Lambda}$Pb. Only the non-spin-flip states are shown for the p_{Λ} states.

5.4 γ Spectroscopy with (K^-, π^0) Reaction (2)

By using the (K^-, π^0) reaction which converts a proton into a Λ, we can produce light neutron-rich hypernuclei and mirror hypernuclei to those produced by the (K^-, π^-) and (π^+, K^+) reactions. A Λ hyperon implanted in a neutron-rich nucleus having a neutron skin or halo may result in a drastic change of the nuclear structure, such as disappearance of neutron skin/halo and large shrinkage. From this viewpoint, measurements of level schemes and $B(E2)$ values of n-rich hypernuclei, such as ${}^{7}_{\Lambda}$He, ${}^{9}_{\Lambda}$Li, and ${}^{11}_{\Lambda}$Be, produced by the (K^-, π^0) reaction on ^{7}Li, ^{9}Be, and ^{11}B targets, are particularly interesting.

The CSB in the ΛN interaction can be studied by precise comparison of level structure between mirror hypernuclei, such as ${}^{4}_{\Lambda}$H – ${}^{4}_{\Lambda}$He, ${}^{12}_{\Lambda}$B – ${}^{12}_{\Lambda}$C, and ${}^{16}_{\Lambda}$N – ${}^{16}_{\Lambda}$O, where ${}^{4}_{\Lambda}$H, ${}^{12}_{\Lambda}$B, and ${}^{16}_{\Lambda}$N hypernuclei are produced by the (K^-, π^0) reaction on ^{4}He, ^{12}C, and ^{16}O targets. These data will play an essential role to understand the origin of the CSB in the ΛN interaction as discussed in Sect. 4.2.

Example of ${}^{7}_{\Lambda}$He

^{6}He is a neutron-rich nucleus with a two-neutron skin. The first excited state of 2^+ is observed as an unbound resonance state (see Fig. 18), but its structure is not well known; the $B(E2; 2^+ \to 0^+)$ has not been experimentally obtained. Figure 18 shows expected level schemes and density distributions of valence neutrons of ^{6}He and ${}^{7}_{\Lambda}$He calculated with a cluster model for $\alpha + n + n$ and ${}^{5}_{\Lambda}\mathrm{He} + n + n$ [58] (E. Hiyama, 2000, private communication) When a Λ is added to ^{6}He, the neutron skin in the ground state is expected to shrink. The ${}^{6}\mathrm{He}(2^+) + s_\Lambda$ state, which has widely spread two valence neutrons, becomes bound by the presence of a Λ, and the core $E2$ transitions $(5/2^+, 3/2^+ \to 1/2^+)$ can be observed. These $B(E2)$ values are predicted to be 10 times smaller than the $B(E2)$ of ${}^{6}\mathrm{He}(2^+ \to 0^+)$, as shown in Fig. 18 (top) (E. Hiyama, 2000, private communication). The predicted $B(E2)$ change is caused by a drastic shrinkage of valence neutron wave functions in ^{6}He, as shown in Fig. 18 (bottom).

In this experiment, we use the ${}^{7}\mathrm{Li}(K^-, \pi^0)$ reaction with 0.8 GeV/c K^- beam from the K1.1 beamline (0.75×10^6/spill with 50 GeV 15 μA proton beam) and employ a high-resolution π^0 spectrometer with an acceptance of $\sim$10 msr and a resolution better than 2 MeV (FWHM). Since these $E2$ transitions compete against weak decay, we can directly measure the lifetimes of the excited states using weak decay products (π^- and proton). Combined with the branching ratios of those $E2$ transitions, the $B(E2)$ values can be derived as described in Sect. 5.5. The yield of γ-weak coincidence events for $5/2^+$; $T = 1$ state is estimated to be 50 counts in 30 days' beam time, and the $B(E2)$ value can be determined within 30% error.

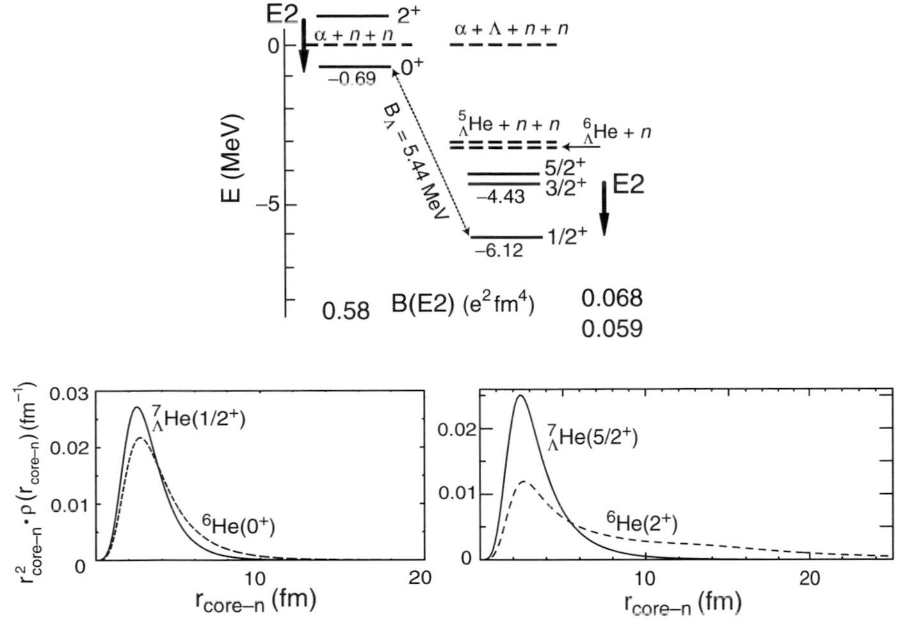

Fig. 18. *Top*: Expected level scheme of ${}^{7}_{\Lambda}$He and $B(E2)$ values calculated by Hiyama et al. with a three-body (α (${}^{5}_{\Lambda}$He) + $n + n$) cluster model [58]. *Bottom*: Calculated density distributions of valence neutrons are compared for ^{6}He(0^+) and ${}^{7}_{\Lambda}$He$(1/2^+)$ and for ^{6}He(2^+) and ${}^{7}_{\Lambda}$He$(5/2^+)$ (E. Hiyama, 2000, private communication).

5.5 $B(M1)$ Measurements with γ-Weak Coincidence Method

If a deviation of the in-medium g_Λ value from the free-space value is once found, we need to extend the study to other hypernuclei in order to understand the mechanism of the deviation. The nuclear density dependence of the g_Λ value obtained from $B(M1)$ measurement for a light to heavy hypernuclei is expected to be sensitive to partial restoration of chiral symmetry. On the other hand, the meson exchange current from the kaon exchange and the two pion exchanges with a Σ intermediate state affects the $B(M1)$ value depending on isospin and structure of the nucleus.

The DSAM can be applied only when the stopping time of the recoiling hypernucleus in the target material (typically, of the order of 1 ps for p-shell hypernuclei) is comparable to the lifetime of the γ-emitting excited state. This condition is not often satisfied. On the other hand, when the lifetime of the excited state is of the same order as the weak decay lifetime ($\sim 10^{-10}$ s), the γ transition competes against the weak decay, and the γ transition rate can be measured by a newly proposed method, "γ–weak coincidence method." In particular, this method is applicable to $B(M1)$ measurement of medium and heavy hypernuclei, in which the doublet spacing energy is usually small (ΔE < 0.1 MeV) and consequently the $M1$ transition rate proportional to

ΔE^3 is small. With this method, we plan to measure $B(M1)$ values for several hypernuclei ranging from light to heavy ones.

γ–Weak Coincidence Method

As explained in Fig. 19, we measure the lifetime of the upper state B directly from the time difference between hypernuclear production and emission of weak decay products (protons and π^- from $\Lambda p \to np$ and $\Lambda \to p\pi^-$ decays) in coincidence with γ transitions of $B{\to}A$. The branching ratio m of the $B{\to}A$ transition is measured from the γ-ray yield of $B{\to}A$ in coincidence with the $C{\to}B$ transition. If there are no γ transitions like $C{\to}B$ from upper states, we need to separate B and A in the (K^-, π) spectrum, which is usually difficult for the doublet states of (A, B).

The time spectrum of weak decay products measured in coincidence with the $B{\to}A$ γ-ray is expressed as follows:

$$P^{B\to A}(t) = \frac{\lambda_A \lambda_B}{\lambda_B - \lambda_A} m N_B^0 (e^{-\lambda_A t} - e^{-\lambda_B t}),$$

where λ_B and λ_A denote the total decay rates of B and A respectively, and N_B^0 denotes the initial population of the state B. From this growth-decay function, we can determine λ_B and λ_A.

When the $B{\to}A$ γ transition is much slower than the weak decay and this transition is suppressed, the number of γ–weak coincidence events decreases, which makes the sensitivity for λ_B worse. In such a case, however, λ_B can be determined from the time spectrum of weak decay products in coincidence with $C \to B$ γ-rays, which is approximately a single exponential decay with λ_B. In general, the time spectrum measured in coincidence with the $C \to B$ γ-ray is

$$P^{C\to B}(t) = \lambda_B N_B^0 \left[\left(1 - m\frac{\lambda_B}{\lambda_B - \lambda_A} \right) e^{-\lambda_B t} + m\frac{\lambda_A}{\lambda_B - \lambda_A} e^{-\lambda_A t} \right].$$

By measuring both $P^{B\to A}(t)$ and $P^{C\to B}(t)$ and fitting them together to the equations above, we can precisely determine λ_B for a wide range of λ_B.

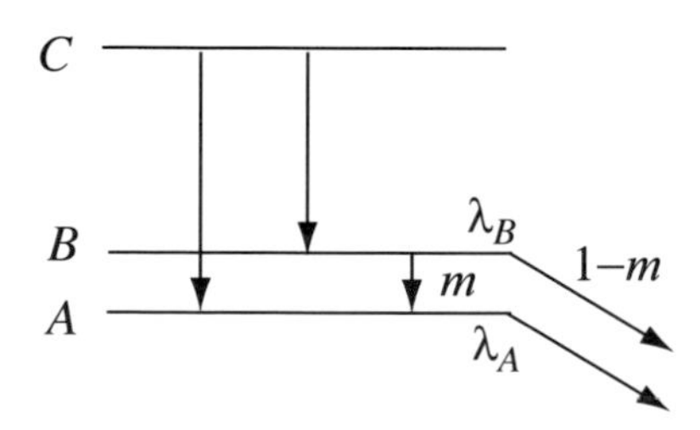

Fig. 19. Method of $B(M1)$ measurement from coincidence events of γ-ray and weak decay products (see text).

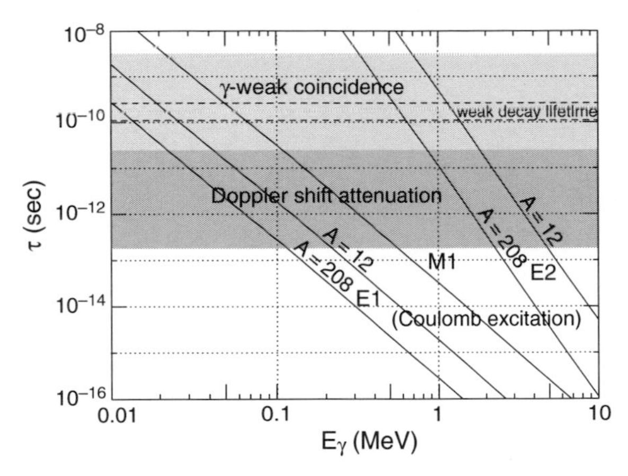

Fig. 20. Lifetime ranges to which the DSAM and the γ–weak coincidence method can be applied for transition probability measurement of hypernuclei, shown together with the $E1/M1/E2$ lifetimes for a Weisskopf unit plotted as a function of transition energy. For lifetimes shorter than 10^{-13} s, Coulomb excitation may be used in some cases using a "hypernuclear beam" produced in the projectile rapidity region from ion beams in the future.

In the γ-weak coincidence measurement a half of the Hyperball-J setup is replaced by a set of decay product detectors (decay arm) similar to those used in our previous hypernuclear weak decay experiments (KEK E307, E462, E508) [46, 59]. The solid angle for the decay product is about 30% of 4π sr. Just upstream of the target we install fine-segmented fast plastic scintillation counters giving start-timing signals. In the decay arm, fast scintillators for stop-timing signals are installed as close as possible to the target together with fine-segmented SSDs for tracking the decay product. Behind them are a stack of scintillation counters for range measurement. Protons and pions are identified and their velocities are measured from range and dE/dx data. According to our previous experience, we will be able to achieve an overall time resolution of 200 ps (FWHM) for the lifetime measurement.

This method is used not only for $B(M1)$ but also for $B(E2)$ measurement. Figure 20 shows the ranges of lifetimes to which the DSAM and the γ–weak coincidence method can be applied, together with the $E1/M1/E2$ lifetimes for a Weisskopf unit plotted as a function of transition energy. The γ–weak coincidence method is applied to $B(M1)$ with energies less than 0.1 MeV and to $B(E2)$ with energies around 1 MeV.

Expected Results – ${}^{12}_{\Lambda}$C Case

Here we consider the case of ${}^{12}_{\Lambda}$C, for example. According to Millener's calculation with the new parameter set (Eq. (1)) the ground state doublet spacing of ${}^{12}_{\Lambda}$C$(2^-, 1^-)$ is predicted to be 0.15 MeV. The $B(M1)(2^- \to 1^-)$ is predicted to be 0.061 μ_N^2 [51], which corresponds to the decay rate of (294 ps)$^{-1}$ for 0.15 MeV. By assuming the weak decay rate of the upper (2^-) state to be the same as the ground (1^-) state (228 ps)$^{-1}$, we estimate the branching ratio of the γ transition to be $m = 0.44$.

We will use the 1.1 GeV/c (K^-, π^-) reaction at the K1.1 beamline to effectively populate the spin-flip 2^- state. Assuming the same beam and spectrometer conditions as in Sect. 5.1 ($1.0 \times 10^6 K^-$/spill, 100 msr), an efficiency for detecting weak decay products of 0.1, the cross sections and the γ-ray branching ratios shown in Fig. 13, and a target thickness of 10 g/cm^2, we expect 15,000 γ-weak coincidence events for $m = 1$ in 70 days with 30 GeV 9 μA (or 17 days with 50 GeV 12 μA). Figure 21 shows the simulated time spectra of the weak decay products in coincidence with the spin-flip $M1(2^- \rightarrow 1^-)$ transition and in coincidence with the upper $1_2^- \rightarrow 2^-$ transition, for various cases of the spin-flip transition rates. The simulation indicates that the $B(M1)$ value can be determined within 5% statistical error for a wide range of $B(M1) = 0.1$–$10\,\mu_N$. Including possible systematic errors of ~5%, we can determine the $B(M1)$ value within 10% error (or $|g_\Lambda - g_c|$ value within 5% error) in total.

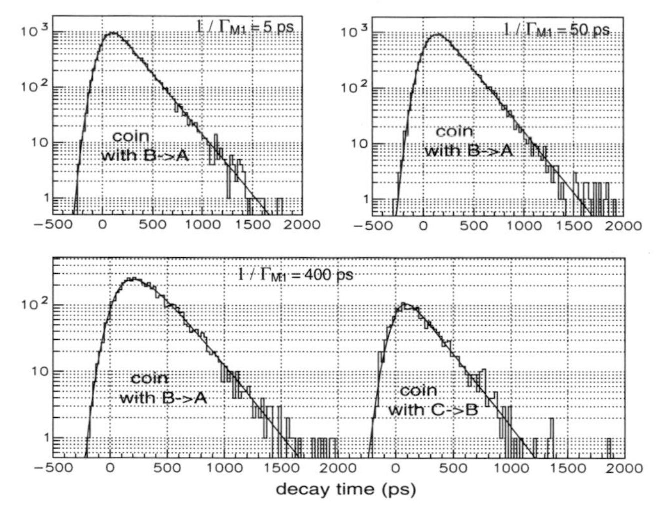

Fig. 21. Simulated time spectrum of weak decay products of $^{12}_{\Lambda}$C measured in coincidence with B→A (the spin-flip $M1(2^- \rightarrow 1^-)$ transition) and with C→B (the $1_1^- \rightarrow 1^-$ transition) for various values of the B→A (spin-flip $M1$) transition rates (Γ_{M1}). A time resolution of 200 ps FWHM is folded. By fitting these histograms to the expected functions (see text), both decay rates λ_A and λ_B can be determined within 5% statistical errors.

6 Summary

The γ-ray spectroscopy of Λ hypernuclei is one of the main subjects in strangeness nuclear physics program at the J-PARC Hadron Hall. By utilizing intense and pure K^- beams available at J-PARC, hypernuclei will be produced via the $(K^-,\ \pi^-)$ or $(K^-,\ \pi^0)$ reactions and γ-rays will be detected in coincidence. The first experiment (E13) will be performed at the K1.8 beamline employing the SKS spectrometer with a modified setup (SksMinus) and a new Ge detector array Hyperball-J. It aims at the investigation of some of s, p and sd-shell hypernuclei ($^{4}_{\Lambda}$He, $^{7}_{\Lambda}$Li, $^{10}_{\Lambda}$B, $^{11}_{\Lambda}$B, and $^{19}_{\Lambda}$F) in order to measure the g-factor of Λ in a nucleus from a spin-flip $B(M1)$ value and to examine various aspects of the hyperon–nucleon interactions (ΛN spin-dependent interactions, ΛN–ΣN coupling, charge symmetry breaking (CSB) in ΛN interaction, etc.). After E13, we will further investigate structures of a wide variety of Λ hypernuclei, such as p-shell, sd-shell, and medium and heavy hypernuclei up to $^{208}_{\Lambda}$Pb, including neutron-rich ones, and also systematically measure $B(E2)$ and $B(M1)$ values for various hypernuclei to explore impurity effect to nuclear structure and nuclear medium effects to baryons, not only by the Doppler shift attenuation method (DSAM) but also by the γ–weak coincidence method.

Acknowledgments

The author thanks all the collaborators in the J-PARC E13 experiment, in particular, K. Tanida, M. Ukai, T. Koike, K. Miwa, K. Shirotori, K. Hosomi, and N. Chiga. The author is also grateful to E. Hiyama, T. Harada, T. Motoba, K. Itonaga, and D. J. Millener for theoretical estimations and valuable discussions.

References

1. Tamura, H., et al.:, Phys. Rev. Lett. **84**, 5963 (2000)
2. Tanida, K., et al.:, Phys. Rev. Lett. **86**, 1982 (2001)
3. Akikawa, H., et al.:, Phys. Rev. Lett. **88**, 082501 (2002)
4. Tanida, K., et al.:, Nucl. Phys. A **721**, 999c (2003)
5. Miwa, K., et al.:, Nucl. Phys. **A754**, 80c (2005)
6. Ukai, M., et al.:, Phys. Rev. Lett. **93**, 232501 (2004)
7. Tamura, H., et al.:, Nucl. Phys. A **754**, 58c (2005)
8. Miura, Y., et al.:, Nucl. Phys. A **754**, 75c (2005)
9. Ukai, M., et al.:, Phys. Rev. C **73**, 012501 (2006)
10. Ma, Y., et al.:, Eur. Phys. J. A **33**, 243 (2007)
11. Ukai, M., et al.:, Phys. Rev. C **77**, 054315 (2008)
12. Ajimura, S., et al.:, Phys. Rev. Lett. **86**, 4255 (2001)
13. Hashimoto, O., Tamura, H.: Prog. Part. Nucl. Phys. **57**, 564 (2006)
14. Hiyama, E., et al.:, Phys. Rev. Lett. **85**, 270 (2000)
15. Dalitz, R.H., Gal, A.: Ann. Phys. **116**, 167 (1978)

16. Millener, D.J., et al.:, Phys. Rev. C **31**, 499 (1985)
17. Millener, D.J., Lect. Notes Phys. **724**, 31 (2007)
18. Gibson, B.F., Lehman, D.R.: Phys. Rev. C **37**, 679 (1988)
19. Akaishi, Y., et al.:, Phys. Rev. Lett. **84**, 3539 (2000)
20. Hiyama, E., et al.:, Phys. Rev. C **65**, 011301 (2002)
21. Nemura, H., et al.:, Phys. Rev. Lett. **89**, 142504 (2002)
22. Bodmar, A.R., Usmani, Q.N.: Phys. Rev. C **31**, 1400 (1985)
23. Nogga, A., Kamada, H., Glöckle, W.: Phys. Rev. Lett. **88**, 172501 (2002)
24. Hiyama, E., et al.:, Phys. Rev. C **59**, 2351 (1999)
25. Motoba, T., Bandō, H., Ikeda, K.: Prog. Theor. Phys. **80**, 189 (1983)
26. Tamura, H., Eur. Phys. J. A **13**, 181 (2002)
27. Brown, G.E., Rho, M.: Phys. Rev. Lett. **66**, 2720 (1991)
28. Hatsuda, T., Lee, S.H.: Phys. Rev. C **46**, R34 (1992)
29. Naruki, M., et al.:, Phys. Rev. Lett. **96**, 092301 (2006)
30. Muto, R., et al.:, Phys. Rev. Lett. **98**, 042501 (2007)
31. Hungerford, E.V., Biedenharn, L.C.: Phys. Lett. **142B**, 232 (1984)
32. Yamazaki, T.: Nucl. Phys. A **446**, 467c (1985)
33. Takeuchi, T., Shimizu, K., Yazaki, K.: Nucl. Phys. A **481**, 693 (1988)
34. Saito, K., Oka, M., Suzuki, T.: Nucl. Phys. A **625**, 95 (1997)
35. Oka, M., Saito, K., Sasaki, K., Inoue, T.: AIP proceedings **594**, 163 (2001)
36. Gopal, G.P., et al.:, Nucl. Phys. B **119**, 362 (1977)
37. Shirotori, K.: Master thesis, Tohoku University (2007)
38. Rusek, A.: Nucl. Phys. A **639**, 111c (1998)
39. Koike, T., et al.:, Proc. 9th Int. Conf. on Hypernuclear and Strange Particle Physics (HYP2006), 10–14 October 2006, Mainz, Ed. by J. Pochodzalla and Th. Walcher, Springer, p.25 (2007)
40. Hull, E., et al.:, IUCF Annual Report **143** (1992–93)
41. Tanida, K., et al.:, "Measurement of X rays from Ξ^- atoms", J-PARC proposal E03 (2006), `http://j-parc.jp/NuclPart/pac_0606/pdf/p03-Tanida.pdf`
42. Nakazawa, K., et al.:, "Systematic study of double strangeness system with an emulsion-counter hybrid method", J-PARC proposal E07 (2006), `http://j-parc.jp/NuclPart/pac_0606/pdf/p07-Nakazawa.pdf`
43. Tamura, H., et al.:, J-PARC proposal E13 (2006), `http://j-parc.jp/NuclPart/pac_0606/pdf/p13-Tamura.pdf`
44. Chrien, R.E., et al.:, Phys. Rev. C **41**, 1062 (1990)
45. Millener, D.J.: Nucl. Phys. A **754**, 48c (2005)
46. Bhang, H., et al.:, Phys. Rev. Lett. **81**, 4321 (1998)
47. Bedjidian, M., et al.:, Phys. Lett. **62B**, 467 (1976)
48. Bedjidian, M., et al.:, Phys. Lett. **83B**, 252 (1979)
49. Motoba, T.: Nucl. Phys. A **639**, 135c (1998)
50. Hotchi, H., et al.:, Phys. Rev. C **64**, 044302 (2001)
51. Itonaga, K., et al.:, Prog. Theor. Phys. Suppl. **117**, 17 (1994)
52. Fetisov, V.N., et al.:, Z. Phys. A **339**, 399 (1991)
53. Millener, D.J.: Nucl. Phys. A **691**, 93c (2001)
54. Myaing Thi Win, Hagino, K.: Phys. Rev. C 78, 054311 (2008)
55. Sakuda, T., Bando, H.: Prog. Theor. Phys. **78**, 1317 (1987)
56. Yamada, T., et al.:, Prog. Theor. Phys. Suppl. **81**, 104 (1985)
57. Hasegawa, T., et al.:, Phys. Rev. C **53**, 1210 (1996)
58. Hiyama, E., et al.:, Phys. Rev. C **53**, 2075 (1996)
59. Kang, B.H., et al.:, Phys. Rev. Lett. **96**, 062301 (2006)

Search for Pentaquark Θ^+

Megumi Naruki

High Energy Accelerator Research Organization, KEK, 1-1 Oho, Tsukuba, Ibaraki 305-0801, Japan
megumi.naruki@kek.jp

1 Introduction

So far, all subatomic particles have been known to be either mesons, which contain a quark and an antiquark, or baryons, which contain three quarks. For example, π^+ is a meson that contains an up-quark and an anti-down-quark ($u\bar{d}$), while proton is a baryon that contains two up-quarks and one down-quark (uud). However, quantum chromodynamics (QCD), the theory of the strong force, allows for any multiquark systems, as far as they are colorless. Therefore, the "exotic hadrons," particles which consist of four or more quarks, have been intensively searched for over 30 years. The spectroscopy of multiquark systems will shed light on the quark dynamics in the nonperturbative regime.

A minimal exotic hadron is a tetraquark which consists of two quarks and two antiquarks ($qq\bar{q}\bar{q}$). The $a_0/f_0(980)$, $D^*_{s0}(2{,}317)$, and $D_{s1}(2{,}460)$ mesons could be candidates for tetraquarks since their masses contradict the expectations from quark models. Another candidate is a series of new particles which were observed at B factories in a decay into two charmed mesons or a charmonium with other mesons [1]. Their masses or branching ratios are far from the theoretical expectations for a normal charmonium. However, the quark composite of all these candidates is reducible to a normal meson ($q\bar{q}$); therefore, they cannot be a direct evidence of tetraquarks. Recently, a narrow resonance Z(4,430) has been observed in the $\pi^+\psi'$-invariant mass spectrum with a significance of 6.5σ by Belle [2]. It has an irreducible quark component and the first charged candidate for tetraquarks. A further study is expected to confirm the existence of tetraquarks.

1.1 First Possible Evidence for Pentaquark Θ^+

In 2003, the LEPS collaboration observed a narrow resonance at 1,540 MeV/c^2 in the K^- missing mass spectrum for the γn $\rightarrow$ K^+K^-n reaction on a

Naruki, M.: *Search for Pentaquark Θ^+*. Lect. Notes Phys. **781**, 139–160 (2009)
DOI 10.1007/978-3-642-00961-7_6

carbon target [3]. The observed width is consistent with the mass resolution of 25 MeV/c^2. This new particle was named pentaquark Θ^+ that contained four quarks and one antiquark. Θ^+ is the first exotic hadron that has irreducible quark component, $uudd\bar{s}$. The LEPS group was motivated by Diakonov et al. who predicted a mass of 1,530 MeV/c^2 and a width of less than 15 MeV/c^2 as a pentaquark Θ^+ in the framework of a chiral soliton model in 1997 [4]. In this model, pentaquarks emerge as rotational excitations of a soliton, treating the known N(1,710) resonance as a member of the anti-decuplet. The agreement on the mass and width arouses a series of pentaquark searches, discussed in the following paragraphs. The striking feature of the Θ^+ is its narrow width, which is very attractive for experimental searches but not trivial for theories. Jaffe and Wilczek proposed a di-quark model, in which the pentaquark consists of two bosonic di-quarks and one antiquark, and its fermionic state demands the orbital angular momentum to be 1 [5]. This feature explains a mechanism to suppress the decay of Θ^+ into one meson and one baryon system. In any framework, the discovery of exotic baryons provides an opportunity to deepen our understanding of the quark dynamics at low energy.

1.2 Positive Results for Θ^+

The first discovery was immediately confirmed by other experiments at various facilities, summarized in Table 1 [3–14]. They observed a narrow peak

Table 1. Positive results for Θ^+. The results are arranged in chronological order.

Experiment		$E_{beam}(\sqrt{s})$	Reaction	Mass [MeV/c^2]	Width [MeV/c^2]	Significance
LPES	[3]	≤2.4 GeV	$\gamma C \to K^- n K^+$	1540±10±5	<25	3.2σ
DIANA	[6]	≤750 MeV/c	$K^+ Xe \to p K^0_s X$	1539 ± 2 ± 3	<9	3.4σ
CLAS(d)	[7]	≤3.1 GeV	$\gamma d \to p K^- n K^+$	1542±2±5	<21	(5.2σ)
SAPHIR	[8]	≤2.8 GeV	$\gamma p \to K^0_s n K^+$	1540±4±2	<25	4.8σ
ITEP(ν)	[9]	57 GeV	$\nu A \to p K^0_s X$	1533±5	<20	4.0σ*
CLAS(p)	[10]	3–5.47 GeV	$\gamma p \to \pi^+ K^- n K^+$	1555±1±10	<26	(7.8σ)
HERMES	[11]	27.6 GeV	$e^+ d \to p K^0_s X$	1528±2.6±2.1	12±9±3	4.2σ
COSY-TOF	[12]	2.95 GeV/c	$pp \to \Sigma^+ p K^0_s$	1530±5	<18	4.7σ
ZEUS	[13]	(300,318 GeV)	$e^+ p \to e' p K^0_s X$	1521.5 1.5$^{+2.8}_{-1.7}$	8±4$^{+2.0}_{-1.4}$	4.6σ
NOMAD	[15]	45.3 GeV	$\nu A \to p K^0_s X$	1528.7±2.5	2∼3	4.3**
SVD	[14]	70 GeV/c	$pA \to p K^0_s X$	1526±3±3	<24	(5.6σ)
	[16]	70 GeV/c	$pA \to p K^0_s X$	1523±2±3	<14	8.0σ
LEPS	[17]	<2.4 GeV	$\gamma d \to \Theta^+ \Lambda(1520)$	1530		4∼5σ

Significance is estimated from $N_S/\sqrt{N_S + N_B}$ or $N_S/\Delta N_S$ or $N_S/\sqrt{N_B}$ (parentheses), where N_S is the number of signals, ΔN_S is the error in N_S, and N_B is the counts of background counts. *Calculated using the fit result shown in [9]. **It is not specified how the significance was estimated.

in the nK^+- or pK_s^0-invariant mass spectra near 1,540 MeV/c^2. However, the statistical significance ranges between 3 and 8. Each result is independent; therefore, it is statistically impossible that all results are statistical fluctuations. Nevertheless it is urgent to establish the existence of the Θ^+ with higher statistics and a well-understood background.

The width is limited by the experimental resolution in each experiment. The HERMES and ZEUS collaborations derived the intrinsic width fitting the spectrum with a Breit–Wigner shape convoluted with the Gaussian resolution obtained through a detector simulation. The HERMES expected that the intrinsic width was $\Gamma = 17 \pm 9$(stat)± 3(syst) MeV/c^2, whereas the extracted width was estimated to be $\Gamma = 8 \pm 4$(stat) MeV/c^2 by ZEUS. On the other hand, it is suggested that the width of Θ^+ is less than a few MeV/c^2 [18] from the analysis of past KN scattering data. Such a narrow width is quite unusual for a hadron whose mass is sufficiently higher than the decay threshold; therefore, there should be some mechanism to forbid the decay of Θ^+. We expect to understand the structure of exotic hadrons by measuring the width of Θ^+ together with its spin and parity.

1.3 Negative Results for Θ^+

Negative results for the Θ^+ search performed so far are summarized in Table 2. In the e^+e^- collider data [19–22], the Θ^+ could be generated through a fragmentation process. For a normal hadron, the BaBar collaboration re-

Table 2. Negative results for Θ^+.

Experiment		$\sqrt{s}(E_{beam})$	Reaction	Upper limit
BES	[19]	3.1 GeV	$e^+e^- \to J/\psi \to \Theta\bar{\Theta}$	1.1×10^{-5} BR (90% CL)
		3.7 GeV	$e^+e^- \to \psi(2S) \to \Theta\bar{\Theta}$	8.4×10^{-6} BR (90% CL)
ALEPH	[20]	91.2 GeV	$e^+e^- \to Z \to pK_s^0X$	6.2×10^{-4} BR (95% CL)
BaBar	[21]	10.58 GeV	$e^+e^- \to \Upsilon(4S) \to pK_s^0X$	1.0×10^{-4} BR (95% CL)
Belle	[22]	11 GeV	$e^+e^- \to BB \to ppK_s^0X$	2.3×10^{-7} BR (90% CL)
CDF	[23]	1.96 TeV	$p\bar{p} \to pK_s^0X$	$0.03 \times \Lambda^*$(90% CL)
SPHINX	[24]	(70 GeV)	$p + C \to \bar{K}^0\Theta^+ + C$	$0.1 \times \Lambda^*$(90% CL)
HERA B	[25]	41.6 GeV	$pA \to pK_s^0X$	$2.7\% \times \Lambda^*$(95% CL)
HyperCP	[26]	(800 GeV)	$pCu \to pK_s^0X$	0.3% K_s^0p
FOCUS	[27]	($\leq$300 GeV)	$\gamma BeO \to pK_s^0X$	$0.02 \times \Sigma^*$(95% CL)
Belle	[28]	($\sim$0.6 GeV/c)	$K^+n \to pK_s^0$	$\Gamma < 0.64$ MeV (90% CL)
PHENIX	[29]	200 GeV	$Au + Au \to nK^-X$	–
BaBar	[30]	9.4 GeV	$eBe \to pK_s^0X$	–
WA89	[31]	25.2 GeV	$\Sigma^+A \to pK_s^0X$	1.8 μb/A (99% CL)
CLAS	[32]	1.6–3.8 GeV	$\gamma p \to \bar{K}_s^0K^+n$	0.8 nb (95% CL)
CLAS	[33]	0.8–3.6 GeV	$\gamma d \to pK^-nK^+$	0.15–3 nb (95% CL)
CLAS	[34]	0.8–3.6 GeV	$\gamma d \to nK^+\Lambda$	5–25 nb (95% CL)
COSY-TOF	[35]	3.0 GeV/c	$pp \to \Sigma^+pK_s^0$	0.15 μb (95% CL)
NOMAD	[36]	45.3 GeV	$\nu A \to pK_s^0X$	2.13×10^{-3} νCC (90% CL)

ported that the production rate in e^+e^- collisions scales with the mass, and the rate for baryons decreases steeper than that of mesons [37]. At this moment, there is no theoretical prediction for the pentaquark production rate, but it is expected that the production rate of pentaquarks via the fragmentation process would be much smaller than that of the normal hadrons.

Further, no evidence for Θ^+ has been observed in high-statistics searches via hadronic production at a number of high-energy facilities [23–27, 29, 31]. The production mechanism of Θ^+ is unknown; however, it is suggested that its production could be suppressed at high energy, if it exists [38].

One of the comparable results with positive evidences is the exclusive measurement in the $K^+n \rightarrow pK_s^0$ reaction by Belle [28]. They estimated the upper limit of the width $\Gamma \leq 0.64\,\mathrm{MeV/c^2}$ at the 90% CL for the mass of $1.539\,\mathrm{MeV/c^2}$. In the same reaction, the positive results were reported with a significance of 5.3σ in the xenon bubble chamber DIANA [6, 39]. The width of $\Gamma(K^+n \rightarrow \Theta^+ \rightarrow pK_s^0)$ is estimated to be $0.36 \pm 0.11\,\mathrm{MeV/c^2}$, which does not contradict with the Belle results.

The positive and negative results are reported also in electro-production experiments (HEREMES, ZUES vs. BaBar). However, in these results, no upper limit has been reported on the width; therefore, it is not possible to compare them directly with other results.

The most recent photo-production data have been analyzed by CLAS. They repeated the search for Θ^+ with higher statistics in the same reaction $\gamma d \rightarrow pK^-nK^+$ for which they formerly reported the positive evidence [33]. The new data do not show any evidence for the narrow Θ^+ resonance. Since the elementary process of the $\gamma d \rightarrow pK^-nK^+$ reaction is $\gamma n \rightarrow K^-\Theta^+$, the CLAS result could be compared with the first LEPS data. The LEPS collaboration continues to accumulate the data; however, owing to the different detector acceptances of the CLAS and LEPS experiments, there is inescapably an uncertainty in the comparison of their results. They plan to construct a new beamline named LEPS2 in order to increase the beam intensity and extend the detector acceptance to a backward region covered by a CLAS detector [40].

In general, the kinematical regions in null experiments are different from those in the experiments reporting positive evidence; therefore, the negative results cannot rule out the positive evidences at the current stage. A further study with higher statistics is being awaited.

2 Past Experiments with Meson Beams

It should be noted that there are only few experiments to search the Θ^+ pentaquarks via hadronic reactions. In particular, a meson-induced reaction using a proton target is still unique to KEK-PS and J-PARC. The proposed experiment will make a unique contribution to understand the production mechanism of Θ^+.

2.1 KEK-PS E522

The search for Θ^+ in the $\pi^- p \to$ K$^-$X reaction was conducted in the E522 experiment for the first time at the K2 beamline of KEK 12-GeV proton synchrotron [41]. The beam pions were irradiated on a polyethylene target at the beam momenta of 1.87 and 1.92 GeV/c. The typical beam intensity was 3.3×10^5 pions/spill. Figure 1 shows the missing mass spectra obtained in the E522 experiment. While no peak attributable to the Θ^+ was observed for 1.87 GeV/c, a hint of a peak structure, whose width was consistent with the experimental resolution of 13.4 MeV/c^2, was observed at a mass of $1530.6^{+2.2}_{-1.9}$(stat.)$^{+1.9}_{-1.3}$(syst.) MeV/c^2. A differential production cross section of Θ^+ is estimated to be 1.9 μb/sr in the laboratory frame. Assuming isotropic Θ^+ production, this value corresponds to a total cross section of 2.9 mb, if the Θ^+ exists.

However, the statistical significance of the above bump is only 2.5–2.7σ, which is insufficient to claim the existence of Θ^+. They derived the upper limits on the production cross section in the laboratory frame to be 1.6 and 2.9 μb/sr at the 90% confidence level at the beam momenta of 1.87 and 1.92 GeV/c, respectively.

2.2 KEK-PS E559

Another search for Θ^+ has been performed at the K6 beamline of KEK in the $K^+p \to \pi^+X$ reaction [42], which is the inverse reaction of $\pi^-p \to K^-X$ in which Θ^+ was searched by E522 collaboration described in the previous section. The data were obtained in 2005 with a K^+ beam of 1.2 GeV/c on a newly developed liquid hydrogen target. The scattered π^+ was measured using a superconducting kaon spectrometer (SKS) with a mass resolution of 2.4 MeV/c^2 (FWHM) for the Θ^+ mass. In the missing mass spectrum of $K^+p \to \pi^+X$ reaction, no significant structure was observed. The upper limit of the differential cross section in the laboratory frame is estimated to be 3.5 μb at the 90% CL.

2.3 Production Mechanism

What are the implications of the results obtained in meson-induced reactions? Although the significance of the bump structure observed in the E522 experiment is insufficient to consider it as a positive evidence, on what basis may these results be consistently explained, if the signal is real?

Figure 2 shows the Feynman diagrams for the $\pi^-p \to K^-\Theta^+$ reaction. The coupling between $NK\Theta^+$ should be sizeable as far as the Θ^+ decays into

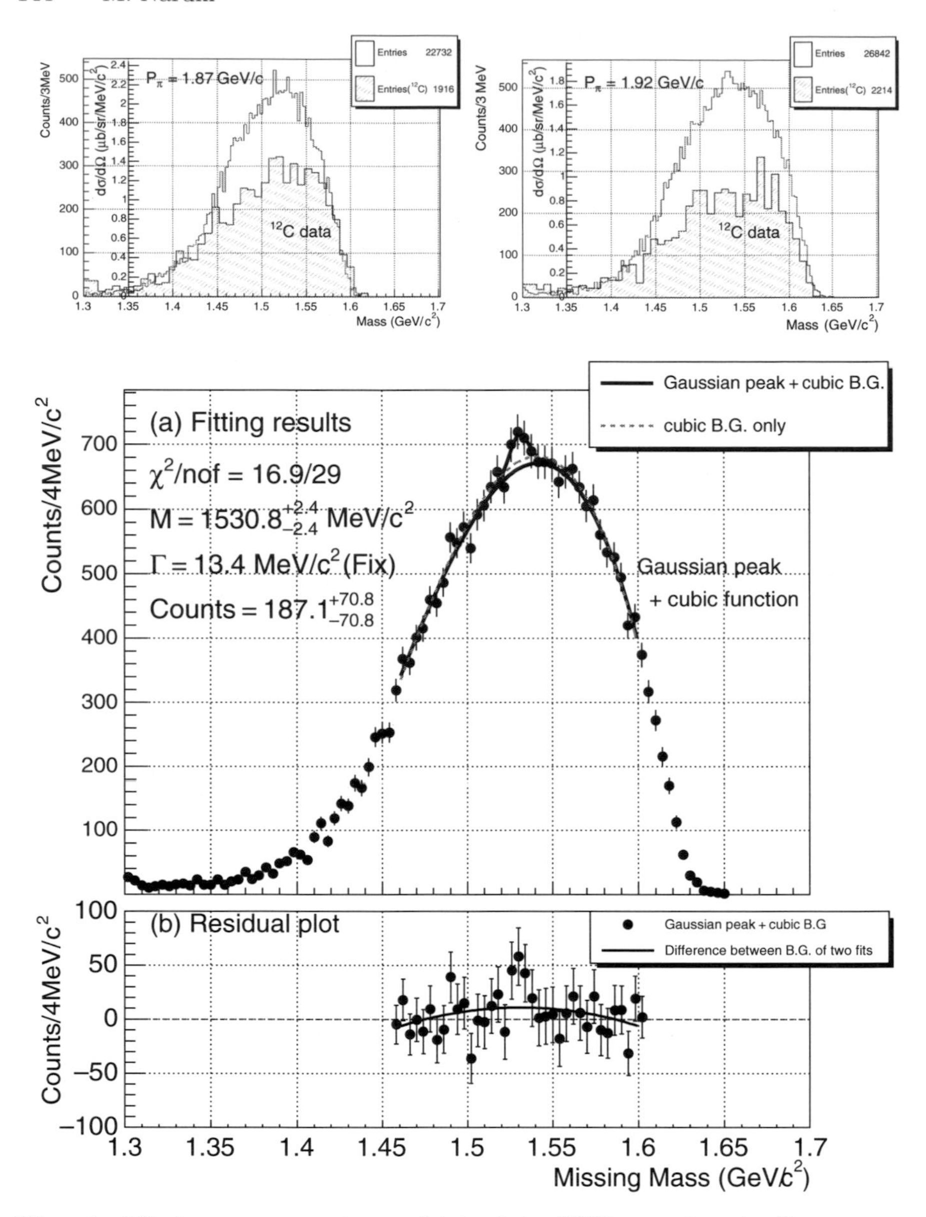

Fig. 1. Missing mass spectrum obtained in E522 experiment. *Top*: $p_{\pi^-} = 1.87\,\mathrm{GeV/c}$. *Bottom*: $p_{\pi^-} = 1.92\,\mathrm{GeV/c}$. A bump structure is seen at $1.53\,\mathrm{GeV/c^2}$ for $p_{\pi^-} = 1.92\,\mathrm{GeV/c}$.

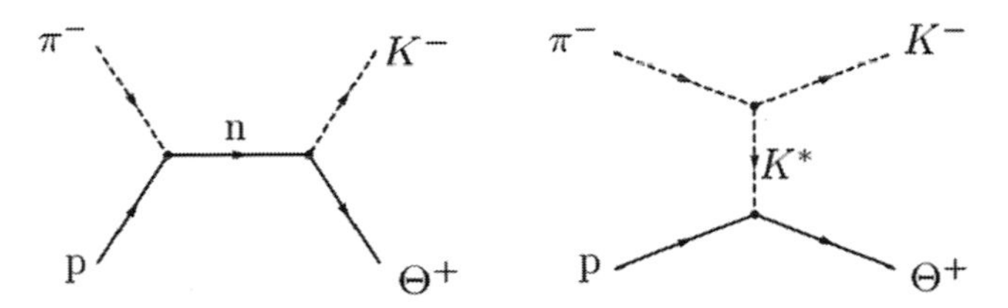

Fig. 2. Feynman diagrams for $\pi^- p \to K^- \Theta^+$ reaction.

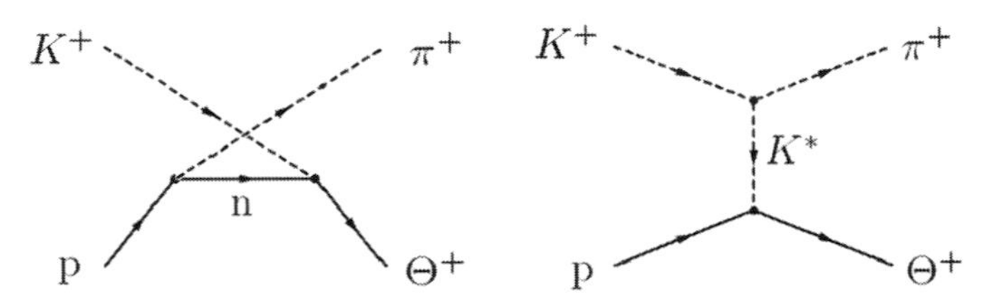

Fig. 3. Feynman diagrams for $K^+ p \to \pi^+ \Theta^+$ reaction.

NK system, whereas the coupling between $K^* N \Theta^+$ is unknown. It should be noted that the recent CLAS measurement in the $\gamma p \to \bar{K}_s^0 K^+ n$ reaction [32] suggests that the contribution to the K^* exchange process is highly suppressed.

In the case of $K^+ p \to \pi^+ \Theta^+$ reaction, there is no s-channel process but a u-channel process, as shown in Fig. 3. If the contribution of a t-channel process is negligible, the differential cross section must be peaked at backward scattering angles. In the E559 experiment, the detector acceptance covers a forward region; therefore, the measurement is not sensitive to observe the Θ^+ produced via u-channel process with the current statistics. This point is minutely inspected in [42].

From the KEK experiments and the CLAS experiment, the K^* exchange process is expected to be negligible compared to the decay probability of Θ^+ into the KN system. Therefore, only the s-channel process will contribute to the $\pi^- p \to K^- \Theta^+$ reaction. Please note that this process is suppressed at higher energies. In the next section, the relation between the pion-induced reaction and photo-production process will be described.

$\gamma p \to \pi^+ K^- n K^+$

The most convincing candidate for Θ^+ is the resonance observed in the $\gamma p \to \pi^+ K^- K^+ n$ reaction reported by the CLAS collaboration [10]. They observed the narrow resonance at 1,555 MeV/c^2 in the nK-invariant mass spectrum with a significance of 7.8σ. They also showed an invariant mass spectrum of the $K^- \Theta^+$ system and found a peak structure with a mass of 2.4 GeV/c^2 as

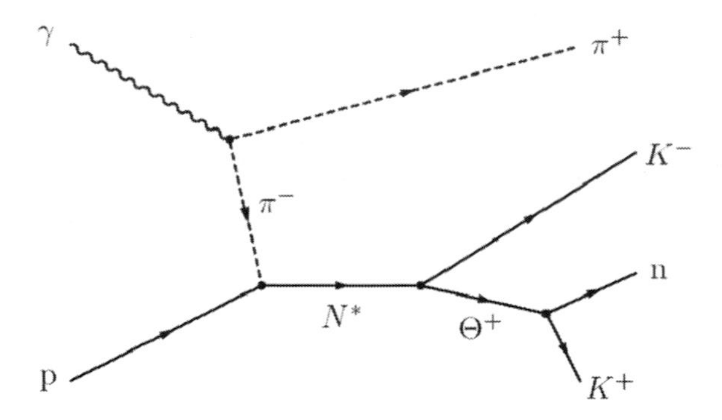

Fig. 4. Feynman diagram for $\gamma p \rightarrow \pi^+ K^- K^+ n$ reaction.

an intermediate state. This suggests that Θ^+ may be produced through the decay of an intermediate N* resonance. Of course any N^*-like baryon state can propagate in the s-channel process provided the energy is sufficient. Please note that the $\gamma p \rightarrow \pi^+ K^- K^+ n$ reaction contains the $\pi^- p \rightarrow K^- \Theta^+$ reaction as an elementary process, as shown in Fig. 4.

Encouraged by the strong positive evidence and the E522 result, the experiment J-PARC E19 has proposed an experimental search for Θ^+ in the $\pi^- p \rightarrow KX$ reaction at J-PARC. The details of the experiment are described in the following sections.

3 J-PARC E19 Experiment

The E19 experiment was accepted as the day-1 experiment at the Nuclear and Particle Physics Facility of J-PARC [43]. The purpose of the experiment is to confirm the existence of Θ^+ produced in the $\pi^- p \rightarrow K^- X$ reaction. This is a natural expansion of the KEK-PS E522 experiment, which was performed at the KEK-PS K2 beamline, as described before. When compared to E522, where the same reaction was employed, we will accumulate ~100 times more statistics with 5 times better mass resolution of 2.5 MeV/c^2 (FWHM) and ~10 times better S/N ratio. With these improvements combined, the expected sensitivity in the laboratory frame reaches 75 nb/sr for a narrow Θ^+ ($\Gamma < 2$ MeV/c^2) and 150 nb/sr for $\Gamma = 10$ MeV/c^2.

The threshold momentum of the reaction is 1.71 GeV/c. Since the peak structure was found only at the beam momentum of 1.92 GeV/c in the E522 experiment, the energy dependence of the cross section is also investigated with the beam momenta of 1.87, 1.92, and 1.97 GeV/c.

With regard to the decay width, if a Θ^+ peak is found, we can determine its width down to 2 MeV/c^2 with the expected missing mass resolution. If it is less than 2 MeV/c^2, this experiment will set a world record for the upper limit of the decay width.

If the bump observed in E522 is not a statistical fluctuation but a real peak, we can detect Θ^+ with a yield of 1.2×10^4 with a beam time of 1 week with an excellent missing mass resolution, as discussed below. This experiment will become the starting point to measure the spin or parity in order to understand its nature.

If Θ^+ is not identified in the proposed experiment, a quite severe upper limit of 75 nb/sr will be obtained. The differential cross section of 75 nb/sr in the laboratory frame corresponds to the total cross section of 100 nb, assuming that Θ^+ is produced isotropically. It is quite unusual that production cross section via hadronic reaction is less than 100 nb; therefore, further theoretical studies should be conducted to understand the production mechanism of Θ^+. The method of estimating the sensitivity of the experiment is described in Sect. 3.6.

Here, we can compare the expected sensitivity with theoretical calculations [44–46]. It is sure that there are unknown parameters such as the coupling constants for $KN\Theta$ and $K^*N\Theta$ interactions together with the various form factors in their calculations; however, the parameters are adjusted to reproduce the known hyperon productions such as $\pi p \to K\Lambda$. The obtained cross sections are ~1 μb or more for the width of ~1 MeV/c^2 and $J^P = 1/2^+$ assignment at the beam momentum of 1.92 GeV/c. In any case, a major contribution is caused by an s-channel process, whose cross section is proportional to the intrinsic width of Θ^+. Therefore, the width can be derived also from the measurement of the cross section for the $\pi^- p \to K^-\Theta^+$ reaction. This is described in Sect. 3.7 in detail.

3.1 Experimental Apparatus

The experiment will be performed at the K1.8 beamline providing a high-intensity secondary beam up to 2 GeV/c. The layout of the beamline is shown in Fig. 5. We will use the 1.87, 1.92, and 1.97 GeV/c π^- beams that are delivered to a 12.5-cm-thick liquid hydrogen target. The target was newly developed for the KEK-PS E559 experiment. The volume and pressure fluctuations of the target are less than 1%. The momentum of scattered K^- is measured in the range of 0.7–0.95 GeV/c by using the SKS spectrometer (Fig. 6). The details of the spectrometer can be found elsewhere [47]. A correlation between the scattering angle and the momentum of K^- is shown in Fig. 7. The momentum of the scattered K^- corresponds to Θ^+ production ranges from 0.7 to 0.9 GeV/c, whereas the background events spread over the entire region. The combination of the K1.8 beamline and the SKS is ideal for the experimental search of Θ^+.

The intensity of the π^- beam is limited by the rate capability of the detector system and is assumed to be 1×10^7 π/spill with a 4-s cycle. The required intensity of the primary proton beam to achieve the 10^7 π/spill is only 4×10^{12} protons/spill and it is about 1% of the designed intensity of

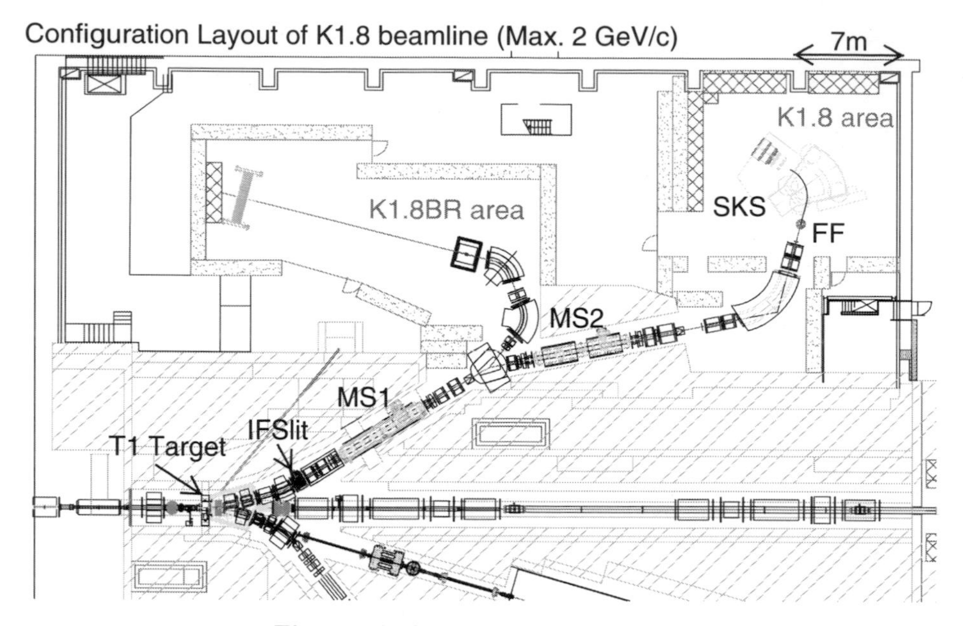

Fig. 5. The layout of K1.8 beamline.

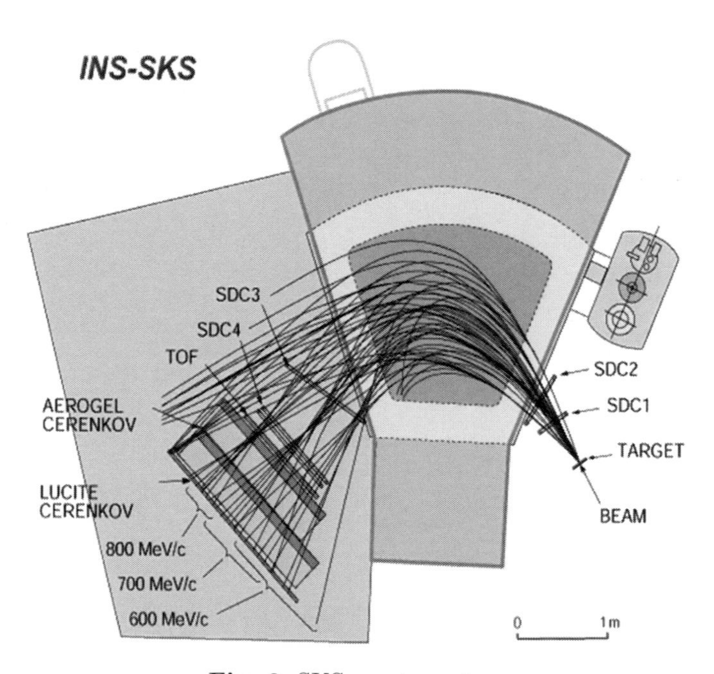

Fig. 6. SKS spectrometer.

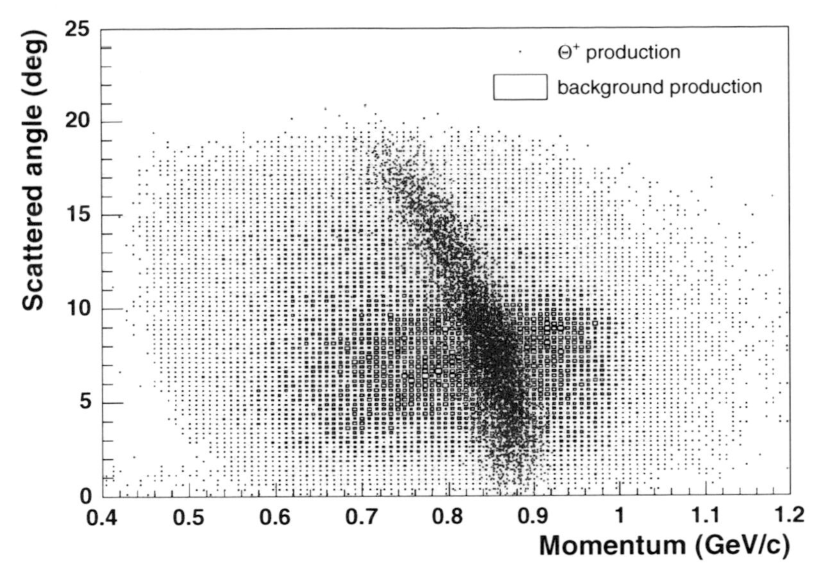

Fig. 7. Correlation between the scattered angle and the momentum of K^-.

J-PARC 50-GeV synchrotron. Thus, the experiment is feasible at a very early stage of day 1.

The SKS spectrometer, together with the beamline spectrometer of K1.8, gives us an excellent mass resolution of 2.5 MeV/c^2 (FWHM), which is five times better than E522. The mass resolution was estimated with the experience on the SKS spectrometer and the design performance of the beamline spectrometer. It is also noted that SKS was long used for hypernuclear experiments using the (π^+, K^+) reactions and can distinguish kaons from pions (and protons). The use of the liquid hydrogen target gives us twice better S/N ratio compared to the polyethylene target.

3.2 K1.8 Beamline

The K1.8 beamline is designed to transport a high-intensity and pure kaon beam with double-stage electrostatic separators. Figure 5 shows the beamline layout together with a K^- spectrometer in the K1.8 experimental area. Table 3 lists the design parameters optimized for a K beam [48]. The first beam has been delivered to the K1.8BR beamline with a 1.2 kW primary proton beam in February, 2009. The construction of the K1.8 beamline will be completed in the summer of 2009, and the first secondary beam will be delivered in the autumn of 2009.

Table 3. Design parameters of K1.8 beamline. The parameters are tuned for K^- beam.

		Phase II (50 GeV, 15 μA)	Phase I (30 GeV, 9 μA)
Max. mom.	[GeV/c]		2.0
Length	[m]		45.853
Acceptance	[msr·%]		1.4
K^- intensity	[/spill]		
	1.8 GeV/c	6.6×10^6	1.4×10^6
	1.1 GeV/c	3.8×10^5	8.0×10^4
K^-/π^- @ FF	@ 1.8 GeV/c	8	6.9
Singles rate @ MS2	@ 1.8 GeV/c	$>3.3 \times 10^7$	$>8 \times 10^6$

3.3 Beam Analyzer

The momentum of a π^- beam is analyzed using the K1.8 beamline spectrometer located after the second mass slit, which is indicated as MS2 in Fig. 5. The beam analyzer consists of *QQDQQ* magnets, tracking detectors, and timing counters. The expected momentum resolution $\Delta p/p$ is 1.4×10^{-4} when a position resolution of 200 μm is achieved with the tracking devices placed before and after the *QQDQQ* magnet system.

The beam rate at the position of tracking chambers determines the acceptable beam intensity. The chambers should be sufficiently thin in order to reduce materials, keeping good resolutions and efficiencies under the high-rate environment. At the beamline chambers, 1 mm pitch multiwire proportional chambers (MWPC) will be newly constructed. The maximum acceptable rate is estimated to be 2.5×10^7/s from an experience on the past experiments performed at the KEK-PS K6 beamline. Since the particle rate at the upstream of the beam spectrometer was estimated to be twice as high as that at the downstream, the rate capability of the detector system after the *QQDQQ* system is expected to be 1.3×10^7/s. The rate per spill is 1×10^7 with a flat top of 0.7 s. It is pointed out that the flat top could be extended to 3 s. In this case, the acceptable rate of the tracking chambers will be 4×10^7/spill.

Segmented plastic scintillator hodoscopes will be located at the upstream and downstream of the beam analyzer being used in a trigger and off-line particle identification with the TOF technique. An aerogel Čerenkov counter with an index of 1.03 is located just upstream of the target in order to separate kaons from pions in the trigger level. Time-of-flight measurement between two beam hodoscopes that are placed 8 m apart can separate kaons from pions at 2 GeV/c in an off-line analysis.

3.4 Spectrometer for Scattered Particles

For the K^- detection, we will use the existing SKS spectrometer. The scattered K^- are identified with the counters located downstream of the spectrometer magnet. The momentum of K^- is analyzed with the SKS magnet and the tracking chambers. A solid angle of 0.1 sr with a wide angular coverage of up to 20° is achieved. The measured momentum resolution is represented in the following equation [49]:

$$\frac{dp}{p} = 0.096 \times p[GeV/c]\% + 0.092\%. \tag{1}$$

We will use four sets of drift chambers (SDC1–4) as tracking detectors of the SKS spectrometer. SDC1 and SDC2 located after the target should be high-rate chambers, since the beam particles will pass through them. These chambers are similar to the beamline chambers, except for the size. Other chambers do not require a high-rate tolerance, since the beam does not pass through these detectors.

SDC3 and SDC4 are located at the exit of the superconducting magnet. For the π/K separation, TOF wall, aerogel Čerenkov counters, and Lucite Čerenkov counter are used. These detectors located downstream are a part of the existing spectrometer.

3.5 Missing Mass Resolution

The SKS spectrometer enables us to measure the width of Θ^+ with the highest resolution in the world. The missing mass resolution is estimated as follows.

The missing mass, M, is expressed as

$$\begin{aligned} M^2 &= (E_{\pi^-} + m_p - E_{K^-})^2 - (\boldsymbol{p}_{\pi^-} - \boldsymbol{p}_{K^-})^2 \qquad (2) \\ &= m_{\pi^-}^2 + m_{K^-}^2 + m_p^2 + 2\,(m_p E_{\pi^-} - m_p E_{K^-} - E_{\pi^-} E_{K^-} + p_{\pi^-} p_{K^-} \cos\theta), \end{aligned}$$

where θ is the scattering angle and m, E, $\boldsymbol{p}$, and p are the mass, energy, momentum, and magnitude of the momentum, respectively. The subscripts π^-, K^-, and p imply the beam, scattered K^-, and the target, respectively. Therefore, the missing mass resolution ΔM is expressed as follows:

$$\Delta M^2 = \left(\frac{\partial M}{\partial p_{\pi^-}}\right)^2 \Delta p_{\pi^-}^2 + \left(\frac{\partial M}{\partial p_{K^-}}\right)^2 \Delta p_{K^-}^2 + \left(\frac{\partial M}{\partial \theta}\right)^2 \Delta\theta^2, \tag{3}$$

$$\frac{\partial M}{\partial p_{\pi^-}} = \frac{1}{M}[\beta_{\pi^-}(m_p - E_{K^-}) + p_{K^-} \cos\theta], \tag{4}$$

$$\frac{\partial M}{\partial p_{K^-}} = -\frac{1}{M}[\beta_{K^-}(m_p + E_{\pi^-}) - p_{\pi^-} \cos\theta], \tag{5}$$

$$\frac{\partial M}{\partial \theta} = -\frac{1}{M} p_{\pi^-} p_{K^-} \sin\theta, \tag{6}$$

where β_{π^-} and β_{K^-} are the velocities of the beam and scattered particles, respectively. Δp_{π^-}, Δp_{K^-}, and $\Delta\theta$ are resolutions of the measurement for momenta of the beam, scattered particles, and the scattering angle, respectively.

The momentum resolution of the beam analyzer at 1.97 GeV/c is estimated to be 0.6 MeV/c (FWHM) by using the beam momentum resolution of 1.4×10^{-4} at 1 GeV/c [48]. The momentum resolution of the scattered K^- is estimated from the measured momentum resolution of the SKS spectrometer described in the previous section. The resolution of the scattered angle is estimated to be 0.26° through a Monte Carlo simulation using Geant4 [50], assuming that the Θ^+ is produced isotropically. The result is shown in Fig. 8 together with the residual distribution of the momentum of the scattered particles. Monte Carlo simulation slightly underestimates the momentum resolution compared with the measured momentum resolution; hence, Eq. (1) was used to calculate the mass resolution for a conservative estimation. Using these values, the missing mass resolution can be estimated to be 2.5 MeV/c^2 (FWHM). With this expected missing mass resolution, we can determine the width of Θ^+. If it is less than 2 MeV/c^2, this experiment can set us a world record for the upper limit of the width.

Figure 9 shows the expected missing mass distribution. The signal yield is estimated using the cross section obtained in the previous experiment E522. The main background originates from ϕ-meson decay, Λ decay, and three-body phase space. The cross sections of these processes were measured and found to be several tenths of microbarns. In the spectrum, the background shapes from the ϕ production, Λ production, and the phase space are shown by blue, red, and green histograms, respectively. Note that the background shapes are

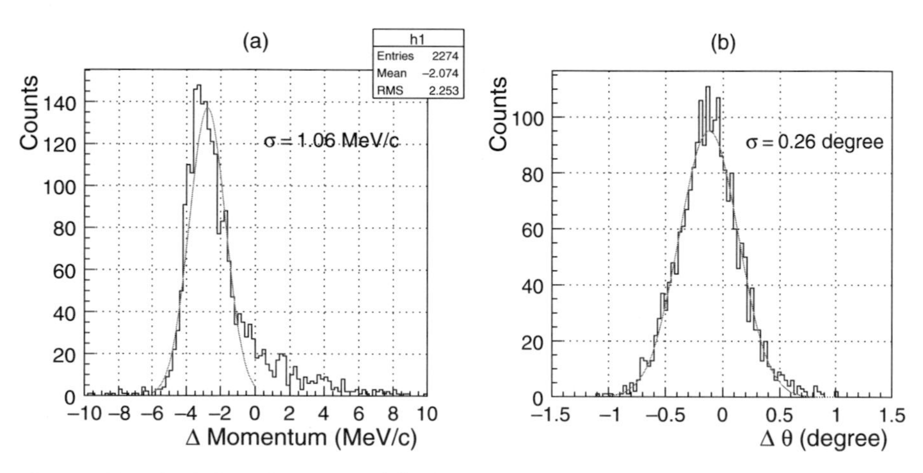

Fig. 8. Residual distribution of (**a**) the momentum of the scattered K^- and (**b**) the scattering angle estimated through detector simulation using Geant4 [50]. The *lines* show the fit results with a Gaussian function.

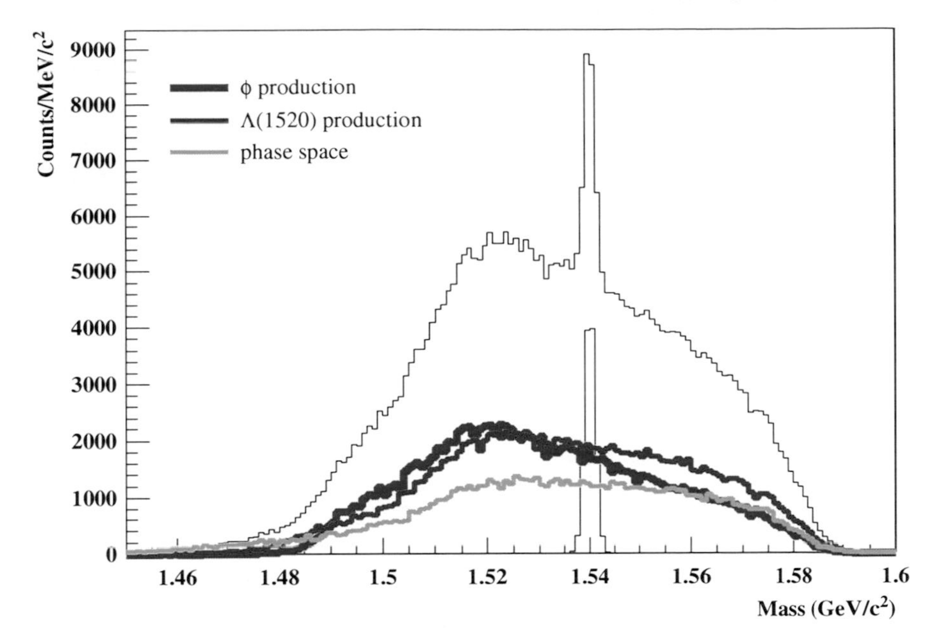

Fig. 9. Expected missing mass spectrum.

scaled by a factor of 2.5, since the statistics of the Monte Carlo simulation are insufficient. If Θ^+ is real in the E522 experiment, we can confirm the existence in the proposed experiment with 62σ significance.

3.6 Expected Sensitivity

With 1 week of beam time, the target will be irradiated with 1.44×10^{12} pions in total, which corresponds to 4.8×10^{11} for each π^- momentum. The number of events is 100 times larger than that of E522 experiment. The expected yield is estimated with the known parameters; the detector acceptance of the SKS is 0.1 sr, the expected analysis efficiency is 0.5, and K^- decay rate in the SKS amounts to 50%. Assuming 1.9 µb/sr for the production cross section of Θ^+, which is based on the E522 result, we will obtain

$$4.8 \times 10^{11} \times 5.3 \times 10^{23} \times 1.9 \times 10^{-30} \times 0.1 \times 0.5 \times 0.5 = 1.2 \times 10^4$$

events for each momentum setting.

The background can also be estimated from the E522 result. In E522, the background cross section was $0.8\,\mu\text{b/sr/MeV/c}^2$ for a proton target[1] at $1530\,\text{MeV/c}^2$. The beam momentum dependence of the background cross section was found to be negligibly small. In this experiment, this corresponds to 5.0×10^3 counts/MeV/c^2 for each momentum setting.

[1] This is obtained by a subtraction method using CH_2 and C targets.

With these estimations combined, we expect to observe a 62σ peak if Θ^+ is a narrow resonance ($\Gamma < 2$ MeV/c^2). Even if it is as wide as $\Gamma = 10$ MeV/c^2, still the significance of the peak is 48σ, which is sufficiently high to claim the existence of the Θ^+. These high statistics enable us to determine the angular distribution of the production cross section in a future experiment.

The expected sensitivity of the experiment is 75 and 150 nb/sr for $\Gamma < 2$ and $\Gamma = 10$ MeV/c^2, respectively. With regard to the width, the sensitivity is down to $\Gamma = 2$ MeV/c^2 or even better because of the high statistics and excellent resolution of the spectrometer system.

3.7 Comparison with Theoretical Predictions

There are some theoretical predictions for the production cross section of Θ^+ in the $\pi^- p \rightarrow K^- \Theta^+$ reaction. Oh et al. calculated the cross section assuming that the width of Θ^+ is 5 MeV/c^2. The result is shown in Fig. 10 [44]. They considered two types of the form factors to take account of the

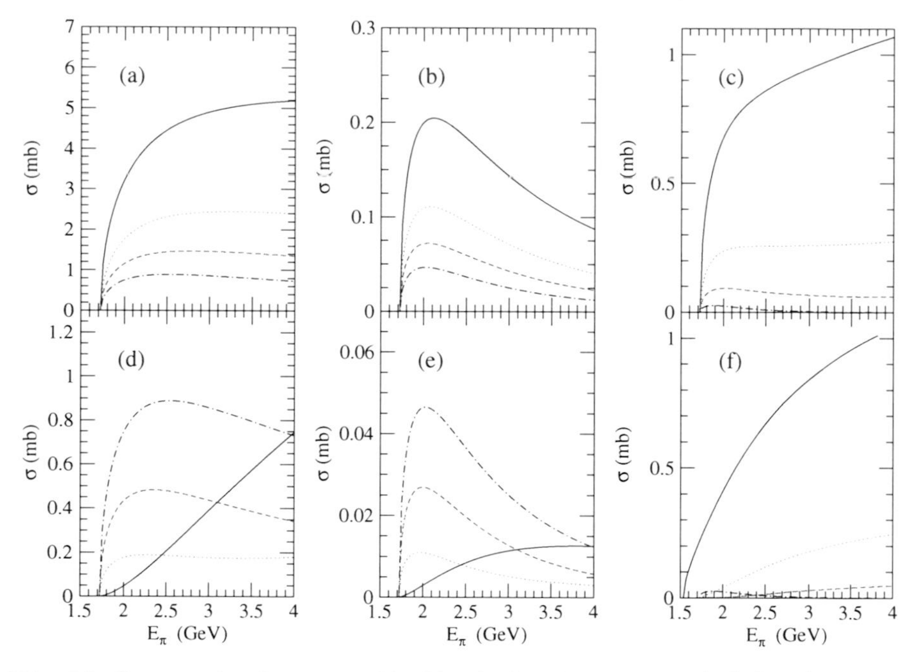

Fig. 10. Cross section for $\pi^- p \rightarrow K^- \Theta^+$ calculated by Oh et al. [44]: (**a, d**) without form factors, (**b, e**) with the form factors of $\Lambda^2/\{\Lambda^2 + q_i^2\}$ where $\Lambda = 0.5$ GeV, and (**c, f**) with the form factors of $\Lambda^4/\{\Lambda^4 + (r - M_{ex}^2)^2\}$ where $\Lambda = 1.8$ GeV. The *solid, dotted, dashed*, and *dot-dashed lines* correspond to the coupling constants for $K^* N \Theta^+$ $g_{K^* N \Theta^+} = -2.2, -1.1, -0.5$, and 0.0 in the *upper figures* and 2.2, 1.1, 0.5, and 0.0 in the *bottom figures*, respectively.

finite size of hadrons: (1) monopole-type form factor $\Lambda^2/\{\Lambda^2 + q_i^2\}$, where q^2 denotes the three moment a of external particles and (2) a form factor of type $\Lambda^4/\{\Lambda^4 + (r - M_{ex}^2)^2\}$, where M_{ex} and r denote the mass of the exchanged particle and the square of the transferred momentum, respectively. In the first form factor, the cut-off parameter Λ was set to 0.5 GeV, the kaon mass. In the second form factor, Λ was set to 1.8 GeV, which was optimized to reproduce the Λ photo-production data. Since the coupling constant for the vertex $K^*N\Theta^+$ is unknown, they considered it as a parameter ranging from −2.2 to 2.2. As a result, the cross section ranges from 2 to 190 μb at the beam momentum of 2 GeV/c. It is sure that there is an uncertainty in the form factor. Nevertheless, their calculation shows that the measured cross section can constrain the coupling constant for the $K^*N\Theta^+$ interaction, and from the energy dependence of the total cross section, we can determine which form factor is appropriate to reproduce the Θ^+ production.

The numerical comparison with the former experimental result obtained by E522 can be more clearly done with the differential cross section, shown in Fig. 11. Here, only the case shown in Fig. 10(c) or (f) is considered. They considered the width of Θ^+ as 5 MeV/c^2. If the K^* exchange process does not exist as suggested by the experiments, the cross section of the s-channel process is proportional to the decay width of Θ^+. Hence, the cross section could be five times smaller if the width of Θ^+ is ∼1 MeV/c^2. However, the cross section increases by about four times at 2 GeV, as shown in Fig. 10(c) and (f).

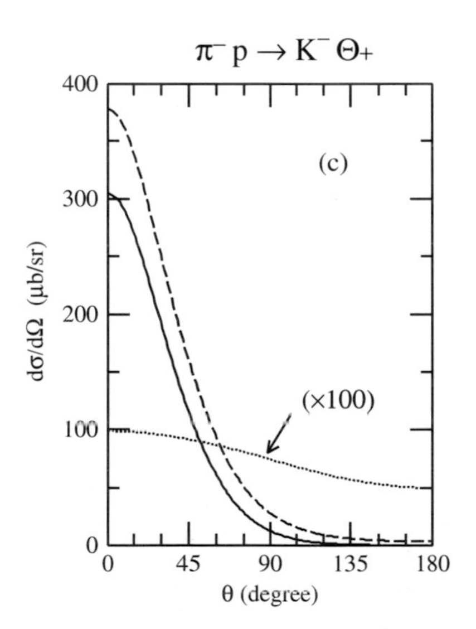

Fig. 11. Differential cross section for $\pi^- p \to K^- \Theta^+$ at $E_\pi = 2.5$ GeV [44]. The *solid, dotted,* and *dashed lines* correspond to $g_{K^*N\Theta^+} = g_{KN\Theta^+}$, $g_{K^*N\Theta^+} = 0$, and $g_{K^*N\Theta^+} = -g_{KN\Theta^+}$, respectively. $g_{K^*N\Theta^+}$ and $g_{KN\Theta^+}$ denote the coupling constants for $K^*N\Theta^+$ and $KN\Theta^+$, respectively.

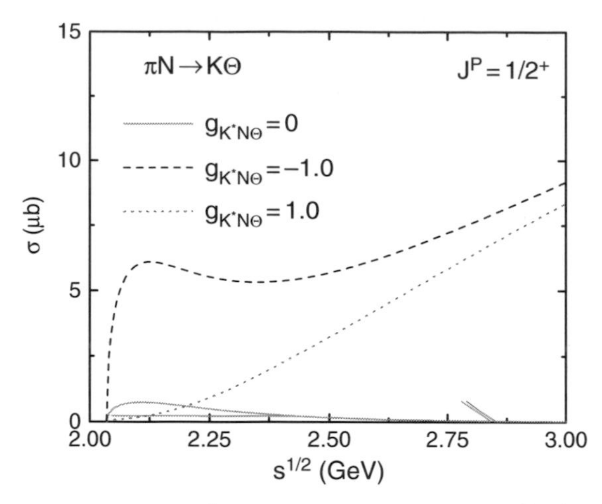

Fig. 12. Cross section for Θ^+ production in the $\pi N \to \bar{K}\Theta^+$ reaction [51].

The rescaled differential cross section for $\Gamma_{\Theta^+} = 1$ amounts to $1\,\mu$b/sr, which is three times smaller than that in the previous experiment. In the future experiment J-PARC E19, the expected sensitivity reaches 75 nb/sr for $\Gamma = 1\,\mathrm{MeV/c^2}$, which is much smaller than the theoretical prediction.

Formerly, Liu and Ko calculated the cross section for the $\pi N \to \bar{K}\Theta^+$ reaction together with the $KN \to \pi\Theta^+$ reaction using an effective Lagrangian [46, 51]. Figure 12 shows their result assuming that the Θ^+ has a width of $1\,\mathrm{MeV/c^2}$ and positive parity. The cross section of the s-channel process is $0.5\,\mu$b for $\pi N \to \bar{K}\Theta^+$ at $\sqrt{s} = 2.25\,\mathrm{GeV}$. They used the monopole-type form factor and pseudo-scalar-type couplings between $KN\Theta^+$ and πNN systems as done by Oh et al., but obtained 20 times smaller cross section. The cause of this inconsistency is unknown; however, the sensitivity of our experiment is still $\sim$4 times better than their prediction.

Hyodo and Hosaka proposed a model introducing two-meson couplings to calculate the cross section for $\pi^- p \to K^-\Theta^+$ and the relevant reaction $K^+p \to \pi^+\Theta^+$ [45]. Figure 13 shows one of their results when the cross section has the smallest value. They used pseudo-vector-type couplings, and the difference between their results and results of Oh et al. is examined in detail in [45] (T. Hyodo, private communication). The result for $\pi^- p \to K^-\Theta^+$ is comparable to the E522 data, whereas that of $K^+p \to \pi^+\Theta^+$ is much larger than the E559 data. Therefore, the effect of the two-meson couplings is expected to be small. In that case the Born terms make a major contribution to the cross section, which amounts to $\sim 0.5\,\mu$b at $\sqrt{s} = 2.2\,\mathrm{GeV}$ (T. Hyodo, private communication).

The sensitivity of our experiment is better by a factor of 4 or more than the theoretical predictions obtained so far, assuming that the width of the Θ^+ is $1\,\mathrm{MeV/c^2}$. Therefore, further theoretical studies are required to understand

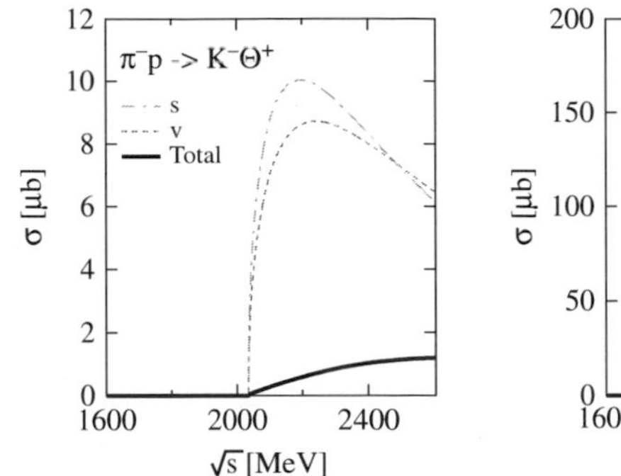

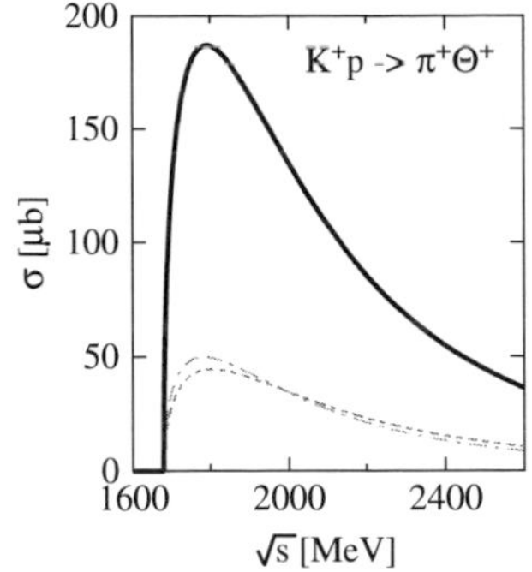

Fig. 13. Total cross section for the $J^P = 1/2^+$ case with a hadronic form factor [45]. The *solid line* shows the total amplitude; *dash-dotted* and *dashed lines* denote the contribution from scalar- and vector-type correlations between two mesons, respectively.

the extraordinary narrow width or explain the suppression of the Θ^+ production, if Θ^+ is not observed in the J-PARC E19 experiment.

4 Future Prospect

Here, I would like to mention the possibilities to advance the research on pentaquarks and other exotic hadrons, which can be performed at J-PARC in future.

If Θ^+ is observed, it is certainly important to determine its spin and parity as a next step. The spin can be determined by measuring the decay angular distribution of Θ^+. New tracking devices and range counters will be installed in a backward region in order to detect decay particles of Θ^+.

If Θ^+ is not observed in the E19 experiment, there is still a possibility that an intermediate N^* resonance plays a role in the Θ^+ production. In order to produce the N^* observed in the CLAS experiment [10], whose mass is 2.4 GeV/c^2, the threshold momentum of π is 2.5 GeV/c. Since the maximum momentum is 2.0 GeV/c for the secondary beamline, which is planned to be constructed in phase 1, a new beamline of a higher momentum is required to verify the CLAS result.

The possibility of producing a Θ^+ hypernuclei has been studied theoretically and experimentally (see [52] or summary in [53]). A letter of intent (LOI) for a search for Θ^+ hypernuclei using (K^+, p) reaction has been submitted to J-PARC PAC [54]. Since the magic momentum of this reaction is 600 MeV/c, K1.1 beamline, which provides the kaon beam whose momentum of 1.1 GeV/c, is favorable to maximize the production probability of Θ^+ hypernuclei. If Θ^+ can be produced in a medium, it will provide new information on the inner structure of the Θ^+.

It is expected that there are other members of the multiplet to which Θ^+ belongs. There have been searches for pentaquarks that have a different quark configuration from that of Θ^+. The NA49 collaboration found a positive evidence for a narrow state at 1.86 GeV/c^2 on the $\Xi^-\pi^-$-invariant mass spectrum [55]. The resonance is called as $\Xi^{--}(1862)$ and has a quark structure of $ddss\bar{u}$. On the other hand, there are null results against this evidence; therefore, the situation is still questionable for the existence of Ξ^{--}. At the J-PARC Hadron Facility, Ξ^{--} can be searched for in the (K^-, K^+) reaction. The threshold momentum is 2.3 GeV/c; therefore, it is necessary to construct a new secondary beamline providing the higher momentum exceeding 2.3 GeV/c.

The H1 collaboration reported a narrow resonance of $\Theta_c^0(3099)$, which has a quark configuration of $uudd\bar{c}$. There is a theoretical prediction that the pentaquark state is more stable when the antiquark is heavy [56]. There is a possibility to produce charmed pentaquarks in pA reactions using a high-intensity primary proton beam. For example, 12 GeV/c proton beam enables us to produce a particle whose mass is less than 3.1 GeV/c^2 in $pp \to ppX$ reaction. It is planned to construct a new beamline that provides primary proton at the hadron facility. The energy of the primary beam is sufficient for the programs that search for heavy pentaquarks. Therefore, it is also expected to realize the construction of the primary beamline.

The most critical measurement for confirming the existence of Θ^+ is to search for Θ^+ in the formation reaction of $K^+N \to \Theta^+$. Since the Θ^+ width is expected to be less than 1 MeV/c^2, the width of the resonance will be determined in the reaction with a high-intensity and low-momentum K^+ beam, if Θ^+ exists. The LOI was already proposed by T. Nakano [43].

There are many theoretical studies concerning the exotic hadrons, particularly for tetraquarks and heptaquarks. For example, the properties of $ud\bar{s}\bar{s}$ tetraquark ϑ^+ have been intensively studied since the configuration is closely related to the structure of Θ^+. It is pointed out that the mass of ϑ^+ is expected to be 1.4–1.5 GeV/c^2 near the K^*K threshold and the width to be O(10–100) MeV/c^2 [57–60]. With the high-intensity kaon beam, it is possible to search for ϑ^+ in the K^+N interaction at J-PARC. For example, there is an exclusive reaction $K^+p \to \Sigma^+\vartheta^+$ whose threshold momentum is 3 GeV/c. Since there should be at least one system with strangeness S = −1 like KN or Λ/Σ or heavier strange baryons in a final state, the beam momentum is required to be greater than 3 GeV/c for the ϑ^+ search. The upgrade plans of the J-PARC Hadron Facility were intensively discussed recently [53, 61]. It is desired to construct the primary beamline and new secondary beamline that has a higher momentum for the search of exotic baryons at J-PARC. The study of multiquark systems will explore the world of nonperturbative QCD and search for exotic hadrons will mark the dawn of a new era in hadron physics.

References

1. Swanson, E.S.: Phys. Rept. **429**, 243–305 (2006) hep-ph/0601110
2. Olsen, S.L., Choi, S.-K., et al.: (2007) hep-ex/0708.1790
3. Nakano, T., et al.: Phys. Rev. Lett. **91**, 012002 (2003)
4. Polyakov, M., Diakonov, D., Petrov, V.: Z. Phys. A **359**, 305–314 (1997)
5. Jaffe, R., Wilczek, F.: Phys. Rev. Lett. **91**(23), 232003 (2003)
6. Barmin, V.V., et al.: Phys. Atom. Nucl. **66**, 1715–1718 (2003)
7. Stepanyan, S., et al.: Phys. Rev. Lett. **91**, 252001 (2003)
8. Barth, J., et al.: Phys. Lett. B **572**, 127–132 (2003)
9. Dolgolenko, A.G., Asratayn, A.E., Kubantsev, M.A.: Phys. Atom. Nucl. **67**(4), 682–687 (2004)
10. Kubarovsky, V., et al.: Phys. Rev. Lett. **92**(3), 032001 (2004)
11. Airapetian, A., et al.: Phys. Lett. B **585**, 213–222 (2004)
12. Abdel-Bary, M., et al.: Phys. Lett. B **595**, 127 (2004)
13. Chekanov, S., et al.: Phys. Lett. B **591**, 7–22 (2004)
14. Aleev, A., et al.: Phys. Atom. Nucl. **68**, 974–981 (2005) hep-ex/0401024
15. Cammileri, L.: Nucl. Phys. B (Proc. Suppl.) **143**, 129–136 (2005)
16. Aleev, A., et al.: hep-ex/0509033
17. Muramatsu, N., et al.: AIP Conf. Proc. **870**, 455–459 (2006)
18. Gibbs, W.R.: Pentaquark in $k^+ - d$ total cross section data. Phys. Rev. C **70**, 045208 (2004)
19. Bai, J.Z., et al.: Phys. Rev. D **70**, 012004 (2004)
20. Schael, S., et al.: Phys. Lett. B **599**, 1–16 (2004)
21. Aubert, B., et al.: hep-ex/0408064
22. Abe, K., et al.: hep-ex/0409010
23. Litvintsev, D.O., et al.: Nucl. Phys. B (Proc. Suppl.) **142**, 374–377 (2005)
24. Yu.M. Antipov, et al.: Eur. Phys. J. **A21**, 455 (2004)
25. Abt, I., et al.: Phys. Rev. Lett. **93**, 212003 (2004)
26. Longo, M.J., et al.: Phys. Rev. D **70**, 111101(R) (2004)
27. Link, J.M., et al.: Phys. Lett. B **639**, 604–611 (2006)
28. Mizuk, R., et al.: Phys. Lett. B **632**, 173 (2006)
29. Pinkerton, C., et al.: J. Phys. **G30**, S1201 (2004)
30. Goetzen, K., et al.: Nucl. Phys. (Proc. Suppl.) **164**, 117–120 (2007)
31. Adamovich, M.I., et al.: Phys. Rev. C **72**, 055201 (2005)
32. Battaglieri, M., et al.: Phys. Rev. Lett. **96**, 042001 (2006)
33. McKinnon, B., et al.: Phys. Rev. Lett. **96**, 212001 (2006) hep-ex/0603028
34. Niccolai, S., et al.: Phys. Rev. Lett. **97**, 032001 (2006)
35. Abdel-Bary, M., et al.: Phys. Lett. B **649**, 252–257 (2007)
36. Samoylov, O., et al.: Eur. Phys. J. C **49**, 499–510 (2007)
37. Halyo, V., et al.: SLAC-PUB-11523
38. daté, S., Titov, A.I., Hosaka, A., Ohashi, Y.: Phys. Rev. C **70**, 042202 (2004)
39. Barmin, V.V., et al.: Phys. Atom. Nucl. **70**, 35–43 (2007)
40. Nakano, T., et al.: AIP Conf. Proc. **915**, 162–167 (2007)
41. Miwa, K., et al.: Phys. Lett. B **635**, 72 (2006)
42. Miwa, K., et al.: Phys. Rev. C **77**, 045203 (2008)
43. http://j-parc.jp/NuclPart/index e.html
44. Oh, Y., et al.: Phys. Rev. D **69**, 014009 (2004)
45. Hyodo, T., Hosaka, A.: Phys. Rev. C **72**, 055202 (2005)

46. Liu, W., Ko C.M.: Phys. Rev. C **68**, 045203 (2003)
47. Fukuda, T., et al.: Nucl. Inst. Meth. A **361**, 485–496 (1995)
48. Noumi, H.: Status of the secondary beam lines in the hadron hall at j-parc (2005)
49. Hotchi, H.: PhD thesis (2000)
50. Agostinelli, S., et al.: Nucl. Inst. Meth. A **506**, 250–303 (2003)
51. Ko, C.M., Liu, W.: nucl-th/0410068
52. http://www.rcnp.osaka-u.ac.jp/penta04/
53. http://www.rcnp.osaka-u.ac.jp/Divisions/plan/kokusai/ws071111.html
54. http://j-parc.jp/NuclPart/PAC for NuclPart e.html
55. Alt, C., et al.: Phys. Rev. Lett. **92**, 042003 (2004)
56. Kim, H., Sarac, Y., Lee, S.H.: Phys. Rev. D **73**, 014009 (2006)
57. Burns, T.J., Close, F.E.: Phys. Rev. D **71**, 014017 (2005)
58. Morimatsu, O., Kanada-En'yo, Y., Nishikawa, T.: Phys. Rev. D **71**, 094005 (2005)
59. Karliner, M., Lipkin, H.J.: Phys. Lett. B **612**, 197–200 (2005)
60. Cui, Y., et al.: Phys. Rev. D **73**, 014018 (2006)
61. http://ag.riken.jp/J-PARC/J-PARC-WS-at-RIKEN-2008/index.html

In-Medium Mass Modification of Vector Mesons

Satoshi Yokkaichi

RIKEN Nishina Center, RIKEN, 2-1, Hirosawa, Wako, Saitama 351-0198, Japan
yokkaich@riken.jp

The mass modification of mesons in medium has been investigated theoretically and experimentally, partly motivated by the interest in the chiral symmetry restoration in medium.

The E325 experiment performed at KEK-PS was one of such experiments. They observed the vector meson mass modification in the e^+e^- decay channel. In particular, the ϕ meson modification was observed for the first time. At J-PARC, a next-generation experiment, E16, has been proposed in order to perform a systematic study of the spectral modification of vector mesons, ρ, ω, and ϕ.

1 Introduction

The spontaneous breaking of the chiral symmetry in quantum chromodynamics (QCD) is considered to play the main role in the mass-generation mechanism of light quarks and hadrons.

In the early universe, about 10^{-11} s after the Big Bang, the Higgs mechanism caused the spontaneous breaking of the electroweak gauge symmetry and the Higgs condensation. The condensate, namely the finite vacuum expectation value of the Higgs field, generated the mass of weak bosons, leptons, and quarks. In the case of light quarks, namely the u, d, and s quarks, the masses given by the Higgs mechanism are much smaller than those of hadrons, which consist of the light quarks. Next, till about 10^{-4} s after the Big Bang, the chiral symmetry was broken spontaneously and the quark–antiquark pairs condensed in vacuum. It is considered that the u/d quarks and s quark acquired the effective masses of about 300 and 500 MeV/c^2, respectively, due to the quark–antiquark pair condensate. Such a situation is schematically shown in the left panel of Fig. 1. At the same time, the QCD phase transition occurred and quarks (and gluons) were confined in the hadrons. The relation of the chiral symmetry breaking and the quark confinement is itself

Yokkaichi, S.: *In-Medium Mass Modification of Vector Mesons.*
Lect. Notes Phys. **781**, 161–193 (2009)
DOI 10.1007/978-3-642-00961-7_7

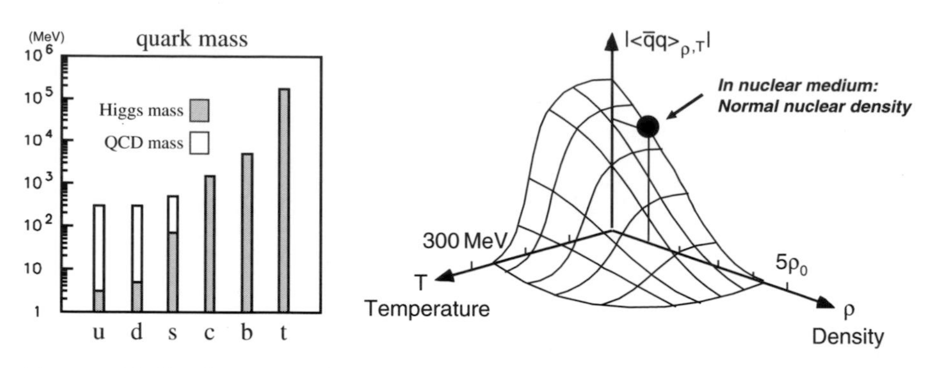

Fig. 1. (*Left*) The origin of quark mass. Light quark masses mainly consist of the QCD mass, namely given by the spontaneous chiral symmetry breaking. (*Right*) The chiral condensate $|\langle\bar{q}q\rangle|$ as a function of the temperature (T) and density (ρ), taken from [9].

an interesting issue itself and is discussed elsewhere [1–3], though we discuss here the breaking and restoration of the chiral symmetry.

In QCD, the chiral symmetry is actually broken explicitly and slightly. It is confirmed by the small quark mass term (current quark mass), generated by the Higgs mechanism, in the QCD Lagrangian. At lower energies, the chiral symmetry is broken spontaneously and largely, and the constituent quark mass of the light quarks and also the hadron mass are generated. The chiral condensate, $\langle\bar{q}q\rangle$, namely the expectation value of the antiquark–quark pairs, is an order parameter of the chiral phase transition. The finite value of the condensate implies that the chiral symmetry is broken. The QCD vacuum is in the broken phase, namely the vacuum expectation value of the pairs, $\langle 0|\bar{q}q|0\rangle = \langle\bar{q}q\rangle_0$, is not zero. The pion is the Nambu–Goldstone (NG) boson associated with the spontaneous chiral symmetry breaking. Because the original chiral symmetry is approximate due to the current quark mass, as mentioned above, the pion is massive while the NG boson should be massless. Such a mass-generation mechanism in QCD was originally proposed by Nambu and Jona-Lasinio [4] for the nucleon mass and is widely accepted now.

As a consequence of the mass-generation mechanism, in the hot and/or dense matter, the chiral symmetry could be restored and the masses of quarks and hadrons are considered to be changed. The chiral condensate in medium as a function of the baryon density ρ, $\langle\bar{q}q\rangle_\rho$, is calculated [5, 6] in the lowest order of the density ρ as follows:

$$\frac{\langle\bar{q}q\rangle_\rho}{\langle\bar{q}q\rangle_0} = 1 - \frac{\sigma_{\pi N}}{m_\pi^2 f_\pi^2}\rho, \tag{1}$$

where $f_\pi \sim 93\,\mathrm{MeV}$ is the pion decay constant and $\sigma_{\pi N} \sim 45\,\mathrm{MeV}$ is the pion–nucleon sigma term (see Ref. [7] for recent further developments). Furthermore, $\langle\bar{q}q\rangle_{\rho,T}$ is calculated by Lutz et al. [8, 9] as shown in the right panel of Fig. 1. In any case, at the normal nuclear density, $\rho_0 \sim 0.17\,\mathrm{fm}^{-3}$, the condensate is decreased to about 2/3 of its vacuum value (ρ=0, T=0), namely an ordinary cold nucleus is expected to behave as a medium in which the hadron properties are modified due to the partially restored chiral symmetry.

In order to investigate such properties of QCD experimentally, several experiments were performed to measure the mesons in medium, at high temperature and/or density. In this chapter, we aim at introducing such theoretical and experimental attempts, particularly at finite density. In Sect. 2, we review the theoretical predictions about the in-medium mass modification of mesons. In Sects. 3 and 4, the invariant mass spectroscopy experiments performed using leptonic and photonic decay modes are reviewed. Taking account of the less possibility of the final state interaction, the leptonic decay modes are considered to be suitable to measure the mass shape modification directly. Unfortunately, it is out of the scope of this chapter to review the important experiments (and related theoretical works) such as the $\rho \to \pi^+\pi^-$[10, 11], $\sigma \to \pi^+\pi^-$[12], $\phi \to K^+K^-$[13] reactions, and the measurements of the deeply bound pionic atoms [14, 15]. In Sects. 5 and 6, we introduce the projects at J-PARC to investigate the in-medium meson properties.

2 Theoretical Predictions

There are many theoretical predictions about the meson mass modification at finite density.

Brown and Rho [16] proposed a scaling law based on an effective Lagrangian. The effective mass of a hadron in medium scales as

$$\frac{m_\sigma^*}{m_\sigma} \approx \frac{m_N^*}{m_N} \approx \frac{m_\rho^*}{m_\rho} \approx \frac{m_\omega^*}{m_\omega} \approx \frac{f_\pi^*}{f_\pi}, \tag{2}$$

where an asterisk denotes the in-medium value. At the normal nuclear density, the ρ and ω masses are reduced by 20%.

On the basis of the QCD sum rule, Hatsuda and Lee [17, 18] showed the decrease in the masses of the ρ, ω, and ϕ mesons at finite density. The in-medium sum rule developed by them deduces the relation between the in-medium chiral condensate $\frac{\langle \bar{q}q\rangle_\rho}{\langle \bar{q}q\rangle_0}$ and the in-medium vector meson mass. By using Eq. (1), the in-medium vector meson masses are numerically obtained as a function of the density ρ and are approximated as

$$\frac{m^*(\rho)}{m(\rho=0)} = 1 - 0.16(\pm 0.06)\frac{\rho}{\rho_0} \tag{3}$$

for ρ and ω mesons, and

$$\frac{m^*(\rho)}{m(\rho=0)} = 1 - 0.15(\pm 0.05)y\frac{\rho}{\rho_0} \tag{4}$$

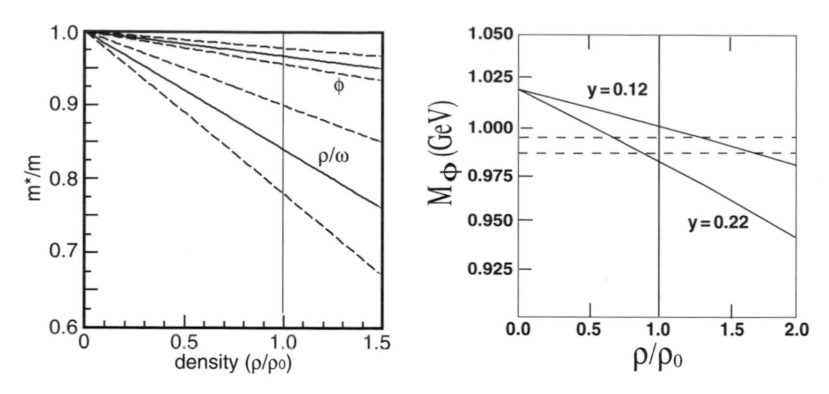

Fig. 2. The mass modification predicted by Hatsuda and Lee [17, 18]. *Left panel* shows the decrease in the masses of ρ, ω, and ϕ. The *solid lines* show the central values for ρ/ω and ϕ ($y = 0.22$) and the *dashed lines* indicate the error bands. In the *right panel*, the decrease in the mass of ϕ for different y values is plotted with the $K\bar{K}$ decay thresholds (*dashed lines*).

for a ϕ meson, where $y = \frac{2\langle \bar{s}s\rangle_N}{\langle \bar{u}u\rangle_N + \langle \bar{d}d\rangle_N}$, the strangeness content of the nucleon, and ρ_0 is the normal nuclear matter density. The range of y is approximately 0.1–0.2, which is expected from experimental observations. These can be plotted as shown in Fig. 2.

Some points should be noted about the calculation. First, m in the formulae is not a pole mass, but the averaged value of the spectral function of the resonance. These two values are very close for narrow resonances such as ω and ϕ, while they could be different for broad resonances such as ρ. Second, the calculation is for the infinite nuclear matter, not a finite nucleus. Third, the meson is at rest in the nuclear matter in this calculation. The momentum dependence of the mass, namely a dispersion relation, possibly exists in the matter. Lastly, the QCD sum rule does not provide the spectral function directly, but gives a constraint on the spectral function as required by QCD. Thus, in principle, spectral functions calculated by using any nuclear model and those predicted on the basis of the QCD sum rule should be consistent with each other.

By using various effective models, the spectral functions of vector mesons in the nuclear matter are calculated [19–23].

There are also predictions about the momentum dependence of the mass modification in hot and/or dense matter. Some predictions in cold and dense nuclear matter are as follows.

On the basis of the QCD sum rule, Lee predicted the momentum dependence [24, 25] of the decrease in vector meson masses for a momentum of less than 1 GeV/c as

$$\frac{m^*(\rho)}{m(\rho = 0)} = 1 - \left\{0.16 + (0.023 \pm 0.007)\left(\frac{p(GeV/c)}{0.5}\right)^2\right\}\frac{\rho}{\rho_0} \qquad (5)$$

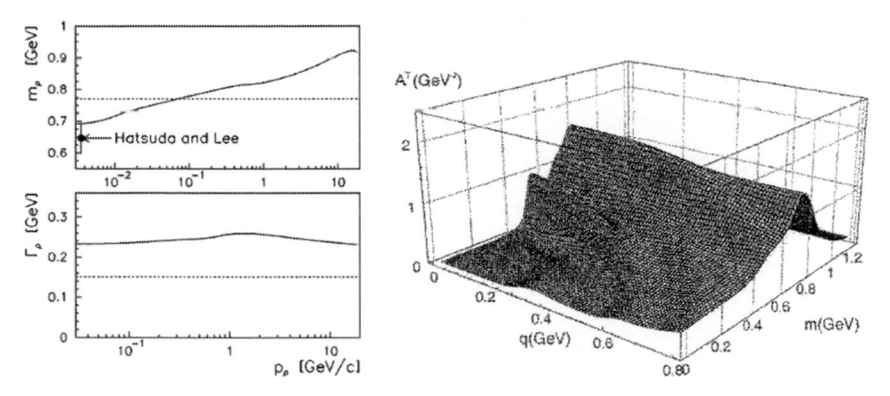

Fig. 3. The predictions of the momentum-dependent mass modification of ρ meson. In the *left panel*, momentum-dependent ρ mass and width in [26] are shown at the normal nuclear density. The consistency with the prediction by the QCD sum rule [17, 18] is checked at zero momentum. In the *right panel*, more complicated ρ spectral function at the normal nuclear density is shown, which is calculated by Post and Mosel [27].

for ρ and ω mesons, and

$$\frac{m^*(\rho)}{m(\rho=0)} = 1 - \left\{ 0.15y + (0.0005 \pm 0.0002) \left(\frac{p(GeV/c)}{0.5} \right)^2 \right\} \frac{\rho}{\rho_0} \tag{6}$$

for ϕ mesons.

Kondratyuk et al. calculated the momentum-dependent mass and width of the ρ meson over a wide range of momentum, as shown in the left panel in Fig. 3 [26]. The calculation is based on the dispersion relation of ρN scattering amplitude in the nuclear matter. Post and Mosel also calculated a ρ spectral function. They obtained more complicated results, e.g., a double peak structure in the low-momentum region [27], as shown in the right panel in Fig. 3.

Lattice QCD is a powerful tool to investigate QCD at high temperatures. The phase structure, equations of state, and even the meson spectral function at high temperature were calculated. On the other hand, at finite density, it is technically difficult in comparison to that at finite temperature because of what is called the "sign problem." Nevertheless, several challenges have been performed recently as reviewed in [28]. On the meson mass, Muroya, Nakamura, and Nonaka calculated using the color SU(2) QCD; the mass dropping of the vector channel is found at finite density system with two flavors [29].

3 KEK-PS E325 Experiment

The experiment 325 at KEK-PS was proposed in 1993. Detector R&D was started in 1994 and construction of spectrometer was started in 1996. The data taking was performed from 1997 to 2002. The first e^+e^- results were published in 2000 [30]. Then the data 100 times as large as the first data were accumulated in the 2001 and 2002 runs in 66 days operation and published in 2006–2007 [31–34]. Mainly, the results of 2001/2002 runs are discussed here.

3.1 Experimental Apparatus

A spectrometer [35] was placed at EP1-B primary beamline in the north counter hall of the KEK 12-GeV proton synchrotron (PS). The beamline was able to provide a primary beam of 4×10^9 proton per pulse (ppp) at a momentum of 12.9 GeV/c, i.e., T = 12 GeV. Asymmetric double slow extraction from the PS was performed [36] to the two primary beamlines, EP1 ($\sim 10^{10}$ ppp) and EP2 ($\sim 10^{12}$ ppp). The latter beamline served some secondary beamline experiments in the east counter hall.

The schematic view of the spectrometer is shown in Figs. 4 and 5. The spectrometer magnet was dipole type and targets were located at the center of the pole pieces. The pole gap was 907 mm and the diameter of each pole was 1,760 mm. The magnetic field strength was 0.71 T vertically at the center and the field integral from the center to the end of the tracking region (R = 1,600 mm) was 0.81 T m. The magnetic field was monitored spill by spill by using an NMR probe on the surface of the lower pole piece. Nuclear targets of thin foils were located in-line along the beam. In the 2001/2002 runs, C and Cu targets were used while CH_2 and Pb were also used in other runs. The target thickness was kept less than 0.6% radiation length in order to suppress the background in the e^+e^- channel from the γ-conversion in the target materials. The typical beam intensity was 6–8 $\times$ 10^8/spill and the target thickness was 0.2–0.4% interaction length in total in the 2001 and 2002 runs.

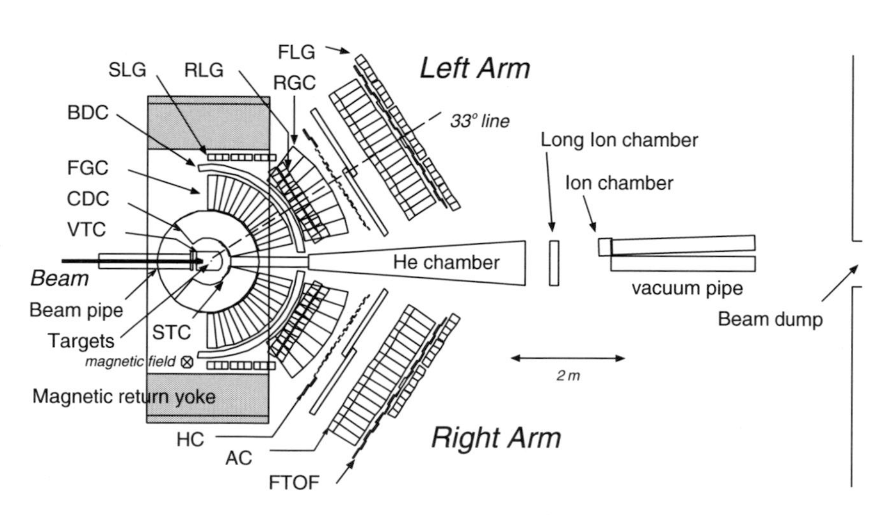

Fig. 4. Schematic plan view of the E325 spectrometer.

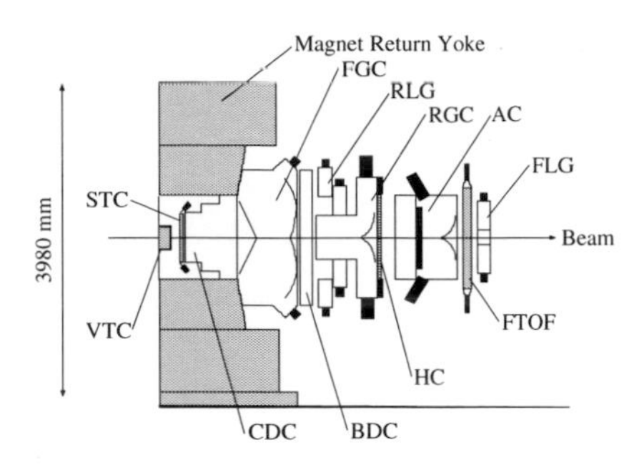

Fig. 5. Schematic cross-sectional view of the E325 spectrometer along the "33° line" shown in Fig. 4.

Three types of tracking devices were operated: vertex drift chamber (VTC), cylindrical drift chamber (CDC), and barrel drift chambers (BDCs). The chamber gas was the mixture of Ar:C_2H_6 = 50:50. The geometrical acceptance of CDC was ±23° vertically and ±12° to ±132° horizontally and that of BDC was ±23° vertically and ±6° to ±96° horizontally. Here, horizontal angles were measured from the beamline. VTC had two super-layers consisting of three drift planes for each, located at the radius of 100 and 200 mm. CDC had three super-layers consisting of three, four, and three drift planes for each, located at the radius of 450, 638, and 825 mm. BDC had a super-layer at 1,605 mm consisting of four drift planes.

The momentum was determined using only CDC and BDCs in order to avoid a possible bias from the limited acceptance of VTC. A candidate of e^+e^- (and K^+K^-) pair was fitted simultaneously with the constraint that the vertex of the pair should be located on the production target. For e^+e^- pairs, the momentum resolution evaluated from the full detector simulation can be approximated as $\sigma_p/p = \sqrt{(1.37 \cdot p(\mathrm{GeV/c}))^2 + 0.41^2}\,\%$. The simulation results were verified using the observed mass peaks for $\Lambda \to p\pi^-$, $K_s^0 \to \pi^+\pi^-$, $\phi \to K^+K^-$, and $\phi \to e^+e^-$. The achieved mass resolution (rms) is 10.7 MeV/c^2 for the $\phi \to e^+e^-$ channel. This is the best value ever achieved in the dilepton measurements in nuclear reactions.

For the electron identification, two types of segmented gas Čerenkov (GC) counters and three types of lead glass EM calorimeters were used. The acceptance of the counters was ±23° vertically and ±12° to ±90° horizontally. Iso-butane gas was used for the radiator of both threshold-type GC counters, namely front and rear GC counters (FGC and RGC). The Čerenkov photons were collected by the convergence mirrors and read out by the phototubes (PMT). The refractive index of the radiator was 1.00127; thus the threshold momentum for pions was 2.7 GeV/c at room temperature. Side, rear, and forward lead glass calorimeters (SLG, RLG, and FLG) were recycled from TOPAZ/TRISTAN [37] and used for the second (or third) stage of the electron identification. In the cascaded operation of the electron identification counters, the achieved rejection factor for pions was 3×10^{-4} with the electron efficiency of 78%.

For the kaon identification, aerogel Čerenkov (AC) counters [38] and a time-of-flight system between the two types of scintillation counters, namely start timing counter (STC) and forward time-of-flight counter (FTOF), were used. The acceptance of the counters was $\pm 6°$ vertically and $\pm 12°$ to $\pm 54°$ horizontally.

3.2 Detector Simulation

The full detector simulation code was constructed using the Geant4 [39] toolkit in order to evaluate the experimental effects on the meson mass spectra, such as the Bethe–Bloch-type energy loss, the Coulomb multiple scattering, and bremsstrahlung. The detector and chamber efficiencies were included. As mentioned above, the validity of the results was confirmed by reproducing the shapes of resonances such as Λ, K^0, and ϕ.

The event generator for the mesons was also developed. The cascade code JAM [40] was used to generate the kinematical distribution of mesons. The observed kinematical distributions of $\omega \to e^+e^-$, $\phi \to e^+e^-$, and $\phi \to K^+K^-$ are well reproduced after filtering by the detector simulator. The results of $\phi \to e^+e^-$ are shown in Fig. 6. The resonances basically have a Breit–Wigner (BW) shape. The ρ meson is a broad resonance and the difficulties to reproduce it are well known; therefore, several special shapes, including that obtained from the relativistic Breit–Wigner (RBW) formula [41], were used for the ρ meson. In the e^+e^- channel, a low-mass tail arising due to the internal radiative effects [42] is also included.

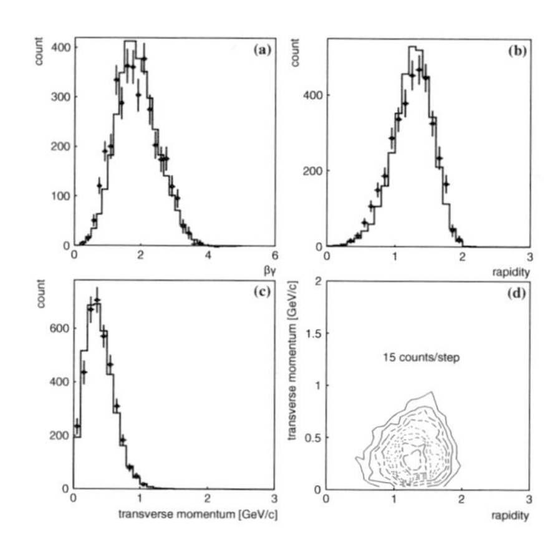

Fig. 6. The observed kinematical distributions of ϕ mesons in the detector acceptance. *Closed circles* show the data and the histograms are the simulation results using JAM and the full detector simulation. The background is subtracted by the sideband method. C and Cu target data are summed. The CM rapidity, y_{CM}, is 1.66.

3.3 Results

$\rho/\omega \to e^+e^-$

The invariant mass spectra in the e^+e^- channel are shown in Fig. 7 [31]. The spectra were fitted with the known hadronic sources, $\rho/\omega/\phi \to e^+e^-$, $\omega \to e^+e^-\pi^0$, and $\eta \to e^+e^-\gamma$, and the combinatorial background, which was evaluated by the event mixing method [41]. As mentioned above, the resonance shapes were generated by the BW/RBW and filtered by the detector simulation. The relative abundance of each component was determined from the fit.

The fit fails for the region of 0.33–2.00 GeV/c^2 for each target, C and Cu, because of the excess just below the ω meson peak for each target data. In addition to the RBW shape for ρ, various shapes such as Gounaris–Sakurai [43], CERES-type RBW without a Boltzmann factor, and even non-relativistic BW were tested, but they failed. However, by excluding the excess region (0.60–0.76 GeV/c^2), the fit reproduces well the data for each target. The interference of ρ and ω was also examined. The shape around the ω peak, particularly the steep fall on the right side, resembles an interference shape whose interference angle is around $2/3\pi$; however, this shape cannot reproduce the data. In fact, any interference shapes in any other interference angle also cannot reproduce the data [31].

$\phi \to e^+e^-$

The e^+e^- invariant mass spectra in the ϕ mass region for the C and Cu targets are shown with the best-fit results in Fig. 8 [32]. The data are divided

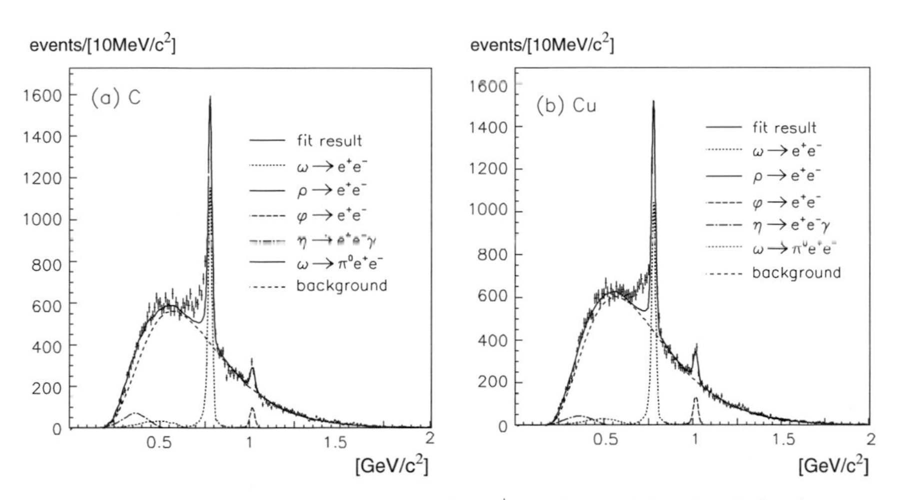

Fig. 7. The invariant mass spectra in the e^+e^- channel for the C (*left*) and Cu (*right*) targets, with the best-fit decomposition.

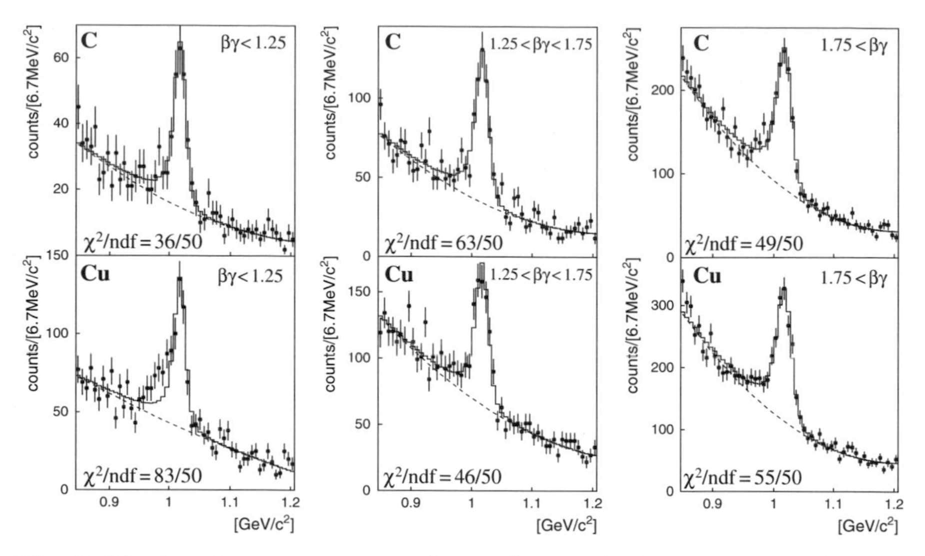

Fig. 8. The invariant mass spectra of $\phi \rightarrow e^+e^-$ in the three $\beta\gamma$ regions for the C and Cu targets. The range of $\beta\gamma$ and target are shown in each panel with the fit quality.

into three $\beta\gamma$ regions: $\beta\gamma < 1.25$, $1.25 < \beta\gamma < 1.75$, and $1.75 < \beta\gamma$. The fit was performed with the ϕ-meson shape generated as described above and a quadratic curve for the background. A cubic curve was also examined and the differences were treated as systematic errors. All the data were reproduced by the fit, except for the slowest data from the Cu target. The fit failed at 99% CL, namely the excess observed on the low-mass side of the peak is statistically significant.

The excess amounts in all the panels are evaluated as follows and shown in Table 1. The excess region was defined as 0.95–1.01 GeV/c^2 and the refit was performed excluding this region for each data. The fit result was subtracted from the spectra and the excess that remained in the region was counted, as

Table 1. The numbers of ϕ mesons (N_ϕ) and excess (N_{excess}) in each data region.

	$\beta\gamma$ range	Mean $\beta\gamma$ of the range	N_{excess}	N_ϕ	$N_{\mathrm{excess}}/(N_{\mathrm{excess}}+N_\phi)$
	<1.25	1.01	$6 \pm 17^{+7}_{-6}$	$257 \pm 26^{+6}_{-7}$	$0.02 \pm 0.06^{+0.02}_{-0.01}$
C	1.25–1.75	1.51	$-4 \pm 26^{+10}_{-12}$	$547 \pm 41^{+13}_{-15}$	$-0.01 \pm 0.05^{+0.02}_{-0.01}$
	1.75<	2.42	$39 \pm 42^{+22}_{-25}$	$1076 \pm 64^{+12}_{-15}$	$0.04 \pm 0.04^{+0.02}_{-0.01}$
	<1.25	0.96	$133 \pm 28^{+5}_{-12}$	$464 \pm 38^{+6}_{-5}$	$0.22 \pm 0.04^{+0.01}_{-0.01}$
Cu	1.25–1.75	1.51	$35 \pm 33^{+9}_{-12}$	$588 \pm 47^{+14}_{-8}$	$0.06 \pm 0.05^{+0.01}_{-0.01}$
	1.75<	2.38	$21 \pm 48^{+25}_{-29}$	$1367 \pm 72^{+24}_{-27}$	$0.02 \pm 0.03^{+0.02}_{-0.01}$

shown in Table 1. Only the slow-Cu data have statistically significant excess, $N_{\rm excess}/(N_{\rm excess}+N_\phi) = 22 \pm 4\%$. If the excess originates from the modified ϕ mesons, this ratio implies the ratio of the number of modified ϕ mesons to the total number of ϕ mesons.

$\phi \to K^+K^-$

The $\phi \to K^+K^-$ invariant mass spectra for the C and Cu targets are shown on the left side of Fig. 9 [33] with the best-fit results. These are also divided into three $\beta\gamma$ regions: $\beta\gamma < 1.7$, $1.7 < \beta\gamma < 2.2$, and $2.2 < \beta\gamma$. In order to fit the data, ϕ meson mass shapes were generated by the detector simulation as well as the $\phi \to e^+e^-$ case mentioned above, except that RBW was the original shape in order to take account of the fact that the K^+K^- threshold is 988 MeV/c^2, which is very close to the peak. The background shape was obtained by using the event mixing method, and normalization factors of the two shapes were determined from the fit.

The fit reproduces all the data regions, namely no statistically significant enhancement is found for the K^+K^- data, which seems different from the e^+e^- data described in the above section, but does not contradict it. It should be noted that the slowest region is for $\beta\gamma < 1.7$ in the K^+K^- data, while it is $\beta\gamma < 1.25$ for the e^+e^- data. These statistics are insufficient for a fit in the region of $\beta\gamma < 1.25$ for the K^+K^- data, as shown on the right side of Fig. 9. Therefore, it is concluded that no enhancement is observed in the region of $1.25 < \beta\gamma$ in both decay channels and in both targets.

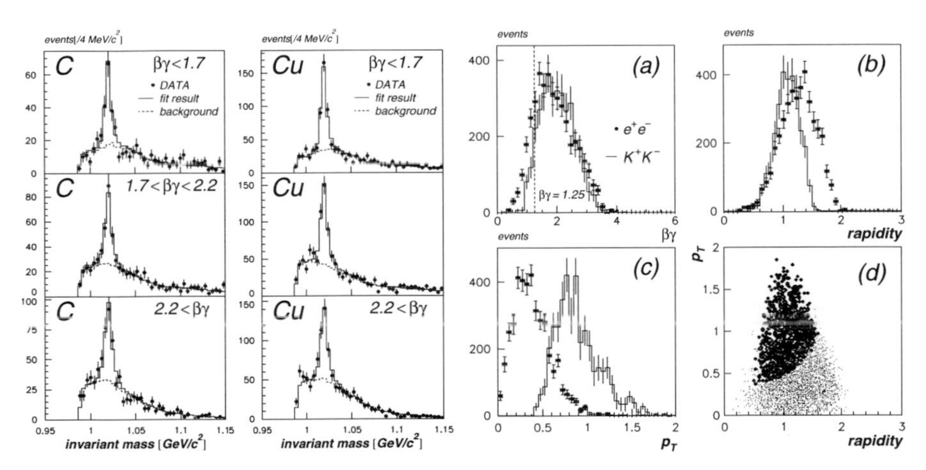

Fig. 9. (*Left*) The invariant mass spectra of $\phi \to K^+K^-$ in the three $\beta\gamma$ regions for the C and Cu targets. (*Right*) The comparison of the kinematical distributions of ϕ mesons measured in the e^+e^- channel and K^+K^- channel.

Production Cross Section

The production cross sections of the ω and ϕ mesons were measured from the e^+e^- channel decays [34] and also from the K^+K^- channel for ϕ, as shown in Fig. 10. Since the spectrometer acceptance mainly covered the backward region in the center-of-mass system, the cross sections were calculated for the backward hemisphere. For the ϕ meson, the cross sections measured from both channels are consistent within errors. The cross sections for the proton were deduced by the CH_2–C subtraction from the data obtained in the 1999 run. The cross section of $p + p \to \omega + X$ is consistent with the data measured in the former experiment $p + p \to \rho^0 + X$ [44], taking account of the fact that the ρ/ω ratio measured simultaneously by them in pp reaction is 1.0 ± 0.2 [44].

The nuclear dependence of the production cross section, α, was also studied. The parameter α is defined as

$$\sigma(A) = \sigma_0 \times A^{\alpha}, \tag{7}$$

where A is a mass number and σ_0 is a proportional constant. We obtained $\alpha_\omega = 0.710 \pm 0.021(\text{stat.}) \pm 0.037(\text{syst.})$ and $\alpha_\phi = 0.937 \pm 0.049 \pm 0.018$ in the e^+e^- decay channel. Naively thinking, the dependence $A^{2/3}$ suggests the meson was produced at the nuclear surface and A^1 suggests the meson was produced at the nuclear volume. The results of JAM reproduce such mass-number dependences of ω and ϕ and are consistent with such a picture, namely ω is produced at the first collision mainly and ϕ is produced in secondary processes in JAM.

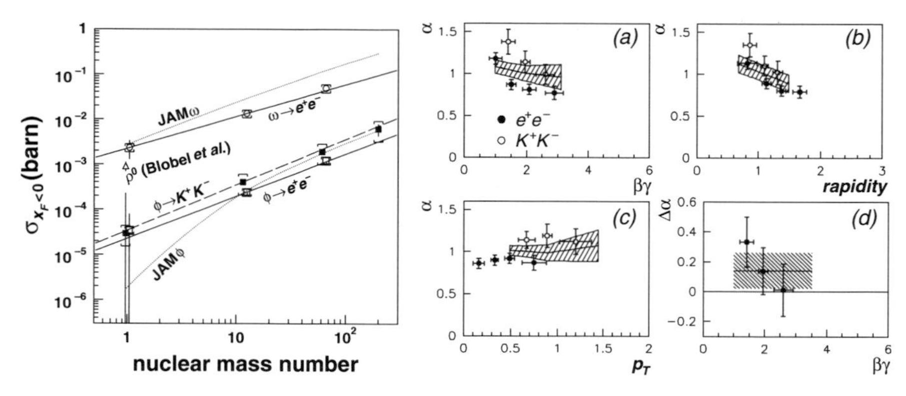

Fig. 10. (*Left*) The production cross sections of ω and ϕ measured in the e^+e^- channel and K^+K^- channel. The results of JAM are overplotted [34]. (*Right*) The dependences of α_{KK} and α_{ee} on (**a**) $\beta\gamma$, (**b**) rapidity, and (**c**) p_T. Panel (**d**) shows the $\beta\gamma$ dependence of the difference $\Delta\alpha = \alpha_{KK} - \alpha_{ee}$[33].

Branching Ratio

In order to investigate the change in the branching ratios of the K^+K^- and e^+e^- channels, the kinematical dependence of α was also studied using the 2001 and 2002 data [33]. The branching ratio of the ϕ meson could be a very sensitive tool to detect the mass modification of not only the ϕ meson itself but also the kaon, due to the small Q-value of the KK decay. Such studies were also performed in the heavy-ion experiments [45].

The relation between the ϕ meson and kaon masses is depicted in the right panel of Fig. 2. When a decrease in the ϕ mass is larger than that in the kaon mass, the Q-value is decreased. Then, the yield of $\phi \to K^+K^-$ (N_{KK}), which is the sum of the in-medium decays and in-vacuum decays, decreases in comparison with that of $\phi \to e^+e^-$ (N_{ee}) (referred to as "KK suppression case"). In the extreme case, the in-medium ϕ mass is less than the in-medium threshold; thus, ϕ cannot decay into K^+K^- inside the nucleus. Furthermore, the "KK enhancement case" can be considered. In this case, the Q-value increases; consequently, N_{KK} is enhanced. Note that such an effect is larger for larger nuclei; thus, for example, R(Cu) is larger than R(C) in the latter case, where the yield ratio R is defined as $\mathrm{R(A)} = \mathrm{N}_{KK}(A)/\mathrm{N}_{ee}(A)$.

The double ratio R(Cu)/R(C) can be correlated with the parameter α defined in Eq. (7) as follows:

$$
\begin{aligned}
\Delta\alpha &= \alpha_{KK} - \alpha_{ee} && (8)\\
&= \left\{ \ln\left[\frac{N_{KK}(A_1)}{N_{KK}(A_2)}\right] - \ln\left[\frac{N_{ee}(A_1)}{N_{ee}(A_2)}\right] \right\} \Big/ \ln(A_1/A_2) \\
&= \ln\left[R(A_1)/R(A_2)\right]/\ln(A_1/A_2). && (9)
\end{aligned}
$$

Thus, the two cases, "KK enhancement case" and "KK suppression case," can be distinguished by the sign of $\Delta\alpha$ and the difference is considered to be larger for the slower ϕ mesons.

The measured $\Delta\alpha$ and its kinematical dependence are shown in the right panel of Fig. 10. On average, $\Delta\alpha = \alpha_{KK} - \alpha_{ee} = 0.14 \pm 0.12$. Unfortunately, it is not statistically significant, namely the significance is less than 3σ, while the tendency favors the "KK enhancement" and the $\beta\gamma$ dependence is reasonable as the slower data have a larger difference. From this result, a limit is set on Γ_{KK} at 90% CL [33].

3.4 Interpretation

Model Calculation

In order to interpret the E325 results, the spectral modification of mesons was simply parametrized using the pole-mass decreasing and width broadening as mentioned in this section, and data were fitted using the modified meson spectra.

It should be noted that the in-medium modification of the meson mass could depend on the density and temperature of the medium. In heavy-ion experiments, the medium is a fireball which consists of the collided nuclei and/or a state of the quark–gluon plasma. Thus an observed mass spectrum is considered to be an integration of various spectra at various densities and temperatures. In E325, the medium is a target nucleus itself. While the temperature is considered to be zero, the density is not unique due to the nuclear surface diffuseness. Furthermore, the dispersion should also be noted, namely the meson mass modification could also depend on the meson momentum itself.

In order to include such effects, a Monte Carlo-type model calculation was adopted to generate the modified meson spectra to fit the data. Basically, the mass shapes were BW type, as mentioned in Sect. 3.3; the pole mass m_0 and the width Γ_0 in the formulae are changed as

$$\frac{m_0^*(\rho)}{m_0} = 1 - k_1 \frac{\rho}{\rho_0} \tag{10}$$

and

$$\frac{\Gamma_0^*(\rho)}{\Gamma_0} = 1 + k_2 \frac{\rho}{\rho_0}, \tag{11}$$

where ρ_0 is the normal nuclear density, namely they are linearly dependent on the density at the decay point of each meson in the calculation. The formula for m_0 is taken from the prediction in [17, 18], though the prediction is not for the pole mass but for the average of the spectrum. As regards Γ_0, no prediction uses such a formula explicitly; however, the predictions in [46, 47] have almost linear dependence on the density around ρ_0. Though the modification could be momentum dependent, namely the parameters k_1 and k_2 could be functions of the meson momentum, they were set to constants.

In the calculation, mesons were generated with a kinematical distribution produced with JAM as same as the usual analysis which is described in Sect. 3.2. The meson production points were distributed uniformly on the whole nucleus for ϕ mesons, and on the surface of incident hemisphere at the half-density radius of the nucleus for ρ and ω mesons. These assumptions were reflected in the observed nuclear dependence of the production cross sections described in Sect. 3.3. Woods–Saxon-type nuclear density distribution was adopted as

$$\rho(r) \propto \frac{1}{1 + \exp[(r - R)/\tau]}, \tag{12}$$

where the "half-density" radii R of the C and Cu nuclei are 2.3 and 4.1 fm and the diffuseness parameters τ are 0.57 and 0.50 fm, respectively [48].

The produced mesons were traced step by step, with the modified mass and width according to the density of the location and decayed according to the probability determined by their own $\beta\gamma$ and width. The formation time of mesons was ignored, namely set to zero.[1] The survived mesons were traced to 100 fm and decayed. The absorption of mesons in nuclei was ignored because the observed kinematical distributions of mesons were composed of only the unabsorbed component of the mesons generated by the proton-induced reactions. The e^+e^- pairs from the decays were input to the detector simulation mentioned in Sect. 3.2 and the output shapes were used to fit the data.

Model Fit

The fits with modified meson shapes were performed to determine the k_1 and k_2 from the data.

In the $\rho/\omega \to e^+e^-$ case, ρ and ω were modified with the same parameter k_1 (or k_2) based on the prediction in [17]. Furthermore the parameters were common to both C and Cu target data. When the k_2 is fixed to zero, the best-fit parameter is obtained as $k_1 = 0.092 \pm 0.002$. It implies that the ρ and ω masses are reduced by 9% at $\rho = \rho_0$ and does not imply the averaged value for the in-medium mesons. This is almost consistent with the prediction of Eq. (3), within the error band. When k_2 is set to 1.0 and 2.0, namely the width at $\rho = \rho_0$ is broadened two and three times the vacuum value, respectively, no mass parameter k_1 is found to reproduce the data shape.

In the $\phi \to e^+e^-$ case, in order to understand the component modified by as much as 22%, it is considered that the decay probability in nuclei could be enhanced; thus, the width could be broadened. The density-dependent widths

$$\frac{\Gamma^{tot}(\rho)}{\Gamma_0^{tot}} = 1 + k_2^{tot}\frac{\rho}{\rho_0} \tag{13}$$

$$\frac{\Gamma^{ee}(\rho)}{\Gamma_0^{ee}} = 1 + k_2^{ee}\frac{\rho}{\rho_0} \tag{14}$$

were introduced and two extreme cases were examined as follows: (i) the branching ratio Γ^{ee}/Γ^{tot} is not changed, i.e., $k_2^{ee} = k_2^{tot}$ and (ii) the partial decay width Γ^{ee} is not changed, i.e., $k_2^{ee} = 0$.

The fits for six data histograms were performed simultaneously with a common set of k_1 and k_2^{tot} and a fitting χ^2/ndf was calculated, namely the parameter is common to both the targets and all the $\beta\gamma$ regions. The parameters were scanned in the ranges $0 < k_1 < 0.06$ and $0 < k_2^{tot} < 10$ for cases (i) and (ii). The confidence ellipsoids were drawn as shown in Fig. 12. As a result, both cases (i) and (ii) can reproduce the data with their best-fit parameters as shown in the left panel of Fig. 11, while the χ^2_{min} for case (i) is smaller than that of (ii).[2] The best-fit parameters, namely for case (i), are $k_1 = 0.034^{+0.006}_{-0.007}$

[1] When a formation time of 1 fm was adopted, the modification parameter was increased by a factor of approximately 1.1 for the ρ/ω case.

[2] Naively thinking, it is strange that the partial width for the leptonic decay is modified, as supposed in case (i), like the hadronic width, which can be changed due to the hadronic effect in medium. More experimental and theoretical studies should be performed on the broadening of the leptonic decay width of ϕ mesons.

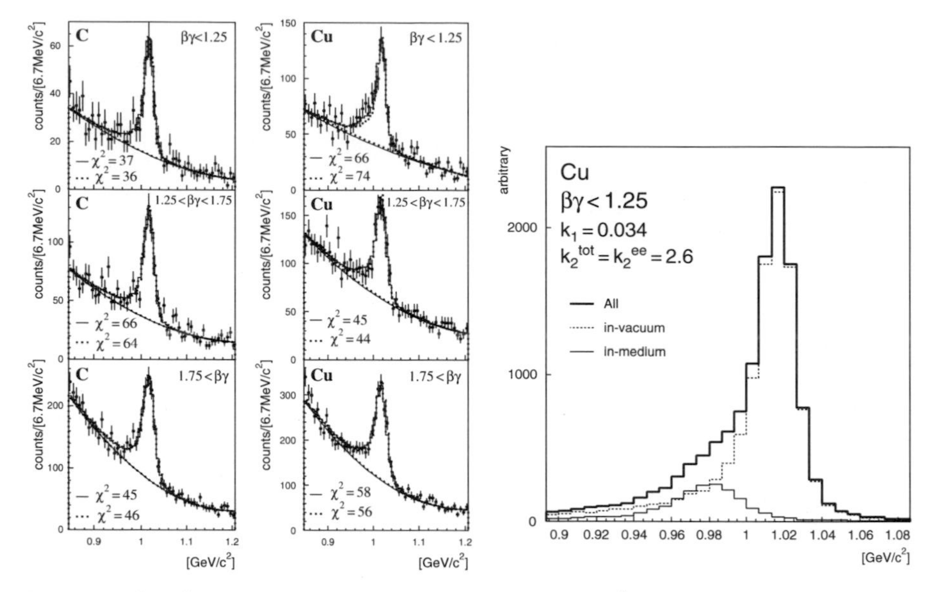

Fig. 11. (*Left*) The invariant mass spectra of $\phi \to e^+e^-$ in the three $\beta\gamma$ regions for the C and Cu targets, with the model fit results. *Solid lines* show the results for case (i) and *dashed* for case (ii) (see text), taken from [32]. (*Right*) An example of the model shape used in the fit for Cu data in the slowest $\beta\gamma$ range. The *thick black line* shows the used shape, the *thin black line* shows the shape of the in-medium decays, and the *dotted line* shows the shape of in-vacuum decays. Here, the boundary between the "in-medium" and "in-vacuum" is set at the half-density radius in the model calculation.

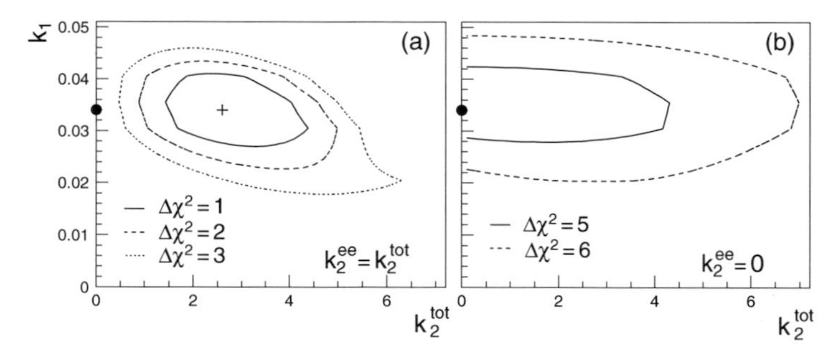

Fig. 12. Confidence ellipsoids according to the shift parameters k_1 and k_2^{tot} for the model fits shown in Fig. 11, taken from [32]. Panel (**a**) is for case (i) and (**b**) is for case (ii). The best-fit point, $\chi^2_{min} = 316.4$, for case (i) is shown by the *cross* in panel (**a**) and that for case (ii), $\chi^2_{min} = 320.8$, is shown by the *closed circle* in panel (**b**). The *circle* is also shown in panel (**a**); in fact, they are the same points in the parameter space. The step for each contour is shown in each panel, commonly measured from the point of $\chi^2_{min} = 316.4$.

and $k_2^{\rm tot} = 2.6^{+1.8}_{-1.2}$. The k_1 value implies that the ϕ meson mass is decreased by 3.4% at $\rho = \rho_0$, i.e., $m^*_\phi(\rho_0) = 985^{+7}_{-6}$ MeV/c^2. This is consistent with the prediction of Eq. (4). Using the center values in both the experiment and theory, y is determined as 0.23. Furthermore, the k_2 implies that the width is broadened by 3.6 times, i.e., $\Gamma^*_\phi(\rho_0) = 16\ ^{+8}_{-5}$ MeV/c^2. It is almost consistent with the predictions [46, 47]. In the former case, the predicted value is 22 MeV/c^2 for the 1,020 MeV/c^2, natural mass of the ϕ meson; however, by using the reduced mass 985 MeV/c^2, the predicted width is around 16 MeV/c^2. In any case, they are within the experimental error.

4 Other Experiments

4.1 Heavy-Ion Collisions

The first remarkable result of the meson mass modification in medium was reported by CERES experiment at SPS, CERN. They measured the e^+e^- invariant mass spectra in 200 A GeV S+Au in 1992 and 158 A GeV Pb+Au reactions in 1995 and 1996 [49, 50]. Significant enhancements were observed in the range of 0.2–0.8 GeV/c^2 in the spectra, which were not observed in the 450 GeV p+Be and p+Au reactions measured by them. Naively thinking, the enhancement could come from the modified ρ mesons in a hot medium, i.e., a fireball formed by the heavy-ion collision. In order to explain the data, various theoretical calculations were performed using the hadronic fireball or quark–gluon plasma, mass modification with the chiral symmetry restoration in hot and dense matter, or pure hadronic calculation, vector meson decays and/or pion annihilation and/or thermal radiation, etc. In the literature [49, 50], the two representative scenarios, the mass dropping (by Brown and Rho [51]) and the width broadening (by Rapp and Wambach [52]) of ρ mesons, are compared with the data, using the same space–time evolution of the fireball by Rapp. However, the data can be explained by both scenarios, namely no definite interpretation can be selected.

The next experiment performed at SPS was NA60. They measured $\mu^+\mu^-$ pairs from the 158 A GeV In+In reaction with a good mass resolution of ~23 MeV/c^2 at 1 GeV/c^2 using the silicon detectors and very large statistics in comparison with CERES. The enhancement is observed again below the clear ω meson peak [53]. They claimed that the enhancement was explained by the width broadening of ρ mesons based on the in-medium many-body calculation by Rapp, and the mass-dropping scenario (Brown–Rho scaling) was excluded. The theoretical approaches to explain the data are reviewed in [54]. According to van Hees and Rapp [55], the in-medium ρ width is 350 MeV/c^2, which is the average value over the space–time evolution of the fireball.

CERES claimed that their new data obtained in 2000, for 158 A·GeV Pb+Au, are also explained by such a width broadening of ρ mesons [56]. The broadening scenario has recently gained predominance; however, the objections still exist [57, 58].

Recently, PHENIX experiment also reported the observation of the enhancement below the ω meson peak in the e^+e^- spectrum measured in Au+Au collisions at $\sqrt{s_{NN}}$ =200 GeV [59].

At a lower energy, HADES experiment measured the e^+e^- spectra in 1–2 A·GeV C+C reactions. They can also measure elementary reactions, namely pion- and proton-induced reactions. They reported the modifications of the e^+e^- spectrum in recent publications [60, 61]. After the subtraction of the combinatorial background, the invariant mass spectrum below 0.15 GeV/c^2 is reproduced by the known π^0 sources, while excesses are observed over 0.15 GeV/c^2. In the range of $0.15 < M < 0.5$ GeV/c^2, the data exceed the expectation, which is based on the known η production, by a factor of 2. In the range of $0.5 < M < 0.8$ GeV/c^2, much more enhancement is observed over the calculation including $\omega \to e^+e^-$, $\rho \to e^+e^-$, and $\Delta \to Ne^+e^-$. Only a thermal source is assumed in their event generator. Furthermore, transport calculations such as HSD, RQMD, and UrQMD also cannot reproduce the data over 0.15 GeV/c^2. The calculations include the collision dynamics, multi-step productions, and so on, but do not include the in-medium mass modification of vector mesons. They also stated that the data are consistent with the result of the 1.0 A GeV C+C reaction performed in DLS experiment at Bevalac [62].

HADES also have 1 A GeV C+C data [63]. The observed enhancement factor in the range of $0.15 < M < 0.5$ GeV/c^2 is 7, and the data are also consistent with the DLS data. They claimed that the Δ and ρ are required to explain the enhancement and their data in elementary processes such as p+p help to understand the enhancement by the transport calculations.

4.2 γ-Induced Reactions

The experiment CBELSA/TAPS [64, 65] was performed using a 0.64–2.53 GeV photon beam at ELSA. Gamma rays from decayed mesons were measured by using TAPS detector, a barrel composed of the BaF_2 and CsI calorimeters. The $\omega \to \pi^0\gamma$ spectrum was reconstructed by three γ-rays, and a comparison between the liquid hydrogen target data and Nb target data was performed after the subtraction of the empirically determined background. A significant excess on the left side of the ω peak was found in the Nb target data [64]. The momentum range of the measured ω mesons are $0.2 < p_\omega < 1.4$ GeV/c and the excess is observed only in the range less than 0.6 GeV/c. By using Giessen BUU transport calculation, the excess was analyzed and found to correspond to a 14% mass decrease at the normal nuclear density ρ_0. New data using a C target also show a similar excess [65].

The nuclear dependence of the production cross section of ω was also measured. The result was analyzed by using the Valencia model and Giessen model, and the obtained in-medium cross section was connected to the in-medium width, 130–150 MeV/c^2 at ρ_0. The width is for the average momentum of 1.1 GeV/c and corresponds to 225–260 MeV/c^2 at rest [66].

The experiment CLAS-G7 [67, 68] was performed using a 0.6–3.8 GeV photon beam at Jefferson Laboratory. Vector meson decays were detected in the e^+e^- channel using CLAS spectrometer. The D_2, C, Fe, and Ti target data are presented. The measured momentum range was $0.8 < p_\rho < 3.0$ GeV/c for the Fe and Ti target data.

The spectra were analyzed using the shape obtained from the Giessen BUU transport calculation after the subtraction of the combinatorial background. The calculation does not include the in-medium mass dropping as predicted by Brown and Rho or Hatsuda and Lee, but includes hadronic interactions that cause the "collisional broadening" of ρ. The broadened widths of ρ in the model were 160, 178, and 203 MeV/c^2 for the D_2, C, and Fe+Ti targets, respectively. By using these shapes, the data were well reproduced, namely the possible modification was explained by the collisional broadening, without mass dropping.

The in-medium mass dropping such as that given by Eq. (10) was also introduced in the model and examined. The upper limit on the modification parameter was deduced as 0.053 at the 95% CL.

4.3 Discussion

The E325, TAPS, and CLAS experiments all claimed that the vector meson mass is modified in nuclei. With the results of the heavy-ion collision experiments, it can be stated that the in-medium modification of vector mesons is established. However, the statements about the nature of modification seem to contradict each other, as one states that the mass is decreased, other states that only the width is broadened. In order to understand the in-medium modification and its relation to the chiral symmetry, we have to solve the confusing situation.

Here, the author would like to make some comments about the "contradiction" in three lower energy experiments.

Both TAPS and CLAS-G7 used the Giessen BUU transport calculation for the evaluation of the modified meson spectra, while E325 used the simple model described in Sect. 3.4. The difference in the mass modifications of 14% by TAPS and 9% by E325 can be interpreted as, for example, to suggest the momentum dependence of the mass modification which is not included by the E325 models or it may simply be due to the different momentum ranges they measured.

Between E325 and CLAS-G7, there is another large difference, i.e., the background normalization, and it should be discussed carefully. However, before suspecting the background estimations, we should note the meson shapes used in their analyses to evaluate the modification. If the data are consistent with each other, different analyses deduce different modification parameters.

For example, the shapes of ω used by CLAS do not appear to be much modified in comparison with that of E325. If CLAS use an ω shape similar to that used by E325, the ρ peak should drop to reproduce the data, and vice versa. Furthermore, the modification parameters themselves are model-dependent results. In order to verify whether the two data contradict each other or not, the same analysis model should be used, at least.

Besides the theoretical models of the meson mass modification, the analyses of $\rho/\omega \rightarrow e^+e^-$ spectra suffer from unknown factors such as the various resonance shapes which ρ mesons could form, unknown ρ/ω production ratio for nuclear target reactions, even the possibility of the $\rho - \omega$ interference. On the other hand, the anomalies observed in the ω spectrum by TAPS and the ϕ spectrum by E325 are found in the comparison between different nuclear targets. In other words, the anomalies, i.e., the excesses observed on the left side of the peaks, are independent of any theoretical models while the background estimations are important to deduce the anomalies. They are clear evidences of the modification of mesons in medium. Of course, theoretical calculations are necessary to connect the data to the chiral symmetry in medium. However, the data can provide the limitation to be satisfied by theoretical predictions.

5 Dielectron Measurement at J-PARC

Several experiments to investigate the meson properties in medium have been proposed to be performed at the J-PARC hadron experimental facility. In this section, dielectron measurement is described and other experiments are described in the next section.

5.1 E16: Dielectron Measurement

The J-PARC E16 experiment [69] was proposed in 2006 as a next step of the e^+e^- invariant mass spectroscopy performed by KEK-PS E325. The spectral modification of ϕ mesons in medium observed by E325 was, in fact, only one data point, i.e., the slowest component of the Cu target data. Even in the ρ/ω case, modifications were found in both targets, C and Cu, while velocity dependence was not analyzed. The existence of the mass modification of vector mesons measured through the e^+e^- decays was established by these data. However, in order to compare it to various theoretical predictions, the modification should be studied systematically and quantitatively, throughout the nuclear size dependence, momentum dependence, and so on.

In the E16 experiment, the goal is to collect $\sim 2 \times 10^5$ ϕ mesons, which is 100 times as large as that collected in E325, for each target. Simultaneously, more statistics are expected for ρ and ω mesons. A factor of 5 comes from the detector acceptance, a factor of 2 comes from the production cross section of ϕ through the increase of incident energy, and a factor of 10 comes from the beam intensity. In order to cope with the higher beam intensity ($\sim 10^{10}$

ppp, i.e., ~10 GHz) and higher interaction rate (10 MHz at the target), new spectrometer should be constructed.

Yield Estimation

The evaluated relative ϕ meson yield is shown in Table 2. The nuclear cascade code JAM [40] was used to evaluate the production cross section and kinematical distribution of ϕ mesons in the p+Cu reaction. The validity of JAM is verified as follows. The kinematical distribution in 12 GeV p+C/Cu reactions are consistent with that observed by E325, as mentioned in Sect. 3.3, while the absolute cross section is slightly overestimated, within a factor of 2 [34]. The energy dependence from 12 to 50 GeV is consistent with the phenomenological scaling for narrow vector mesons in the literature [70]. Thus, the obtained cross sections at 30 and 50 GeV are scaled to the measured one at 12 GeV. The generated ϕ mesons are decayed to e^+e^- pairs and used as an input of a fast simulation to calculate the acceptances. Two detector geometries, both the E325 spectrometer and a new large spectrometer, are used and compared with each other, as shown in Fig. 13.

For a larger incident energy, the larger production cross section of ϕ mesons and the smaller detector acceptance due to the Lorentz boost are obtained. As a result, the figures of merit for the 30/50-GeV cases are larger than that for the 12-GeV case. When the energy is fixed, the yield obtained by the new spectrometer is five times as large as that obtained by the E325 spectrometer. Furthermore, for each spectrometer, the increase in the yield with the incident energy is a factor of 2–2.5, including the Lorentz boost.

In addition, ρ, ω, and even J/ψ are expected to be measured simultaneously in the e^+e^- channel. Both the ρ and ω yields are 5–10 times as large as the ϕ yield. At 50 GeV, the yield of J/ψ is roughly estimated as 1/50 times as the ϕ yield, as summarized in Table 3. The inclusive production cross section of J/ψ in the p+p reaction is estimated to be 0.01 μb at 50 GeV ($\sqrt{s} = 10$) from Fig. 4(b) of the literature [71]. The cross section for the inclusive production of ϕ in p+Cu reaction is measured at 12 GeV, and it is scaled to 50 GeV as described above. The detector acceptance for $J/\psi \to e^+e^-$ is assumed to

Table 2. The ϕ meson yields in p + Cu reactions for two detector geometries.

Beam energy		12 GeV	30 GeV	50 GeV
Production cross section (scaled to 12 GeV)		1	3.0	5.1
Detector acceptance	E325	8.8%	(6.0%)	(4.5%)
	J-PARC	45%	31%	23%
Normalized yield by E325	E325	1.0	(2.0)	(2.6)
	J-PARC	5.1	10.0	12.7

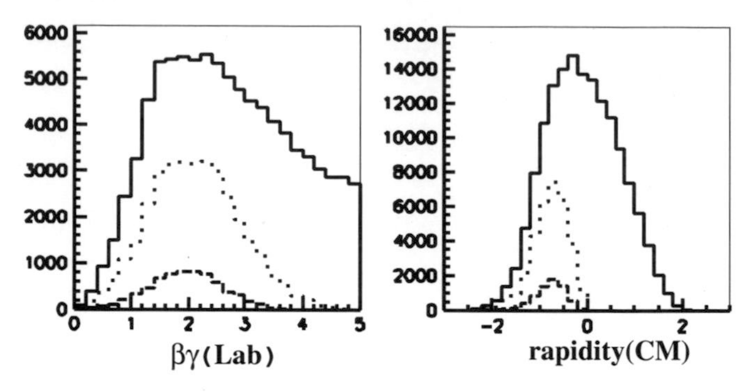

Fig. 13. The acceptance for the $\phi \to e^+e^-$ channel in 50-GeV p + Cu reaction for two detector geometries. *Left panel* shows the $\beta\gamma$ distribution of ϕ mesons in the laboratory system and *right panel* shows the rapidity distribution in the CM system. *Solid lines* show the ϕ meson distribution generated by JAM. *Dotted* and *dashed lines* show the ϕ mesons accepted geometrically by the proposed new spectrometer and the E325 spectrometer, respectively.

Table 3. The estimated yields of ϕ and J/ψ at 50 GeV.

	ϕ	J/ψ	Ratio
p+p cross section		0.01 μb	
p+Cu cross section	∼ 5 mb	∼ 0.5 μb	1/10,000
Branch to e^+e^-	0.03%	6%	200
Yield	2×10^5	4000	1/50

be the same as that for $\phi \to e^+e^-$ and the nuclear mass number dependence of the production cross section is considered to be A^1 for J/ψ.

5.2 Systematic Study of the Modification

In the E16 experiment, systematic studies of the mass modification will be performed as follows using the higher statistics.

Nuclear Size Dependence

The planned target configuration is summarized in Table 4.

In general, a heavy nuclear target is difficult to use for dielectron measurements, because the ratio of the interaction length to the radiation length is worse. In order to suppress the γ-conversion in the target material, the radiation length should be kept less than 0.5%; then, the interaction length becomes shorter for a heavier target and the data are statistically limited. Nevertheless, for the high statistics expected in the E16 experiment, a heavy target such as lead can also be used.

On the other hand, $CH_2 - C$ subtraction effectively provides information of the proton target. The subtraction method also requires high statistics.

Table 4. Targets planned to use.

Nuclei	Interaction length (%)	Radiation length (%)	Thickness (μm)
C	0.05	0.1	200
CH_2	0.05	0.1	400
Cu	0.05	0.5	80
Pb	0.01	0.3	20

The four targets listed in Table 4 are used in an in-line manner, as in the E325 experiment. The expected velocity (momentum) binning and statistics are schematically shown in Fig. 14. In the E325 results, which are shown in the left panel, only one point has significant excess and both model curves with different modification parameters are consistent with the data, as shown in the figure. In E16, as schematically shown in the right panel, a model-selection capability could be increased. The excesses can be detected significantly in almost all the velocity bins and nuclear size dependence is also clearly found. It should be noted that each data point plotted in the figures is deduced from the fit result of each invariant mass spectrum. In place of 6 spectra, 32 spectra with higher statistics will be available for the systematic studies.

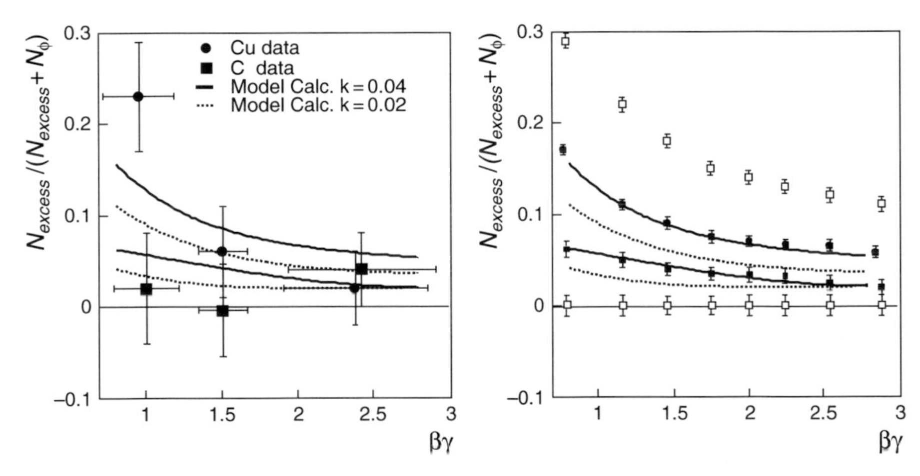

Fig. 14. *Left panel* shows the $\beta\gamma$ dependence of the ratio $N_{\text{excess}}/(N_{\text{excess}}+N_\phi)$ observed in e^+e^- spectra measured by E325, shown in Table 1. *Solid circles* and *squares* are for the Cu and C target data, respectively. *Solid* and *dashed curves* show the results of model calculation, the two cases of the parameter in the prediction. *Right panel* shows the effect of statistics expected in the E16 experiment. *Curves* are the same as in the *left panel.* Statistical error bars are shrunk by the statistics 100 times as high as that of E325. Expected proton and Pb data points are also shown with *open squares.*

Momentum Dependence

By using the high statistics, the momentum dependence of the modification can be extracted. In the E325 analysis for the ϕ meson, the momentum independence of the modification (parameters k_1 and k_2) is assumed. Because there are only two data points (two target nuclei) for a momentum bin and each bin covers a wide momentum range, the determination is thought to be less reliable.

In E16, the momentum bins can be divided more finely and the number of data points (target nuclei) in each bin is four or more; hence, reliability will be increased. The expected plot of the momentum dependence of the mass dropping of ϕ is shown in Fig. 15 with the prediction by Lee [24, 25].

Collision Geometry

As another approach to systematic studies, the dependence on the collision geometry ("centrality") of the p+A reaction can be studied for larger nuclear targets such as Pb. Of course the impact parameter of the interaction could be related to the production point and the in-medium flight length of mesons.

The information of the collision centrality is provided by the charged particle multiplicity and/or the number of slow protons in the target rapidity region [72]. The centrality resolution of the measurements is studied using JAM. The results are shown in Fig. 16. The numbers of charged particles and

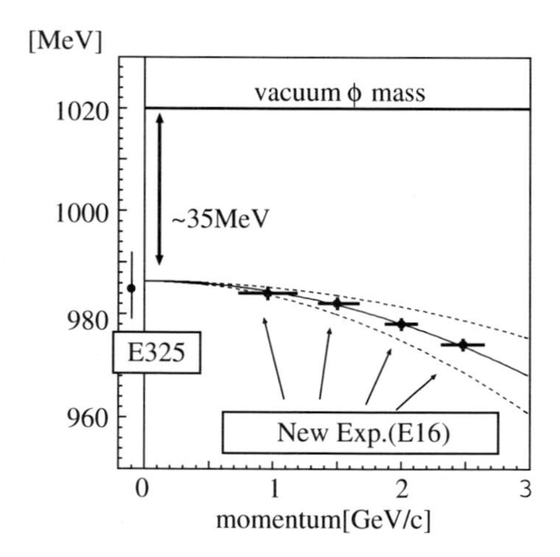

Fig. 15. Momentum-dependent ϕ mass predicted by Lee [24, 25] at the normal nuclear matter density is plotted, where $y = 0.23$, which is selected to reproduce the E325 results. The *solid curve* shows a central value and the *dashed curves* show errors of the momentum-dependent term. Original prediction, which is valid only below 1 GeV/c, is extrapolated to 3 GeV/c. Four observation points expected by E16 are plotted with the errors, which are very small. The E325 result, which neglects the possible momentum dependence, is also plotted.

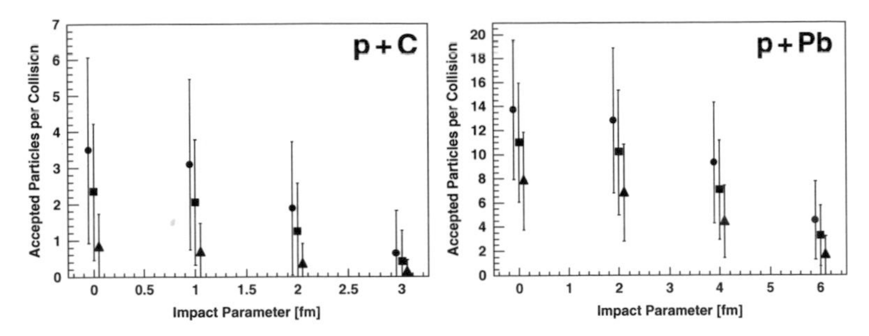

Fig. 16. Evaluation of collision geometry resolution in p+C (*left*) and p+Pb (*right*) collisions at 50 GeV. *Solid circles* show the number of charged particles emitted into the aperture of the spectrometer with a low-momentum cutoff at 0.3 GeV/c; *solid squares* are with an additional high-momentum cutoff at 1.0 GeV/c; *solid triangles* show the number of protons alone with the acceptance and momentum cutoffs.

slow protons that enter the proposed spectrometer are plotted as a function of the impact parameter used in the simulation for the C and Pb targets. In the heavy (Pb) nucleus case, two or three divisions of impact parameter can be determined by the measured particle multiplicity, while such a clear division is impossible for the light (C) nucleus case.

5.3 Experimental Apparatus

The experiment E16 will be performed using a newly built spectrometer at the "high-momentum beamline," which will be available around 2011 or later at J-PARC hadron experimental hall.

Beamline

As shown in Fig. 17, the high-momentum beamline is branched from the main primary beamline (A-line) at the SM1 separation point located in the switch yard to be utilized for the primary beam experiments. Up to $\sim 1 \times 10^{12}$ ppp of primary beam is available. The point is designed to allow an energy loss

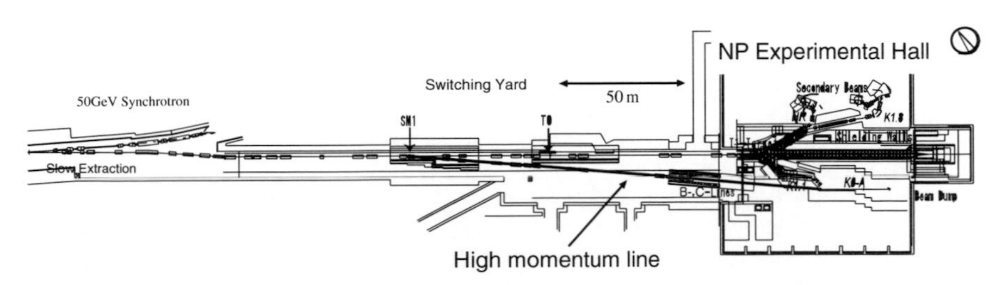

Fig. 17. The plan view of the extraction point from the J-PARC MR (50-GeV PS), the switch yard, the hadron experimental hall, and the high-momentum beamline.

of 15 kW, namely a 2% of the planned full power (750 kW) of the J-PARC 50-GeV PS (MR, main ring). By using a thin production target at SM1, the line is also used as a secondary beamline. The length of the beamline from SM1 to the final focus is approximately 120 m; thus, the low-momentum kaons almost decay in the beamline.

New Spectrometer

As mentioned above, the E325 spectrometer should be replaced to cope with the higher interaction rate, which is 10 MHz at the target, and to cover larger acceptance. The conceptual design of the proposed new spectrometer is as follows.

The schematic plan and front views of the spectrometer are shown in Fig. 18. The main detector components and their locations are summarized in Table 5. The spectrometer covers $\pm 12°$ to $132°$ horizontally and $\pm 45°$ vertically. As summarized in Table 2, the acceptance for the e^+e^- decays of ϕ mesons is increased by a factor of 5 in comparison with that of E325 spectrometer.

As for a spectrometer magnet, the E325 spectrometer magnet can be recycled. In order to increase the vertical acceptance of detectors, the pole pieces are reshaped and the coils are newly built.

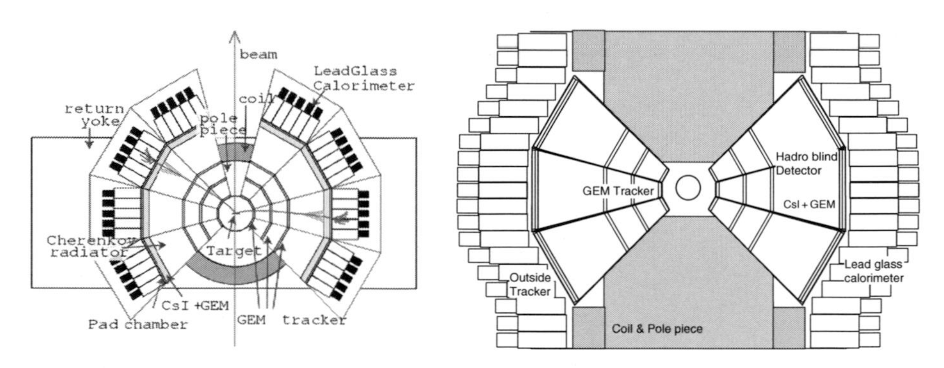

Fig. 18. Schematic plan (*left*) and front (*right*) views of the proposed E16 spectrometer.

Table 5. List of detectors.

Name	Radius (mm)
GEM tracker 1	200
GEM tracker 2	400
GEM tracker 3	600
Čerenkov counter (HBD)	600–1100
Outside tracker	1120
EM calorimeter	1140–1700

The GEM trackers are adopted to cope with high particle rates. The expected particle rates, which are scaled from the E325 experience, are summarized in Table 6. The tracking planes are located at the radius of 200, 400, and 600 mm from the center of the spectrometer magnet. A position resolution of 200 μm in rms is required to achieve a mass resolution of ~10 MeV/c^2 at the ϕ meson mass region. It is achieved with the readout strip of 700-μm pitch. Basically the momenta of tracks are determined by these three planes; however, an outside tracker is also located at a radius of 1,120 mm to help track finding and refining. A cathode strip chamber is a candidate for the outside tracker to cover a large area at a reasonable cost.

Electron identification is performed using two stages of identification counters. As the first stage, a threshold-type gas Čerenkov counter is used. A GEM detector with a CsI photocathode and CF_4 gas is operated as a photon sensor. Since CF_4 is also a radiator gas, a window between the radiator and the photon sensor is not required. The refractive index of CF_4 is 1.000620 (thus a threshold momentum for pions is 3.95 GeV/c) and the cutoff wavelength is approximately 110 nm. The working principle of the counter is schematically described in Fig. 19. Such a type of gas Čerenkov counter is called a hadron blind detector (HBD). Its merits over the segmented gas Čerenkov counter

Table 6. Expected particle rates at various locations. The radius is measured from the target. The angle and distance are measured from the beamline. The rate is calculated from the E325 experience, just scaled by 10.

Radius	Angle	Distance	Rate (mm^{-2})	E325 experience
200 mm	6°	20 mm	5 kHz	350 KHz in 3.5 mm × 220 mm (DC cell)
400 mm	12°	80 mm	1.1 kHz	1.8 MHz in 40 mm × 400 mm (scintillator)
400 mm	60°	350 mm	0.25 kHz	0.4 MHz in 40 mm × 400 mm (scintillator)

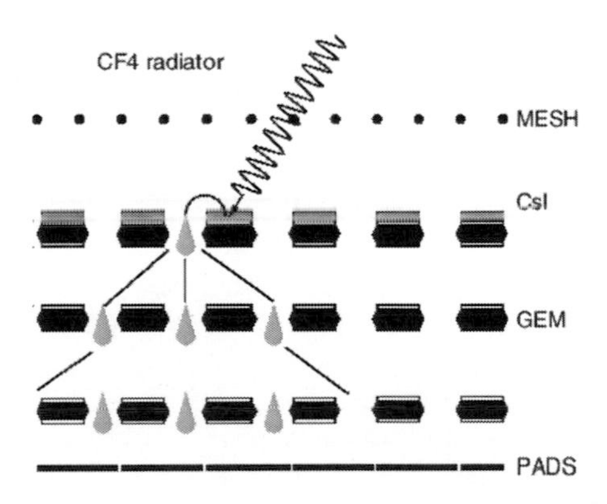

Fig. 19. A schematic view of the working principle of the hadron blind detector (HBD), taken from [73].

read by PMT and adopted in E325 are the position sensitivity for the trigger and no photon loss at the PMT window and the converging mirror.

As a second stage of the electron identification, lead glass EM calorimeters (LG) are used. They are recycled from the TOPAZ experiment at TRISTAN/KEK. About 650 modules are required to cover the acceptance. The energy resolution for electrons is expected to be $8\%/\sqrt{E(\mathrm{GeV})}$ [37]. In the combination of the two stages of counters, a rejection factor of 1.0×10^{-4} for pions is expected.

5.4 Trigger and Background

The online and offline backgrounds were estimated on the basis of the E325 experience.

The main physical background sources in the e^+e^- channel are the Dalitz decays of π^0 ($\pi^0 \rightarrow e^+e^-\gamma$) and the γ-conversion, namely the e^+e^- pair creation on the detector materials by the γ-rays, which mainly arise from the $\pi^0 \rightarrow \gamma\gamma$ decays. In general, these pairs can be eliminated using their small opening angle. Thus, in offline analysis, the background mainly consists of such e^+e^- pairs as they have accidentally large opening angle because each e^+ and e^- in a pair belongs to different Dalitz or conversion pairs. They are called as a combinatorial background. The misidentified pions could be background sources in both online and offline analyses. It is essential to achieve a high rejection factor, about 10^{-3}–10^{-4}, for pions with the electron identification counters.

In the E325 experiment, the main trigger background was the accidental coincidence of the electron ID counters, namely no track from the target was associated with the counters. The singles rates of the counters are expected to be proportional to the interaction rate on targets and also proportional to the amount of beam halo. Both contributions to the trigger rate were almost of the same order of magnitude. In the E16 experiment, as mentioned in the above section, the beam intensity is increased by a factor of 10 while the target thickness is the same in comparison with the E325 experiment. Thus both the interaction rate and the beam halo are expected to be 10 times larger; therefore, the singles rates of counters are also expected to be 10 times larger than that of E325. In order to cope with the high particle rate, the granularity of the counters is increased in the design of the proposed spectrometer.

To trigger the e^+e^- decays of vector mesons, two electron candidates that have large opening angle, for example, more than 60°, are required. An electron candidate is defined as the coincidence of hits in the two stages of electron ID counters, HBD and LG, and also, a hit in the third plane of the GEM tracker. The expected position resolutions for the counters in the trigger logic are 12 cm × 12 cm for HBD and LG, and 10 cm × 10 cm for the tracker. The fast signal is extracted from the last GEM plane of the tracker, which has a size of 30 cm × 30 cm, and the electrode pattern is divided into 10 cm × 10 cm blocks. It is required in the trigger logic that these rough positions are geometrically consistent with a track from the target. With these segmentations, accidental coincidence can be reduced by a factor of 100 with respect to the E325 case; thus, the accidental trigger rate will be kept the same as in E325, namely about 1 kHz, in spite of the fact that the interaction rate is increased to 10 MHz.

The combinatorial e^+e^- pairs were studied by the fast detector simulation in which only the γ-conversion at the target (0.5% X_0) is included. The inputs are only the π^0 and $\pi^\pm$ from the 50-GeV p+Cu reaction produced by JAM. For the 10 MHz interaction rate on the target, a rejection power of 10^{-4} for pions, and an opening angle of 60°, the false trigger rate by combinatorial pairs was 200 Hz over the 1 kHz of expected accidental trigger rate, as discussed above. In offline analysis, the S/N ratio around the ϕ mass region is kept within a factor of 1.5 in comparison with that of E325. These online and offline conditions are acceptable. Furthermore, in the J-PARC phase I, the energy of primary protons is 30 GeV and the conditions are expected to be better than those in the case of 50 GeV, as used for the above estimation.

Summary

The experiment E16 is planned to measure the invariant mass spectra of the low-mass vector mesons (ρ, ω, ϕ) in the e^+e^- decay channel. The goal is to accumulate approximately 10^5 of ϕ mesons for each target (C, CH_2, Cu, and Pb) with the mass resolution of 10 MeV/c^2 or better. The nuclear size dependence, collision geometry dependence, and meson momentum dependence of the modification of mesons will be studied systematically and precisely. Comparison of various theoretical predictions can be performed with the high statistics. The data will provide severe limitations on theoretical predictions, lead the development of interpretation models and help the development of an approach to investigate the chiral symmetry in medium.

The proposal was given the physics approval ("stage 1 approval") by J-PARC PAC in March 2007. Detector development is undergoing as follows. Domestic GEM products of 100 mm × 100 mm size are available and a larger one is under development. CsI-evaporated GEM of 100 mm × 100 mm size is also available. The measurements of the quantum efficiency and the beam test using CF_4 were already performed. The prototypes of the HBD and GEM tracker, which consists of three GEM chambers, will be delivered by the 2008 summer.

6 Production of Mesic Nuclei at J-PARC

There are three letters of intent submitted to the J-PARC PAC to investigate the meson bound states in a nucleus, η-, ω-, and ϕ-mesic nuclei. As summarized in Table 7, all the three letters have a plan to use the (π^-, n) reaction

Table 7. The specifications of mesic nuclei experiments at J-PARC.

	η	ω	ϕ	
Beam (GeV/c)	π^- (0.7–1)	π^- (2–3)	π^- (∼2)	$\bar{p}$ (∼1.3)
Beam intensity(/spill)	1×10^6	1×10^7	1×10^7	2×10^6
Momentum transfer	0	0	400 MeV/c	200 MeV/c
Target	^{7}Li, ^{12}C	^{12}C	^{12}C	^{12}C
Yield(/month)	—	9000	140	240
Resolution (MeV/c^2)				
Missing mass	20–25	9	12	∼15
Invariant mass	—	40	—	—

to produce the mesic nuclei, detect the forward neutrons at zero degree, and measure the missing mass to detect the bound states. The ϕ-mesic nuclei experiment also has a plan to use the $\bar{p}$ beam, in which a better S/N ratio is expected.

6.1 η-Mesic Nuclei

This experiment is planned to investigate the N*(1535) (S$_{11}$(1535)) property in medium. The resonance is a candidate of the chiral partner of the nucleon [74]. The η-nucleus optical potential is dominated by the in-medium property of N*(1535) because of the strong coupling of ηN–N*(1535). For example, two theoretical models, the chiral doublet model and the chiral unitary model, predict different properties of N*(1535) in medium and also different bound states of η [75]. By using the missing-mass spectroscopy in the η-nucleus formation, the two models are distinguishable. Through the chiral doublet model, the information of the chiral restoration in nuclear matter could be obtained.

As a detector system, J-PARC E15 [76] spectrometer is available and the expected missing-mass resolution, which is 20 MeV/c^2 achieved by the time-of-flight length of 12 m for neutrons, is sufficient to distinguish the two models.

6.2 ω-Mesic Nuclei

In this experiment, the ω bound state is exclusively measured. Behind a sweeping magnet, forward-scattered neutron is identified by using the time-of-flight method with plastic scintillation counters and a missing mass is measured [77]. The goal of the missing-mass resolution is 9 MeV/c^2, which could be realized by employing a flight length of 20 m and a time resolution of 80 ps. Simultaneously, three γ-rays from the $\omega \to \pi^0\gamma$ decay are measured using CsI calorimeters with a resolution of $3\%/\sqrt{E}$ and an invariant mass resolution of ∼40 MeV/c^2 could be achieved. The expected yield is ∼9000 in 33 days operation using a ^{12}C target 1 g/cm^2 in thickness. Taking account of the prediction by Nagahiro [78], in which the binding energy of ω is 50 MeV, an incident momentum of 2 GeV/c is suitable.

This is the first attempt of simultaneous measurements of the in-medium mesons, through the initial channel, i.e., the missing-mass spectrum, and the final channel, i.e., the invariant mass spectrum of decay products. The results of this experiment will help to understand the relation between the mesic nuclei spectroscopy and the invariant mass spectroscopy.

6.3 ϕ-Mesic Nuclei

This project plans to use a π^- beam of momentum around 2 GeV/c or a $\bar{p}$ beam of momentum around 1.3 GeV/c for producing a ϕ meson in a nucleus [79]. In each case, p(π^-, n)ϕ or p($\bar{p}$, ϕ)ϕ, the produced ϕ mesons are not at rest but have a low momentum of ~0.4 and ~0.2 GeV/c, respectively, and they have a significant probability to be bound by the target nucleus. The incident energies in both the reactions are selected to maximize the production cross section of ϕ. The total cross section of each elementary reaction is 20 and 4 μb, respectively.

In the case of $p(\pi^-, n)\phi$ reaction, the forward neutron should be detected, as in the above experiments. In addition, the captured ϕ could interact with a proton in the nucleus and decay to $\Lambda + K^+$ with a branching ratio of 37%. Detecting $\Lambda(\rightarrow p\pi^-)$ and K^+ helps the event selection. In the case of $p(\bar{p}, \phi)\phi$ reaction, a ϕ meson produced in the forward direction can be detected through the K^+K^- decay channel and another ϕ could form a bound state in the nucleus. The $\Lambda + K^+$ tagging is also used in this reaction to select the event.

Unfortunately, the yield is so small that the dilepton measurements from the $\phi \rightarrow e^+e^-$ decay are not feasible.

Acknowledgments

First, the author wishes to express his gratitude to all the colleagues in E325 and E16 collaborations.

The experiment E325 was performed successfully with the help of all the staff members of KEK, particularly the PS beam channel group. The data analysis was performed mainly at the RIKEN Super Combined Cluster and RIKEN–CCJ. The author is most grateful for their support. The experiment is partly supported by JSPS, RIKEN SPDR Program, and a Grant-in-Aid by Japan MEXT.

The author also acknowledges the support to the E16 project by the RIKEN challenging fund and the Grant-in-Aid (No. 19340075).

References

1. Hatta, Y., Fukushima, K.: Phys. Rev. D **69**, 097502 (2004)
2. Mocsy, Á., Sannino, F., Tuominen, K.: Phys. Rev. Lett. **92**, 182302 (2004)
3. McLerran, L., Pisarski, R.D.: Nucl. Phys. A **796**, 83 (2007)
4. Nambu, Y., Jona-Lasinio, G.: Phys. Rev. **122**, 345 (1961)
5. Drukarev, E.G., Levin, E.M.: Prog. Part. Nucl. Phys. **27**, 77 (1991)
6. Cohen, T.D., Furnstahl, R.J., Griegel, D.K.: Phys. Rev. C **45**, 1881 (1992)
7. Kaiser, N., de Homont, P., Weise, W.: Phys. Rev. C **77**, 025204 (2008)
8. Lutz, M., Klimt, S., Weise, W.: Nucl. Phys. A **542**, 521 (1992)
9. Weise, W.: Nucl. Phys. A **553**, 59 (1993)
10. Huber, G.M., et al. (TAGX Collaboration): Phys. Rev. C **68**, 065202 (2003)
11. Adams, J., et al. (STAR Collaboration): Phys. Rev. Lett. **92**, 092301 (2004)
12. Grion, N., et al. (CHAOS Collaboration): Nucl. Phys. A **763**, 80 (2005)
13. Ishikawa, T., et al. (LEPS Collaboration): Phys. Lett. B **608**, 215 (2005)
14. Itahashi, K., et al.: Phys. Rev. C **62**, 025202 (2000)
15. Suzuki, K., et al.: Phys. Rev. Lett. **92**, 072302 (2004)
16. Brown, G., Rho, M.: Phys. Rev. Lett. **66**, 2720 (1991)
17. Hatsuda, T., Lee, S.H.: Phys. Rev. C **46**, R34 (1992)
18. Hatsuda, T., Lee, S.H., Shiomi, H.: Phys. Rev. C **52**, 3364 (1995)
19. Klingl, F., Kaiser, N., Weise, W.: Nucl. Phys. A **624**, 527 (1997)
20. Klingl, F., Waas, T., Weise, W.: Phys. Lett. B **431**, 254 (1998)
21. Lutz, M.F., Wolf, Gy., Friman, B.: Nucl. Phys. A **706**, 431 (2002)
22. Post, M., Leupold, S., Mosel, U.: Nucl. Phys. A **741**, 81 (2004)
23. Muehich, P., et al.: Nucl. Phys. A **780**, 187 (2006)
24. Lee, S.H.: Phys. Rev. C **57**, 927 (1998)
25. Lee, S.H.: Phys. Rev. C **58**, 3771 (1998)
26. Kondratyuk, L.A., et al.: Phys. Rev. C **58**, 1078 (1998)
27. Post, M., Mosel, U.: Nucl. Phys. A **699**, 169 (2002)
28. Muroya, S., et al.: Prog. Theor. Phys. **110**, 615 (2003)
29. Muroya, S., Nakamura, A., Nonaka, C.: Phys. Lett. B **551**, 305 (2003)
30. Ozawa, K., et al. (KEK–PS E325 Collaboration): Phys. Rev. Lett. **86**, 5019 (2001)
31. Naruki, M., et al. (KEK–PS E325 Collaboration): Phys. Rev. Lett. **96**, 092301 (2006)
32. Muto, R., et al. (KEK–PS E325 Collaboration): Phys. Rev. Lett. **98**, 042501 (2007)
33. Sakuma, F., et al. (KEK–PS E325 Collaboration): Phys. Rev. Lett. **98**, 152302 (2007)
34. Tabaru, T., et al. (KEK–PS E325 Collaboration): Phys. Rev. C **74**, 025201 (2006)
35. Sekimoto, M., et al. (KEK–PS E325 Collaboration): Nucl. Instrum. Meth. A **516**, 390 (2004)
36. Tanaka, K.H., et al.: Proceedings of the First Asian Particle Accelerator Conference (APAC98), pp. 576–578, KEK, Tsukuba, 23–27 March 1998
37. Kawabata, S., et al.: Nucl. Instrum. Meth. A **270**, 11 (1988)
38. Ishino, M., et al.: Nucl. Instrum. Meth. A **457**, 581 (2001)
39. Agostinelli, S., et al.: Nucl. Instrum. Meth. A **506**, 250 (2003)
40. Nara, Y., et al.: Phys. Rev. C **61**, 024901 (2000)

41. Yokkaichi, S., et al. (KEK–PS E325 Collaboration): Int. J. Mod. Phys. A **22**, 397 (2007)
42. Spiridonov, A.: hep-ex/0510076
43. Gounaris, G.J., Sakurai, J.J.: Phys. Lev. Lett. **21**, 244 (1968)
44. Brobel, V., et al.: Phys. Lett. B **48**, 73 (1974)
45. Adamova, D., et al. (CERES Collaboration): Phys. Rev. Lett. **96**, 152301 (2006)
46. Oset, E., Ramos, A.: Nucl. Phys. A **679**, 616 (2001)
47. Chung, W.S., Ko, C.M., Li, G.Q.: Nucl. Phys. A **641**, 357 (1998)
48. Glauber, R.J., Matthiae, G.: Nucl. Phys. B **21**, 135 (1970)
49. Agakichiev, G., et al. (CERES Collaboration): Phys. Rev. Lett. **75**, 1272 (1995)
50. Agakichiev, G., et al. (CERES Collaboration): Euro. Phys. J. C **41**, 475 (2005)
51. Brown, G., Rho, M.: Phys. Rep. **363**, 85 (2002)
52. Rapp, R., Wambach, J.: Adv. Nucl. Phys. **25**, 1 (2000)
53. Arnaldi, R., et al. (NA60 Collaboration): Phys. Rev. Lett. **96**, 162302 (2006)
54. Damjanovic, S. (NA60 Collaboration): Nucl. Phys. A **783**, 327 (2007)
55. van Hees, H., Rapp, R.: Phys. Rev. Lett. **97**, 102301 (2006)
56. Adamova, D., et al. (CERES Collaboration): nucl-ex/0611022
57. Brown, G., Rho, M.: arXiv:nucl-th/0509001, nucl-th/0509002
58. Brown, G., et al.: Phys. Rep. **439**, 161 (2007)
59. Afanasiev, S., et al. (PHENIX Collaboration): arXiv:0706.3034[nucl-ex]
60. Agakichiev, G., et al. (HADES Collaboration): Phys. Rev. Lett. **98**, 052302 (2007)
61. Muntz, C.: arXiv:0710.3274[nucl-ex]
62. Porter, R.J., et al. (DLS Collaboration): Phys. Rev. Lett. **79**, 1229 (1997)
63. Agakichiev, G., et al. (HADES Collaboration): arXiv:0711.4281[nucl-ex]
64. Trnka, D., et al. (CBELSA/TAPS Collaboration): Phys. Rev. Lett. **94**, 192303 (2005)
65. Metag, V.: Prog. Part. Nucl. Phys. **61**, 245 (2008)
66. Kotulla, M., et al. (CBELSA/TAPS Collaboration): Phys. Rev. Lett. **100**, 192302 (2008)
67. Nasseripour, R., et al. (CLAS Collaboration): Phys. Rev. Lett. **99**, 262302, (2007)
68. Wood, M.H., et al. (CLAS Collaboration): arXiv:0803.0492v1[nucl-ex]
69. Yokkaichi, S., et al.: J-PARC proposal no. 16 (2006)
70. Drijard, D., et al.: Z. Phys. C **9**, 293 (1981)
71. Adler, S.S., et al. (PHENIX Collaboration): Phys. Rev. Lett. **92**, 051802 (2004)
72. Chemakin, I., et al.: Phys. Rev. C **60**, 024902 (1999)
73. Tserruya, I., et al.: Nucl. Instrum. Meth. A **563**, 333 (2006)
74. Itahashi, K., et al.: J-PARC LoI (2007)
75. Nagahiro, H., Jido, D., Hirenzaki, S.: Phys. Rev. C **68**, 035205 (2003)
76. Iwasaki, M., et al.: J-PARC proposal no. 15 (2006)
77. Ozawa, K., et al.: J-PARC LoI (2007)
78. Nagahiro, H., Jido, D., Hirenzaki, S.: Nucl. Phys. A **761**, 92 (2005)
79. Iwasaki, M., Ohnishi, H., et al.: J-PARC LoI (2007)

Experimental Investigations of Mesonic Bound States in Nuclei

Masahiko Iwasaki

Advanced Meson Science Laboratory, RIKEN Nishina Center for Accelerator-Based Science, RIKEN, Saitama 351-0198, Japan
advanced_meson@riken.jp

1 Introduction

Properties of particles in matter/media attract interest in many ways. It is believed that these properties are influenced by the surroundings through the interaction between the two. The present scenario of how the mass of the basic particles is formed is based on the idea that the vacuum is also a kind of medium with a condensate of Higgs bosons. This condensation is realized during the cooling down of the universe at around $T \sim 100\,\text{GeV}$ after the Big Bang. Therefore, to detect the Higgs boson is one of the major goals of the LHC project at CERN.

To explain the origin of the mass of the hadrons, this Higgs mechanism is not sufficient to account for the so-called constituent quark mass $m_q \sim 300\,\text{MeV}/\text{c}^2$ and the baryon formation. To explain that, one needs another phase transition of the vacuum called quark–antiquark pair condensation at around $T \sim 200\,\text{MeV}$. Thus, the vacuum expectation value of $\langle \overline{q}q \rangle$ is non-zero due to the spontaneous chiral symmetry breaking of the vacuum, and this $\langle \overline{q}q \rangle$-condensation is the major source of masses of low-lying hadrons such as protons, neutrons, pions, etc.

The present scenario of the mass formation to the particles depends on how particles interact with the surrounding space where Higgs and quark–antiquark pairs are condensed. Thus, in the hadron sector, in-medium particle properties are fundamentally related to chiral symmetry breaking and its restoration in the nuclear medium. The $\langle \overline{q}q \rangle$ expectation value (chiral order parameter) is a function of temperature and chemical potential (density). Therefore, there is currently great experimental interest to study the effect of chiral symmetry breaking and its partial restoration in the nuclear media.

In a nucleus, it is known that mesons, especially pions, play an important role to bind nucleons. Existence of the mesons was originally predicted by Yukawa as boson which forms the nuclear field as a solution of the Klein–Gordon equation of the total energy zero ($\exp(-m_\pi r)/r$). In the standard model, the basic gauge fields of the strong interaction are the gluons, but

Iwasaki, M.: *Experimental Investigations of Mesonic Bound States in Nuclei.* Lect. Notes Phys. **781**, 195–229 (2009)
DOI 10.1007/978-3-642-00961-7_8

gluons cannot propagate directly between two nucleons in a nucleus due to the confinement nature. The meson field approach is still valid for the nuclear interaction at intermediate and long distances. In this manner, nuclei consist of nucleons and virtual meson fields.

Triggered by the series of recent experimental studies of mesonic atoms, the importance of the experimental search for nuclear bound states of mesons was pointed out. Extensive experimental studies were started only very recently. It is still controversial whether there exist deeply bound mesonic states (in other words, whether one can detect them experimentally) inside the nuclei. This experimental search is one of the important subjects to be explored on 50-GeV proton synchrotron of the J-PARC facility, which will start its operation in 2009.

2 Observation of Deeply Bound Pionic Atoms

One milestone of the study using mesons is the observation of deeply bound pionic atoms at GSI, as shown in Fig. 1 [1, 2]. The pionic atoms had been studied by using x-ray transition of a negatively charged pion captured by a

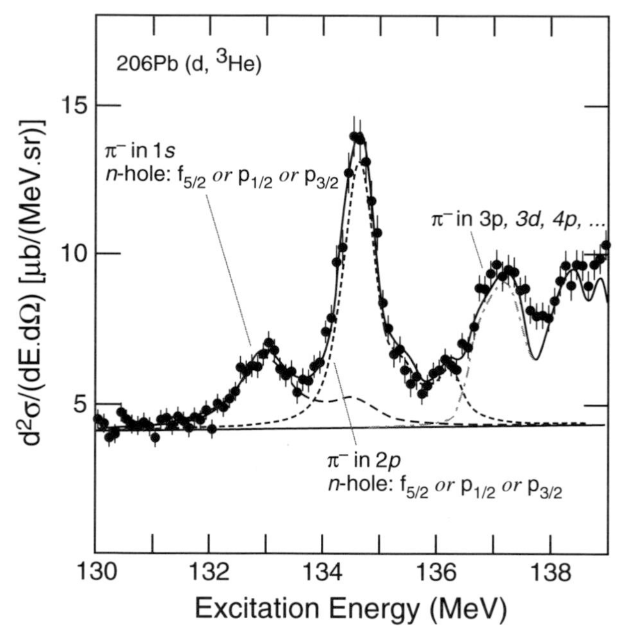

Fig. 1. Double differential cross section measured in the ^{206}Pb(d, ^{3}He) reaction at $T_d = 604.3\,\text{MeV}$ as a function of the excitation energy above the ^{205}Pb ground state covering the bound region of π^- [2]. The curves are the fitted functions to the experimental spectrum. In the fitting, relative yields of the neutron–hole state contributions were fixed as calculated in Ref. [3].

target atom. However, the deeply bound state of the pionic atom in heavy nuclei, such as Pb, cannot be formed by this cascade process. This is because the timescale of the absorption width of the pion due to the strong interaction becomes much shorter than that of the radiative transition in the low-lying level of the atomic states. Until the observation, these deeply bound pionic states are accepted to be experimentally inaccessible.

To overcome this situation, the pion was produced in the atomic orbits directly, via strong interaction, using the "neutron" pickup reaction, ${}^{A}_{Z}X(d,{}^{3}He)\ \{\pi^- \oplus {}^{A-1}_{Z}X\}$ reaction, making a neutron–hole in the daughter nucleus and leaving a π^- in an atomic orbit. If this reaction is realized, the missing mass spectrum should have a structure at the excitation energy of about m_π above the ground state of the daughter nucleus, as shown in the figure. The overall structure agrees remarkably well with the theoretical calculation of the deeply bound pionic state formation [3]. In this reaction, the recoilless condition for the produced π^- is realized at the incident deuteron kinetic energy of ~600 MeV, namely the momentum transfer to the pion is zero, so the atomic state formation can be realized efficiently, in spite of the quite small phase space available for the nuclear reaction.

As shown in Fig. 2, the Bohr radius of the pionic atom in heavy nuclei lies within the nucleus, but the strong interaction of the pion is repulsive in s-wave, so the major part of the wave function is pushed away from the

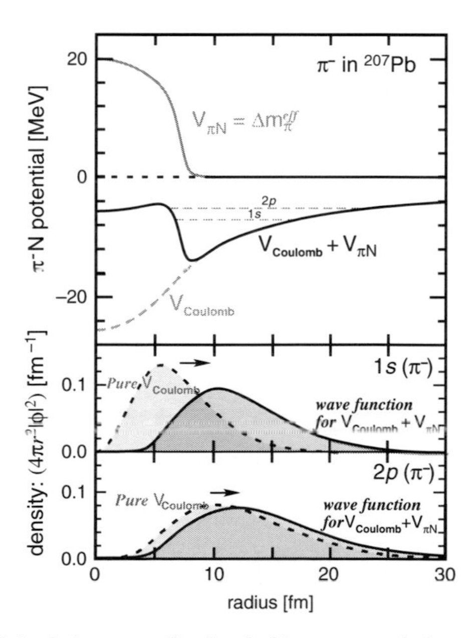

Fig. 2. The potential of the pure Coulomb V_{Coulomb} and the πN optical potential $V_{\pi N}$ is plotted in the case of ^{208}Pb target. Radial distributions of the $1s$ and $2p$ π^- densities for the pure Coulomb case (*dashed*) and that for the total optical potential (*solid*) are shown in the bottom [1].

nucleus, and the pions are bound to the nuclear surface by the Coulomb force. On the other hand, the observed deeply bound pionic states are not in pure atomic state in a common sense, whose size and binding energy are comparable to the nuclear scale, so the states lie in between nuclear and atomic states.

Through the partial overlapping of the wave functions between the pion and the nucleus, one can study the in-medium modification of the s-wave pion–nucleon interaction at threshold through the energy shift and width of the atomic state. This interaction involves the pion decay constant f_π (=92.4 MeV in vacuum) which is, in turn, connected to the quark condensate through the Gell-Mann–Oakes–Renner (GOR) relation [4]. The series of deeply bound pionic atom studies in Sn isotopes can be interpreted as a 35% reduction of f_π^2 in nuclei as reported in Ref. [5],[1] and equivalently, as a signature for the tendency toward chiral symmetry restoration in nuclei [6–9].

The existence of relatively narrow pionic states would not be extremely surprising, because it is in an atomic state (bound by the Coulomb force), and the πN strong interaction is repulsive in s-wave. However, the existence of these deeply bound pionic states themselves was not widely accepted because of the inaccessibility by the cascade process before the clear experiment producing pionic atom directly via strong interaction.

How about other mesons? If the interaction between mesons and nuclei is repulsive, as it is the case of pion, it is natural that the on-shell meson cannot exist at rest in nuclei. If it is attractive, then one may detect meson bound state inside the nuclei, when the absorption width is small enough compared to the attraction.

3 Study of the Kaonic Atom

The second lightest meson is the kaon. Let us first overview the recent studies of the $\overline{K}N$ interaction by the x-ray measurement of the kaonic atoms. As it is discussed in the pionic atom study, the $\overline{K}N$ interaction can be studied by the level shifts and widths of the kaonic atom x-rays. Present kaonic atomic data are still insufficient for a detailed understanding of the $\overline{K}N$ interaction, so it is a good field to be explored in the forthcoming facility such as J-PARC.

In general, if the interaction is attractive, the effect of the strong interaction to the atomic level is much more complicated compared to the repulsive case. To demonstrate the situation, simple calculation result of the level shift of the kaonic hydrogen atom is plotted in Fig. 3, as a function of the real part of the potential between kaon and proton. The s-wave atomic levels are calculated by solving the Schrödinger equation in a Coulomb field and a schematic Yukawa potential as a local part, $(V + iW)\exp(-r/\lambda)/r$, where λ is the range parameter to be 1 fm for the simplicity, and V and W are the constants representing real and imaginal parts of the local part.

[1] This result leads to a proposed systematic study at RIBF (RIKEN Nishina Center).

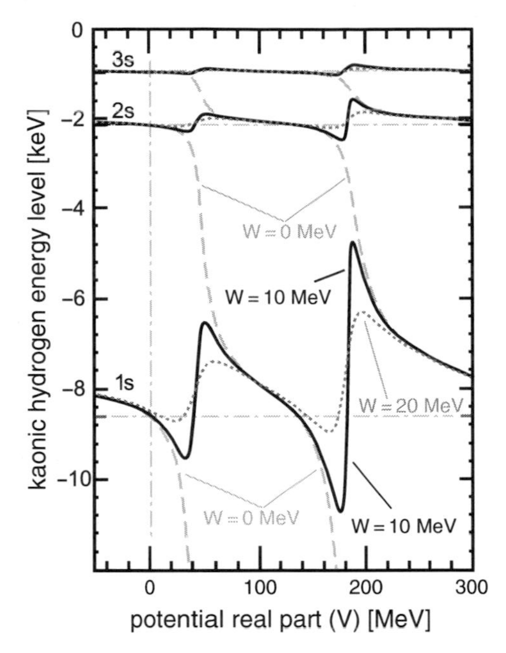

Fig. 3. A diagram of the level shift of the s-wave kaonic-hydrogen atom as a function of real part of the local potential. The level shift is calculated by the simple Schrödinger equation in a Coulomb field together with the Yukawa potential as a local part, $(V + iW)\exp(-r/\lambda)/r$. Three curves of the imaginary part, $W = 0$ (*dashed*), 10 (*solid*), and 20 (*dotted line*), are plotted. The range parameter λ is fixed to be 1 fm.

As shown in the figure, the atomic level shifts have beat pattern when the attractive force is strong enough compared to the imaginary part. This can be understood in terms of an interference between nuclear pole and atomic one. At the node, a bound state of the local potential is formed as a nuclear state, and the atomic level shift changes its sign and moves upward for more attractive interaction. This upward shift is not the result of the repulsion. The atomic ground state changes its nature to the nuclear (local) one, and the wave function is confined in the local potential. It can be understood from the curve of zero imaginary part, $W = 0$. To substitute the vacancy of the phase volume of the atomic ground level, upper level ($2s$ in this case) reduced its energy at around the node.

In the figure, only the s-wave atomic states are plotted, but this is a generic feature to all the partial waves. For example, similar beat pattern is formed in p-wave atomic states when a p-wave nuclear state is formed (cf. Fig. 6 (left)). Thus, the strong upward energy shift of the atomic level can be expected when the attractive interaction is strong enough to form a nuclear state below the atomic state. This means that the atomic level shift is sensitive to the K^-N strong interaction. However, to deduce the interaction in a definitive way including isospin dependence of the interaction, a compilation of the data of many atomic data is needed.

It is also shown that the imaginary part is ineffective to the level shift, when the interaction is repulsive. This is because the overlap between the atomic wave function and the local potential is reduced as in the case of the pionic atom.

3.1 Kaonic Hydrogen Puzzle

Kaonic hydrogen x-ray data have been quite ambiguous for long time. It is known that the K^-p s-wave interaction at the threshold energy can be determined in a fairly model-independent way [10, 11] from the kaonic hydrogen $2p$ to $1s$ atomic data. However, previous atomic data [12–14] sharply contradicted the scattering data even with respect to the sign of the interaction, until first reliable data [15] were obtained by the KpX(KEK PS-E228) group as shown in Fig. 4.

As shown in Fig. 4 (left), most of the theoretical calculations, based on the low-energy $\overline{K}N$ scattering data, give upward shift for the $1s$ level, but the old three experimental values locate in the downward side. There is another difficulty caused by the existence of a sub-threshold resonance ~20 MeV below K^- and p (energy below the mass of the two $m_{K^-} + m_p$), namely $\Lambda(1405)$. Analogous to the energy level shift described above, if there exist a pole below

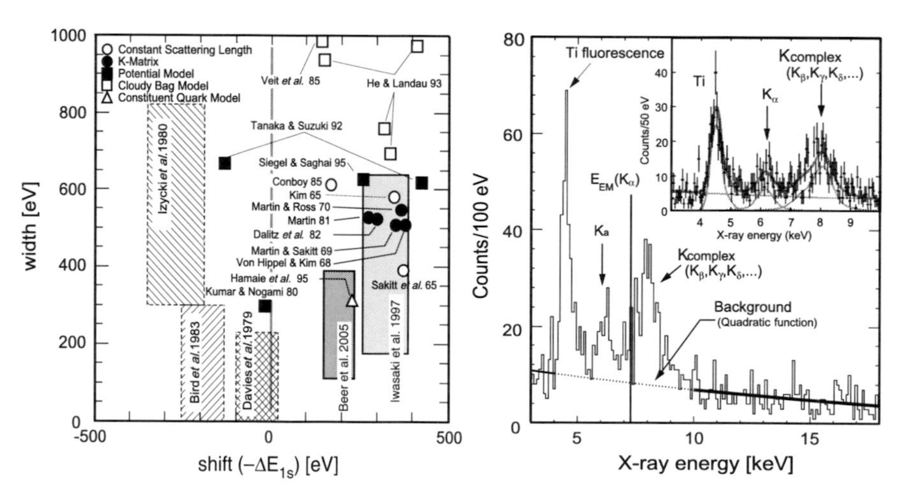

Fig. 4. *Left panel* shows a compilation of the shifts, $-\Delta E_{1s} = -(E^{obs}_{2p\to 1s} - E^{EM}_{2p\to 1s})$, and widths of the kaonic hydrogen atomic $1s$ level. The boxes correspond to the experimental values, where the box size represents the one standard deviation of the experimental error, and the *marks* represent the theoretical one. *Right panel* shows the x-ray spectrum observed at KEK [15]. If the energy level shifts upward, K_α line (energy difference between $2p$ and $1s$) should be smaller than the pure electro-magnetic value $E_{EM}(K_\alpha)$.

the atomic levels, then the atomic level should move upward. These difficulties were known as the "kaonic hydrogen puzzle."

This puzzle was resolved by the KEK experiment, *KEK PS-E228 (KpX)*, in 1997 [15] as shown in Fig. 4 (right). It is worthwhile to describe the KEK experiment briefly to browse experiments that can be realized in the future. Experimentally, the observation of the kaonic hydrogen x-rays is quite difficult, because (i) huge background due to the γ-ray from K^-p atom decays via strong interaction, (ii) extremely low x-ray yield due to the Stark effect, and (iii) difficulty of the absolute energy calibration of the x-ray detector.

For the background reduction, an event tagging method was applied to select the γ-ray free events. Table 1 shows the number of γ-rays produced by the K^-p atom decay by the strong interaction. As shown in the table, two charged-pion events cannot emit γ-rays in the final state, and the branching ratio of these events is quite large $\sim$60%. To overcome other experimental difficulties, (i) gaseous hydrogen target was applied and (ii) titanium foil was installed in the target cell to utilize the fluorescence line as in situ energy calibration source, for the first time.

More recently, the K^-p atom study was extended by the *DEAR* group at INFN using DAΦNE machine [16]. At the ϕ-factory, kaons are produced at very low momentum so the kaon can be stopped in the gaseous target efficiently, and resulting higher statistics for the observed kaonic hydrogen x-rays. However, signal-to-noise ratio (S/N) of the spectrum is not as good as in the previous KEK experiment. The results of the shift and width are also plotted in Fig. 4 (left). These two recent experimental results, *KpX* and *DEAR*, are consistent within the error, so the upward energy level shift is confirmed.

The limited S/N in *DEAR* experiment comes from the insufficient time resolution of their x-ray detector, CCD. To study the isospin dependence of the $\overline{K}N$ interaction using both hydrogen and deuterium targets, an improved experiment is presently running by the *SIDDARTA* group at INFN, using newly developed silicon drift x-ray detector (SDD).

The tagging method applied in the *KpX* experiment can be extended to the further study for the kaonic atom study of light nuclei. The validity of

Table 1. Final states of the K^-p reaction.

Primary reaction	Final state	Branch	$\mathcal{N}^o$ of $\pi^\pm$	$\mathcal{N}^o$ of γ
$\Sigma^+\pi^-$	$p\pi^- + 2\gamma$	0.10	1	2
$\Sigma^+\pi^-$	$n\pi^+\pi^-$	0.10	2	0
$\Sigma^-\pi^+$	$n\pi^-\pi^+$	0.46	2	0
$\Sigma^0\pi^0$	$p\pi^- + 3\gamma$	0.18	0	3
$\Sigma^0\pi^0$	$n + 5\gamma$	0.10	0	5
$\Lambda\pi^0$	$p\pi^- + 2\gamma$	0.04	1	2
$\Lambda\pi^0$	$n + 4\gamma$	0.02	0	4

this method to suppress the kaon-originated background was clearly demonstrated in their spectrum, so this method should be powerful for extending atomic study to the light nuclei when the x-ray emission yield per kaonic atom is expected to be quite weak. To be free from γ-rays in the final state in light nuclei, however, one need to be sensitive to the charge of the pions. For example, $K^-n \rightarrow \Sigma^0\pi^-$ primary reaction can emit γ-ray as $\Sigma^0 \rightarrow \Lambda\gamma$ and then $\Lambda \rightarrow p\pi^-$, in which two π^- are produced together with a γ in the final state. However, $\pi^+\pi^-$ pair tagging is still valid even in the light kaonic atom experiments. Therefore, further light kaonic atom study will be a good candidate for the further study of the $\overline{K}N$ interaction at J-PARC, such as $2p$–$1s$ transition of kaonic deuteron and helium.

3.2 Kaonic Helium Puzzle

In the x-ray study of the kaonic atom, there was another problem called "kaonic helium puzzle." Figure 5 shows the compilation of the kaonic atom data [17–19] (symbol) and the theoretical calculation (curve) based on the optical model [20]. As shown in the figure, there are two data points quite distant from the theoretical curves, one is the helium and the other is the oxygen.

Especially, calculated shift of the kaonic helium $2p$ level should be almost zero (well below 1 eV) while three data points are consistent and having an average value as large as ~40 eV (upward). In the x-ray spectra of these experiments, kaonic helium x-rays were clearly observed. Therefore, a possibility of existence of the deeply bound kaonic states in nuclei is discussed by Wycech as early as 1986 [21], based on the kaonic ^{4}He atom data. However, the downward shift of the atomic $1s$ level, the kaonic hydrogen puzzle was not resolved yet at that time, does not seem to be consistent with the existence of deeply

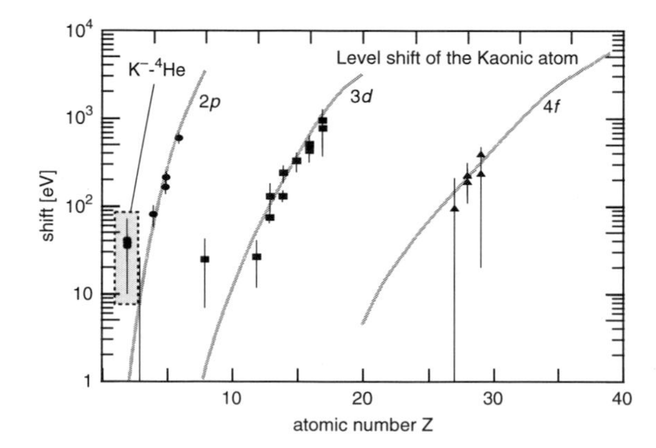

Fig. 5. A compilation of the kaonic atom shifts of the atomic number (Z) from 2 to 40 [20]. Data points of kaonic helium and oxygen are strongly deviating from the optical model theoretical calculation.

bound kaonic states in nuclei. Thus, extensive study of the deeply bound kaonic states was not triggered at that time unfortunately.

Recent extensive experimental study to search for the deeply bound state after the "kaonic hydrogen puzzle" had been settled, a new light was shed on this "helium puzzle" problem again.

According to Ref. [22], the absorption strength of the kaonic helium $2p$ level can be small due to the repulsive nature of the pion in the decay channel (e.g., $K^-p \to \Sigma^+\pi^-$), and as a consequence, the atomic level could have large shift, as shown in Fig. 6 (left). The situation is analogous to the amplitude of the beat pattern shown in Fig. 3, namely the amplitude becomes larger for smaller imaginary potential. Although the predicted shift is within the theoretically favored region of the real part of the potential (thin-hatched region), the upmost shift can be only up to ∼10 eV, which would still not be sufficient to explain the previous experimental result of 40 eV.

This prediction triggered a new precision x-ray experiment of the kaonic helium atom at KEK, *KEK PS-E570*. This experiment was proposed to study the previous experimental result again with much improved accuracy, because if it is confirmed then the existence of the deeply bound kaonic state is indicated very strongly. As it is described, strong upward atomic $2p$ shifts suggesting an existence of nuclear p-wave pole below the atomic one.

They observed the kaonic helium x-ray transition very clearly with good resolution, which is achieved by using SDD as x-ray detectors. An x-ray spectrum is shown in Fig. 6 (right). In the experiment, the fluorescence x-ray peaks were recorded simultaneously using the self-trigger mode of the SDD

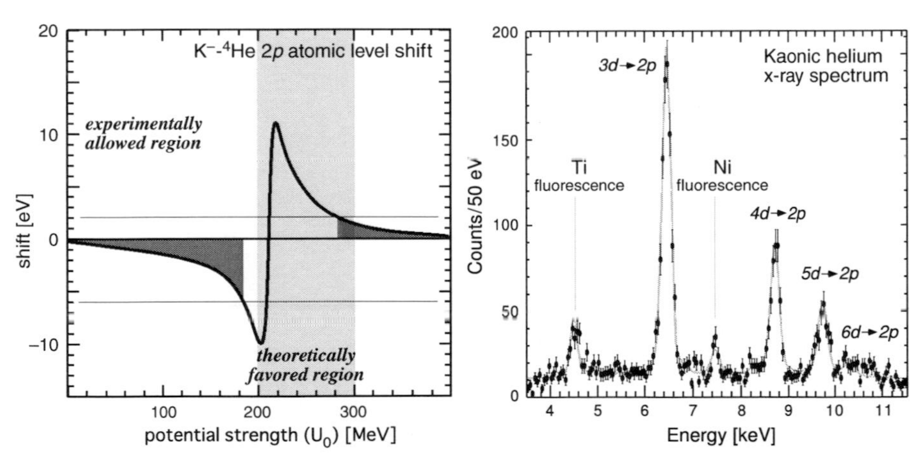

Fig. 6. *Left panel* shows a calculated level shift of the kaonic helium atom by Akaishi [22]. Theoretically favored region of the potential strength is indicated by the *thin-hatched* region. *Right panel* shows an x-ray spectrum obtained by the most recent KEK experiment KEK PS-E570 [23] (*right*), whose observed shift is $-\Delta E_{2p} = -(E^{obs}_{3d\to 2p} - E^{EM}_{3d\to 2p}) = -\ 2 \pm 2\ (stat.) \pm 2\ (syst.)$ eV. The allowed region of the potential strength is indicated by a *thick-hatched area* in the *left panel.*

signal. They sandwiched kaonic helium $3d$ to $2p$ transition line between fluorescence x-ray peaks of Ti and Ni excited by charged pion contaminated in the kaon beam. These two fluorescence lines were used as in situ absolute energy-calibration sources to reduce the systematic error. As shown in Fig. 6 (right), a part of the fluorescence peak can be seen also in their kaon trigger events in the final spectrum.

In contrast to the experimental motivation, their shift result, $-\Delta E_{2p} = -(E^{obs}_{3d\to 2p} - E^{EM}_{3d\to 2p}) = -\ 2 \pm 2\ (stat.) \pm 2\ (syst.)$ eV, is consistent to be zero within the experimental accuracy, and the shift locates slightly below zero to the downward side. The allowed region of the potential strength by the experimental result is indicated by a thick-hatched area in the left panel of Fig. 6 (left).

The S/N ratio of the x-ray spectrum was improved in the experiment using modern technique, but the discrepancy of the shift between the KEK result and previous one is not easy to understand. One of the possible reasons would be the existence of the in situ calibration of the semiconductor device, SDD. Actually, it is known that the peak position shifts depending on the charge deposition rate on the device.

From Fig. 6 (left), the narrow region around the node seems to be allowed. However, it is known that the width of the $2p$ state should be very broad, which is inconsistent with the observed narrow peak in the x-ray spectrum. Therefore, most of the theoretically favored regions suggested by Akaishi are excluded from the experiment. As shown in the figure, though, the K^--^{4}He atom data is sensitive only around the node. To study the $\overline{K}N$ interaction furthermore, the present atomic data are not satisfactory yet, and one needs more atomic data.

One of the most important data to be studied would be the K^--^{3}He atom. It is pointed out in Ref. [22] that the node position of K^--^{3}He atom $2p$ level moves to the deeper side by about 70 MeV, so there is a chance to deduce the potential quite accurately by the observation of K^--^{3}He atom $2p$ level shift and width. This experiment will be performed as *J-PARC E17* [24] at the beginning of the slow extraction mode operation of J-PARC 50-GeV proton synchrotron (PS).

4 Search for Kaon Bound States

A recent hot topic concerns the possible existence of deeply bound kaonic nuclear states. After the so-called "kaonic hydrogen puzzle" was resolved, the attractive nature of the $\overline{K}N$ interaction is undeniable. Assuming the $\Lambda(1405)$ to be a K^-p bound state, Akaishi and Yamazaki predicted meta-stable kaon bound state formation in light nuclei [25], using their coupled-channel calculation (see Fig. 7).

While this prediction triggered many extensive studies, both experimental and theoretical, it is at the same time under lively dispute. Several more

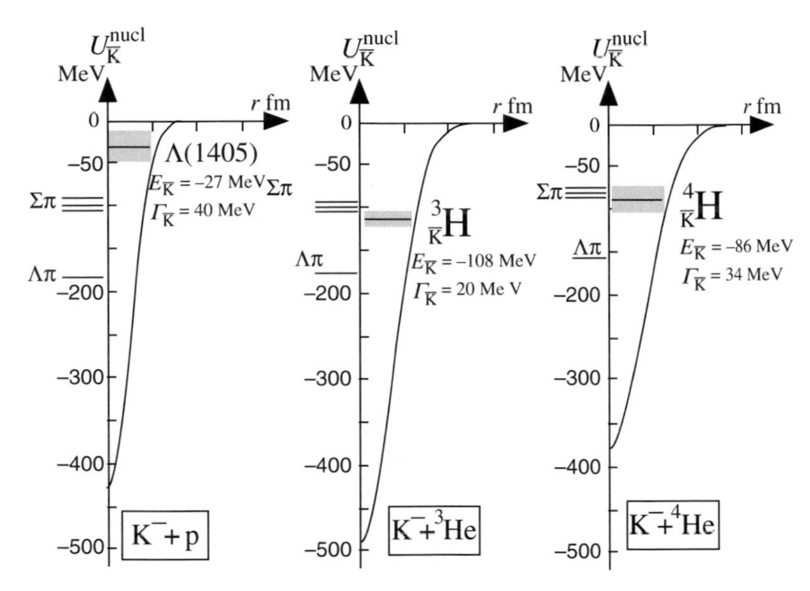

Fig. 7. A theoretically predicted deeply bound kaonic state in light nuclei by Akaishi and Yamazaki [25] by asserting that the $\Lambda(1405)$ is an isospin $I = 0$ bound state of $\overline{K}N$.

detailed coupled-channel calculations of $\overline{K}N$ and $\overline{K}NN$ systems [26–30] find significantly weaker attraction and binding than suggested by the simple local potential approach of Ref. [25]. In the case of atomic study, the shift and width is also sensitive to the nuclear surface structure especially for higher transition lines. Thus, one may study more direct information of the $\overline{K}N$ interaction and the property change of K-mesons in nuclei, if it exists. Actually, potential strength of the theoretical study on K^- atom x-rays has two favored interaction regions, one is as deep as ~200 MeV and another is shallow at around ~80 MeV, depending on the different treatments of the nuclear surface [31].

At present, there are several reports of possible candidates interpreted as deeply bound kaonic states, but very conclusive data are still missing. In a sense, existence of narrow and deep kaonic states seems to be negative, both experimental and theoretical ways, and at least the width of the state would be much larger than the original prediction. This could be understood because the multi-nucleon absorption process is quite large. However, the importance of the experimental study is unchanged. Many experimental studies are planned to be sensitive even for the wider states. The experimental data will provide rich information of the $\overline{K}N$ interaction and will help to understand the nature of $\Lambda(1405)$ in the sub-threshold region.

4.1 KEK Study via Stopped K^- Reaction

Akaishi and Yamazaki obtained a set of two-body coupled-channel potentials ($\overline{K}N$, $\overline{K}N - \Sigma\pi$, and $\overline{K}N - \Lambda\pi$) [25] to reproduce the interaction strengths and their range, using the kaonic hydrogen data, the KN scattering data, as well as the energy and width of $\Lambda(1405)$. They calculated the in-medium $\overline{K}N$ interactions and obtained $\overline{K}$-nucleus potentials. Using this potential, they predicted deeply bound $\overline{K}$ states, whose binding energy can be as large as 100 MeV for several light nuclei. If the state is located below the $\Sigma\pi$ threshold, then the $I = 0$ single-nucleon absorption channel is energetically closed, so the states are expected to have narrower decay widths compared with the width of the $\Lambda(1405)$ (40 MeV). In their calculation, the most stable one is predicted to have I = 0 and be formed by K^- with ^{3}He core as shown in Fig. 7.

It should be noted that this potential is not applicable to the analysis of the atomic level shift, because the size of the nuclei is shrunk due to the existence of the kaonic nuclear state and forms a higher nuclear matter density, which is the one of the reasons why the local potential of the kaon is much deeper than the other theoretical works.

Right after the prediction, a dedicative experiment was proposed to form the state via the neutron Auger effect using kaon at-rest reaction in a liquid helium target, aiming at *neutron* spectroscopy of the ^{4}He(*stopped* K^-, n) reaction [32]. A search experiment for the deeply bound kaonic state was performed by *KEK PS-E471* experimental group at KEK. Their first publication was the spectrum of the ^{4}He(*stopped* K^-, p) reaction, and discussed about the distinct peak observed in the spectrum, and the state corresponding to the peak was denoted as "tribaryon" $S^0(3115)$ [33]. If one assumes that the peak observed in proton spectrum in *KEK PS-E471* corresponds to the kaonic nuclear bound state, then it means that the non-zero isospin state is formed with extremely large (~200 MeV) binding energy far beyond the theoretical prediction, while the predicted state has isospin zero with binding energy about 100 MeV. These are very outstanding features. If the assumption is true, the formed system could be denser than the original theoretical prediction. Therefore, experimental confirmation is strongly required. To qualify this requirement, *KEK PS-E549* experiment was performed by substantially upgrading the experimental setup based on *KEK PS-E471* setup, so as to obtain more detailed information with improved accuracy and much higher statistics both for proton and neutron spectra shown in Fig. 8.

Resolution and the statistics were improved substantially compared to the spectra obtained by the *KEK PS-E471* experiment. The bump at around 500 MeV/c region is caused by the two-nucleon absorption, $K^-NN \rightarrow YN$, which is the major background in the stopped kaon experiment. There is no clear structure shown in both spectra, including the location of $S^0(3115)$, where *KEK PS-E471* once claimed the existence of the narrow peak. Consequently, it must be concluded that the previously reported $S^0(3115)$ structure and its interpretation as a deeply bound, narrow antikaon–nuclear cluster are not confirmed.

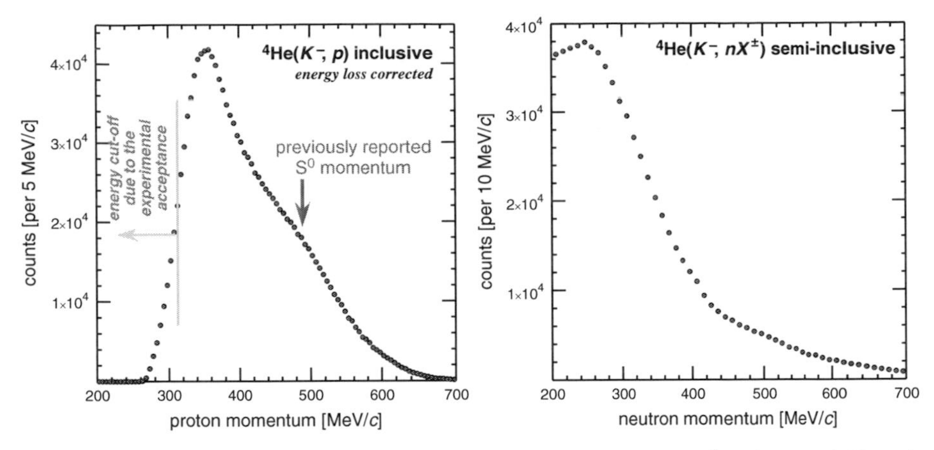

Fig. 8. Momentum spectra of the proton inclusive spectrum of ^{4}He(*stopped* K^-, p) reaction (*left*) and neutron semi-inclusive spectrum of ^{4}He(*stopped* K^-, $nX^\pm$) reaction (*right*). The momentum of the proton and neutron was measured by means of time of flight (TOF). The peak position corresponding to the previously reported $S^0(3115)$ is marked with an *arrow* in the proton spectrum. The neutron inclusive spectrum cannot be realized because at least one charged-particle emission from the kaon absorption reaction is needed to identify the reaction vertex, which is needed for the neutron TOF analysis.

Figure 9 (left) shows the missing mass spectrum of protons after the acceptance correction. In this figure, the vertical axis is normalized by the number of stopped kaons. The evaluated upper limit for the peak formation is given in Fig. 9 (right) by assuming a smooth proton spectral function (third polynomial function) together with a Gaussian centered at 3115 MeV/c^2. The upper limit of the narrow ($\Gamma < 20$ MeV/c^2) peak formation at this mass region is obtained to be well below 10^{-3} at the 95% confidence level [34]. This upper

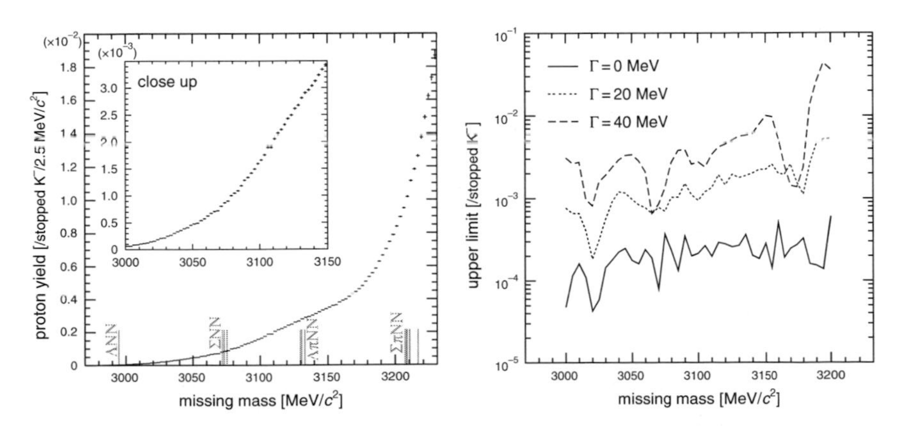

Fig. 9. Acceptance corrected proton missing mass spectrum of ^{4}He(*stopped* K^-, p) reaction (*left*), and upper limit of the formation branching ratio for the kaonic bound state (*right*) [34].

limit is quite severe compared to the at-rest kaon-induced hypernuclear formation probability in the order of percentage. Therefore, the existence of such a narrow state is not very likely according to Ref. [25]. Detailed upper limit as a function of the missing mass is given in Ref. [34]. The peak in the proton spectrum of the *KEK PS-E471* experiment is concluded that it is most probably produced as an experimental artifact in Ref. [35].

4.2 Invariant Mass Analysis of the Stopped K^- Reaction

The difficulty to search kaonic nuclear state via kaon reaction at rest arises from huge background due to the high-momentum nucleon emission caused by the non-mesonic hyperon (Y) formation in kaon two-nucleon absorption $K^-NN \to YN$ and the successive decay $Y \to N\pi$. The strong back-to-back correlation between correlations Y and N is naturally expected. On the other hand, if Y and d are the decay products of the $\overline{K}NNN$ bound state, then the Y and d will be produced back to back instead. The *KEK PS-E549* group performed an Λd-invariant mass study using Λdn final state [36]. Figure 10 (left) shows the Λd-invariant mass of the Λdn final state in which Λ and d are produced in the back-to-back direction.

There are two components: (1) the narrow peak just below the $m_{^4\mathrm{He}} + m_{K^-} - m_n$ threshold and (2) a broad component around 3100–3220 MeV/c^2.

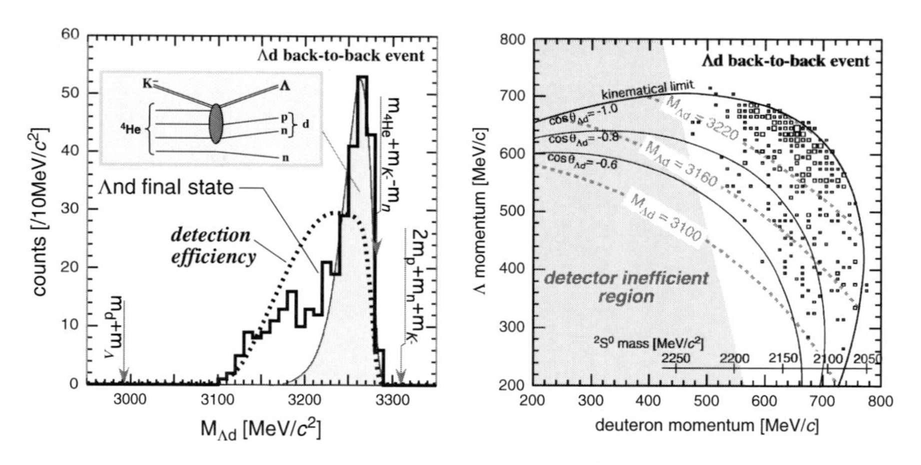

Fig. 10. Λd-invariant mass and momentum correlation between Λ and d [36]. *Left panel* shows a Λd-invariant mass ($M_{\Lambda d}$) spectrum of the Λdn final state produced by the kaon reaction at rest on helium target. The spectrum is for the event set in which Λ and d are recorded in the back-to-back direction. The *dotted line* is the relative detection efficiency of the experiment and the *hatched curve* is for the crude simulation for the Λdn final state requiring the neutron to be a spectator as it is indicated (the rough detector efficiency is taken into account). *Right panel* shows the momentum correlation between Λ and d for Λnd events. Kinematical constraint for invariant mass and opening angle of Λ and deuteron are overlaid.

Component (1) can be attributed to the three-nucleon absorption process $K^- ppn + (n) \to \Lambda d + (n)$, where the neutron (n) is a spectator of the reaction. Actually, very crude simulation of three-body phase space of Λdn together with nuclear form factor to the spectator neutron supports this interpretation. In this simulation, kaon energies are shared between Λ and d, and the neutron behaves as a spectator. Same event set in Λ and d momentum correlation is shown in Fig. 10 (right). The momentum correlation between Λ and d of the three-nucleon absorption process (1) is clearly seen close to the kinematical boundary with strong back-to-back correlation. The similar process $K^- ppn + (n) \to \Sigma^0 d + (n)$ is also detected in their analysis just below the threshold energy [36].

The Λd-invariant mass spectrum from the stopped K^- reaction on p-shell nuclear targets has been reported recently as a possible candidate signal of kaonic nuclear bound state formation [37]. The structure of the spectrum is quite similar to the process (1) described above. Since there exist nearly the same peak structures just below the threshold on both isospin $T = 0$ and $T = 1$ channels (Λd and $\Sigma^0 d$), the interpretation as the formation of kaonic nuclear bound state in Ref. [37] is unlikely for the ^{4}He case, because one should assume two kaonic bound states with the similar masses and widths for $^3S^+{}_{T=0}$ and $^3S^+{}_{T=1}$ states (the multi-baryonic states with strangeness is denoted as $^AS^Z$ for the baryon number A and the charge Z).

On the other hand, the component (2) cannot be explained in a simple way. The low $M_{\Lambda d}$ region distributes over wide region, and the back-to-back correlation between Λ and d is weaker than that of the process (1) as shown in the figure. The events in the low Λd-invariant mass component $M_{\Lambda d} < 3220$ MeV/c^2, Q-value is shared by three particles, namely Λ, n and d, thus these particle's momenta exceed the Fermi motion. Therefore, none of these can be attributed to be a spectator of the reaction. One possible interpretation of these events could be a candidate of the kaonic nuclear state, such as

$$K^-\ {}^4\mathrm{He} \to {}^2\mathrm{S}^0_{T=1/2} + d \tag{1}$$

or

$$\begin{aligned} &K^-\ {}^4\mathrm{He} \to {}^3\mathrm{S}^+_{T=0} + n \text{ and} \\ &{}^3\mathrm{S}^+_{T=0} \to \Lambda + d. \end{aligned} \tag{2}$$

However, there remains a possibility of the two-step schemes of the conventional processes, namely

$$\begin{aligned} &K^-\ {}^4\mathrm{He} \to \Sigma^0 + n + (d) \text{ and} \\ &\Sigma^0 + (d) \to \Lambda + d, \end{aligned} \tag{3}$$

where (d) is a spectator deuteron in ^{4}He nuclei. This cascade process is one of the Σ–Λ conversions, in which mass difference between Σ and Λ is transferred to the spectator deuteron without breaking up the d.

The ΛN-invariant mass analysis is also important. The *FINUDA* experimental group reported data which they interpreted as the possible existence of the bound state of K^-pp (or $^2S^+$ in the above notation) in the Λp-invariant mass spectrum [38]. The non-mesonic multi-nucleon absorption process is expected to be the primary source of the imaginary part of the $\overline{K}N$ potential in the deeply bound region; the presently available experimental information for that process is limited to the total non-pionic absorption rate of stopped K^- [39]. It is indispensable to clarify whether the two-nucleon absorption processes exist as well-separable processes from the possible signature of the kaonic bound states. The two-nucleon absorption process, $K^-NN \rightarrow YN$, must be most cleanly observed when one detects back-to-back correlated YN pairs from the stopped K^- reactions. The *KEK PS-E549* group also reported Λp- and Λn-invariant mass analysis results of 3,000 Λp-pairs and 10,000 Λn-pairs [40]. Figure 11 (left) shows the missing mass spectrum of Λp back-to-back correlated pairs ($\cos\theta < -0.6$) from stopped K^- on ^{4}He target. The spectrum shape is different from the *FINUDA* result.

To understand the structure, the Λp-invariant mass spectrum is divided into three components. The criterion for the classification is given in the missing mass spectrum of Fig. 11 (right). In the figure, the missing mass is calculated from the observed four momenta of Λp pairs by the four-momentum balance. The missing particles in the unhatched region ($M_{miss.} < 1920$ MeV) should be two neutrons and are consistent with the spectator picture whose internal energy is small (below 40 MeV). Therefore, the component around $M_{\Lambda p}{\sim}2310$ MeV in the left figure is most likely the two-nucleon absorption process, $K^-pp \rightarrow \Lambda p$. The observed invariant mass width is consistent with the broadening effect caused by the Fermi motion.

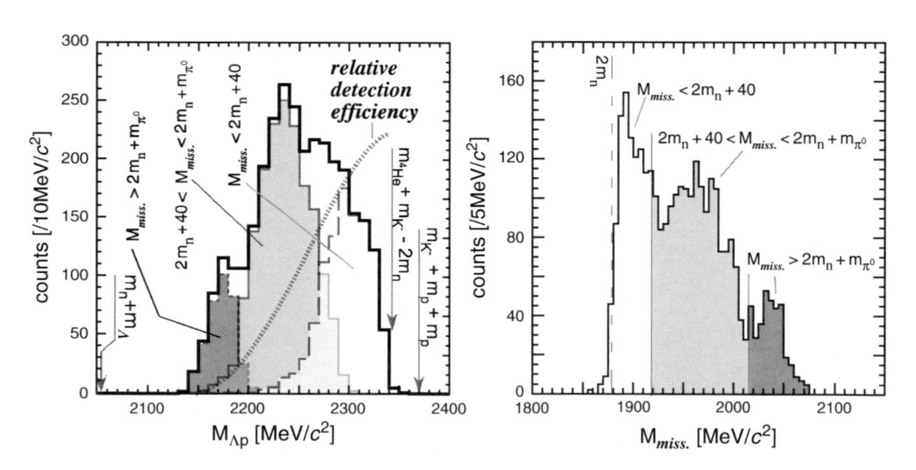

Fig. 11. The Λp-invariant mass spectrum of the K^- reaction at rest on helium target (*left*) and the missing mass distribution calculated by the observed Λp-pair's four momenta (*right*). The two spectra were divided into three sub-components by the missing mass as shown in the right panel [40].

On the other hand, the missing particles in the thick-hatched region ($M_{miss.} > 2015$ MeV) mainly come from two neutrons associated with a pion emission, as shown in the missing mass spectrum (right). Therefore, the component $M_{\Lambda p} < 2190$ MeV is mainly due to the quasi-free hyperon production, such as $K^- pn \to \Lambda p\pi^-$. With respect to the quite limited sensitivity shown in the left figure, this component should have large fraction compared to the other two components.

The non-mesonic Λ production predominantly occurs via the unassigned component (thin-hatched region where $1920 < M_{miss.} < 2015$ MeV). The observed momentum distributions of both Λ and proton of this component distribute around 450 MeV/c having full width of about 200 MeV/c (cf. Ref. [40] for more details). This component is also a candidate caused by the broad structure of kaonic bound state either

$$\begin{aligned} K^-\ {}^4\text{He} &\to {}^2S^+ + (nn) \text{ or} \\ K^-\ {}^4\text{He} &\to {}^3S^+ + (n) \end{aligned} \tag{4}$$

and their decay. However, it is difficult to rule out the possibility of conventional multi-step processes similar to reaction (3) also for this case.

4.3 KEK Study via In-Flight (K^-, N) Reaction

In KEK, there is another approach to search for the deeply bound kaonic state using in-flight (K^-, N) reaction by the *KEK PS-E548* experimental group. This kinematics has several good features to the study. The elementary cross section of the kaon scattering on single nucleon has a peak at around 1 GeV/c as shown in Fig. 12 (left). The forward nucleon from the in-flight reaction has higher momentum than the incident kaon momentum when the backward kaon is trapped in a nuclear state. The momentum transfer to the kaon of this

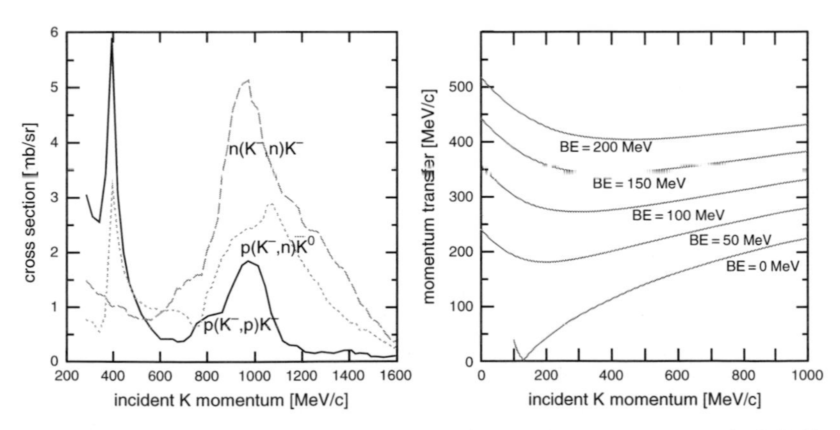

Fig. 12. The elementary cross section of the kaon scattering reaction [41] (*left*) and momentum transfer of the (K^-, N) reaction (*right*).

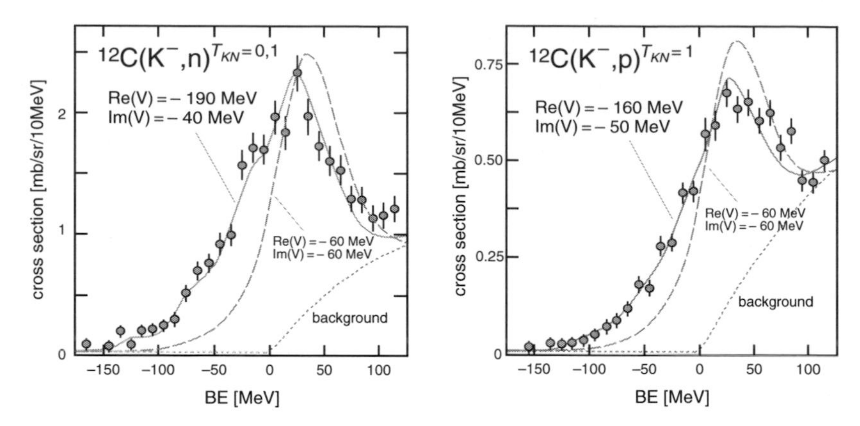

Fig. 13. The (K^-, n) and (K^-, p) spectra obtained by *KEK PS-E548* experiment [42]. The best fit to the spectroscopic function is plotted in *solid line* and the *dotted line* shows the quadratic function to represent for the background in the unbound region. The *dashed line* shows the shallow and absorptive spectroscopic function for the comparison.

process is in the moderate range depending on the binding energy as shown in the figure (right). At the high incident momentum of 1 GeV/c, multi-nucleon reaction process is expected to be naturally suppressed at the formation stage.

Their obtained spectra on the carbon target as a function of the binding energy (BE) are shown in Fig. 13. It is quite remarkable that the spectra is free from the background due to the two-nucleon absorption process, because it should have maximum at around $BE = 300$ MeV region.

The spectra were fitted using spectroscopic function delivered from Green's function method [43, 44]. A strong attractive potential with relatively small imaginary part was found as shown in the figure. They concluded that the attractive potential of about 200 MeV is indispensable to fit the spectra having substantial yield below threshold, and the sub-threshold shape could not be explained by the leakage of quasi-free component by the existence of the imaginary part. They concluded that this result supports the strong attractive potential solution of the atomic x-ray study given in Ref. [31]. On the other hand, the fact that the given potential strength is similar to that derived by the atomic data indicates that the nuclear density is not strongly modified due to the attraction and the potential strength is weaker than that predicted by Akaishi and Yamazaki [25]. It is obviously quite important to pursue further studies in order to deduce the $\overline{K}N$ interaction precisely and check the consistency with other nuclear target data.

As shown in Fig. 13 (left), weak shoulder-like structures below the threshold energy due to the higher partial waves are indicated; however, the existence of these structures is not very strong due to the statistical error. Therefore, more intensive and detailed experimental study is needed to have conclusive understanding of the $\overline{K}N$ interaction.

4.4 Exclusive Experiment Planned at J-PARC

As it is discussed, experimental evidence for the deeply bound kaonic state formation is still missing. Several candidates of the deeply bound kaonic state formation are reported and under discussion, but it is difficult to understand all the data in a consistent way.

To clarify the situation, an exclusive experiment to study the formation in the lightest nuclear system, K^-pp is planned using $^3\text{He}(K^-,n)$"K^-pp" reaction by *J-PARC E15* experimental group [45]. As discussed in the previous section, *KEK PS-E548* data demonstrated that in-flight reaction has several good features compared to the at-rest experiment; (1) smaller momentum transfer especially for the shallow binding energy region; (2) the multi-nucleon absorption process in the formation reaction should be much weaker which will substantially reduce the background process; and (3) this multi-nucleon absorption background component should be well separated from the signal region in energy due to the kinematics as it is demonstrated in the previous KEK experiment, *KEK PS-E548*.

There are two reasons why the lightest K^-pp system is selected as a target. First, it would be ideal if one can analyze the event kinematics in both the missing mass and the invariant mass at the same time. The $^3\text{He}(K^-,n)$"K^-pp" reaction missing mass can be analyzed by the forward neutron, the decay branch "K^-pp" $\rightarrow \Lambda p$ can be utilized for the invariant mass study at the same time, and one can cross-check the consistency between the formation and decay. To trigger the event efficiently and to perform two analysis methods described above, *J-PARC E15* experiment is equipped with cylindrical detector system (CDS) around the liquid ^{3}He target and neutron time-of-flight counter wall in the forward direction after the sweeping magnet of the charged particles. As it is demonstrated in Fig. 10 (left), a peak-like structure could be formed in the partial invariant mass due to the kinematics and the nuclear form factor. Actually, only from the experimental point of view, this peak-like structure of the Λd-invariant mass cannot be discriminated from very shallow kaonic bound state. Therefore, the full understanding of the reaction mechanism is extremely important to understand the reaction both from formation and the decay channel.

Second, the decay width of the kaon due to the multi-nucleon absorption process would be minimum in this system, although the binding energy should be smaller compared to the heavier target nuclei. The binding level is naturally expected to be limited to the ground state for the lightest system which makes the analysis simpler, although the binding energy would be smaller than the heavier target case. Therefore one may observe the sub-threshold structure even more clearly, if there exists substantial reduction of the width due to the smaller absorption channels described above.

Figure 14 shows a recent theoretical calculation of the spectroscopic function by Koike and Harada. The experimental setup is designed to be insensitive to the scattered kaon, the spectroscopic function only for the kaon conversion

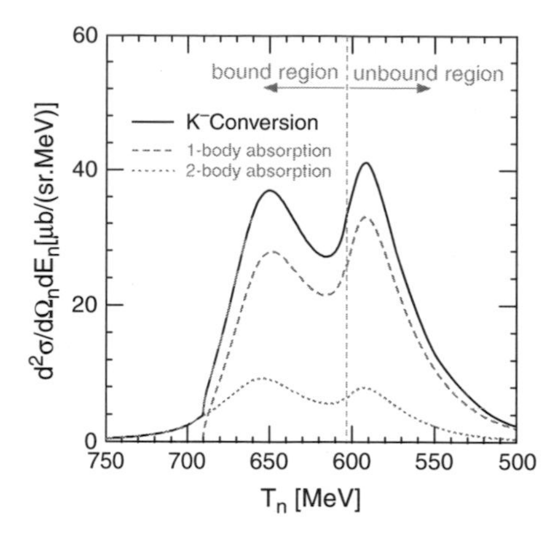

Fig. 14. One of the spectroscopic functions calculated by Koike and Harada [46] plotted as a function of the kinetic energy of the forward neutron. In this example, assumed potential is essentially the same as the deep potential given by Ref. [25].

part is plotted. It is seen that the clear structure can be expected. If this is the case, one can expect a definitive spectrum in the forthcoming experiment. Thus, we definitely need more intensive theoretical work on the spectroscopic function to understand the experimental spectrum.

Above example, however, assumed that the potential is deep. In Ref. [46], shallow potential case is also discussed. If the binding energy is small enough and consequently the "K^-pp" $\rightarrow$ $\Sigma\pi p$ decay channel has relatively large Q-value, then this decay channel can be triggered selectively by $\pi^+\pi^-$ pair in the final state. By this selection, signal component in the bound region will be enhanced substantially, because the channel is selective to the isospin zero channel. Thus, *J-PARC E15* experiment could be sensitive also for the shallow potential case. It is needed to have more detailed theoretical work of the expected spectroscopic function, in which the decay channels are taken into account.

In summary, there is a good chance to observe a structure in the missing mass spectrum even if the $\overline{K}N$ strength is much shallower than Akaishi and Yamazaki's original prediction [25]. Future experimental plan beyond *J-PARC E15* is much dependent on the result. If a spectrum with physically significant information is obtained, then one should proceed to the heavier targets such as helium-4 and carbon, for the detailed understanding of the $\overline{K}N$ interaction utilizing the intense kaon beam available at J-PARC.

5 Search for the ϕ-Meson Bound States

Vector mesons in the nuclear medium have also been extensively studied experimentally. There are two outstanding features of the study: (1) the mass (or energy) shift of the meson in the nuclear medium can be calculated from the QCD sum rule and it is predicted to be about 1.5–3% of the rest mass [47] and (2) the formation and the decay can be obtained by an invariant mass study [48, 49]. It is inconclusive whether the mass shift has been observed in invariant mass spectra; however, the corresponding peaks such as ρ and ϕ have low-energy tails on peaks at the free mass. Their simple model calculation shows that the tail formation results from about a 3% mass reduction in the nuclear medium, which is consistent with the calculated value given by the QCD sum rule. An experimental study, J-PARC E16 [50], is planning to accumulate much higher statistics to obtain a more conclusive result using the dispersion relation between the total energy and the momentum.

Recently, a new experimental approach to search for the ϕ-meson bound state was submitted to the J-PARC PAC as a letter of intent (LOI) [51].

5.1 KEK ϕ-Invariant Mass Study Using e^+e^- Decay Channel

For the case of ρ and ω mesons in hot or dense matter, several experiments reported evidence of the mass modification in a medium [52–57]. On the other hand, experimental information on the ϕ meson modification is very limited compared to ρ and ω mesons. Recently, the invariant mass spectrum of $\phi \to e^+e^-$ was observed using $p + A$ reactions with 12-GeV protons by the *KEK PS-E325* experiment [49], which has an extremely interesting energy dependence. Figure 15 shows the invariant mass spectrum of $\phi \to e^+e^-$, and an excess on the low-mass side of the ϕ meson peak was observed in the low $\beta\gamma$ $(= p/m)$ region of ϕ mesons $(\beta\gamma < 1.25)$ with copper targets. However, in the high $\beta\gamma$ region $(\beta\gamma > 1.25)$, spectral shapes of ϕ mesons were consistent with the Breit–Wigner shape, i.e., spectrum shape in vacuum. Since the mass modification of the ϕ mesons in a target nucleus is expected to be visible only for slow ϕ mesons produced in a heavy target nucleus, they concluded that the excess is considered to be the signal of the mass modification of the ϕ mesons in a target nucleus.

To interpret the invariant mass spectra of ϕ mesons measured by E325 experiment, the simplest approach is to assume that the spectrum shape of ϕ meson is modified in nuclear medium. It allows us to compare the spectrum directly to the theoretical predictions. There are a number of predictions about the mass shift of ϕ mesons in nuclei, from QCD calculations such as the Brown–Rho scaling [58], the QCD sum rule [47, 59], the effective chiral Lagrangian [60], and the renormalization of the kaon [61]. The comparison based on a simple model calculation gives about 3% mass reduction and 3.4 times width broadening of the ϕ in nuclear media [49]. On the other hand, there is criticism of such a straightforward comparison of the invariant mass [62]. Therefore the conclusion that properties of the in-medium ϕ meson have been modified is still controversial.

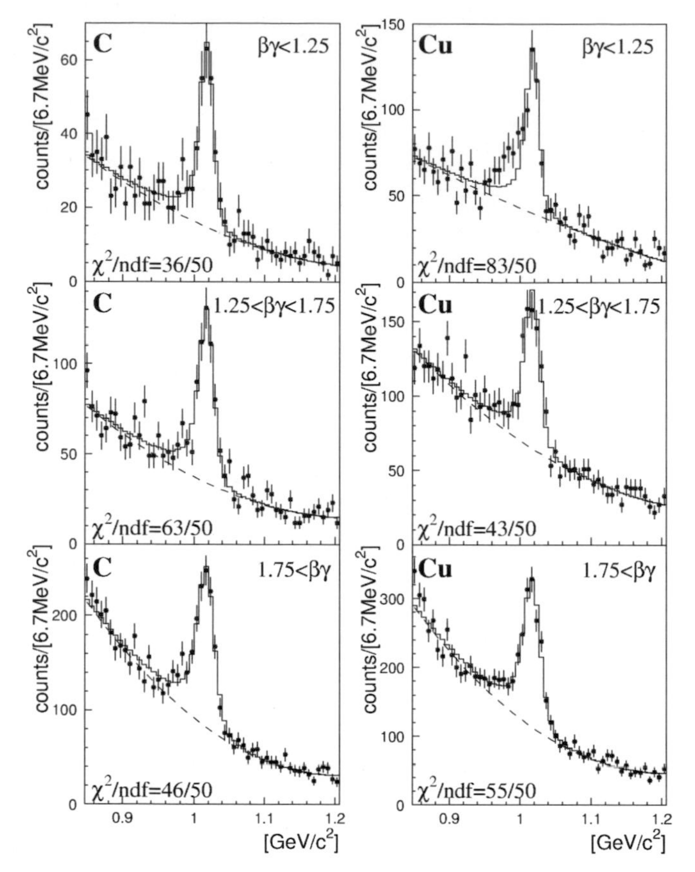

Fig. 15. The invariant mass spectra of $\phi \rightarrow e^+e^-$ at KEK PS E325 [49].

5.2 ϕ in Nuclei

What can one expect when a ϕ is in a nucleus? As described, the in-medium mass shift of the ϕ is predicted to be $\Delta m_\phi/m_\phi = 1.5$–$2.6\%$ by the QCD sum rule as shown in Fig. 16 (left) [59]. The width broadening is naturally expected because several decay channels are open in a nucleus. There are several theoretical predictions and the predicted widths are quite narrow. Klingl–Waas–Weise predicted that the width is below $\Gamma_\phi < 10$ MeV [63], while Oset–Ramos reported that the width can be bigger, $\Gamma_\phi < 16$ MeV (for $\Delta m_\phi \sim 30$ MeV/c^2), taking into account Σ^* and the vertex correction in the chiral unitary model [64] (Fig. 16 (right)). All the currently available theoretical predictions give a quite narrow natural decay width of the ϕ even in the nuclear medium.

On the other hand, experimental ϕ-invariant mass studies in the pA reaction reported that the ϕ mass shift in medium-heavy nuclei (Cu) is about 3% and the natural width broadening of $\Gamma_\phi/\Gamma_\phi^{free} \approx 3.4$ [49]. These numbers are

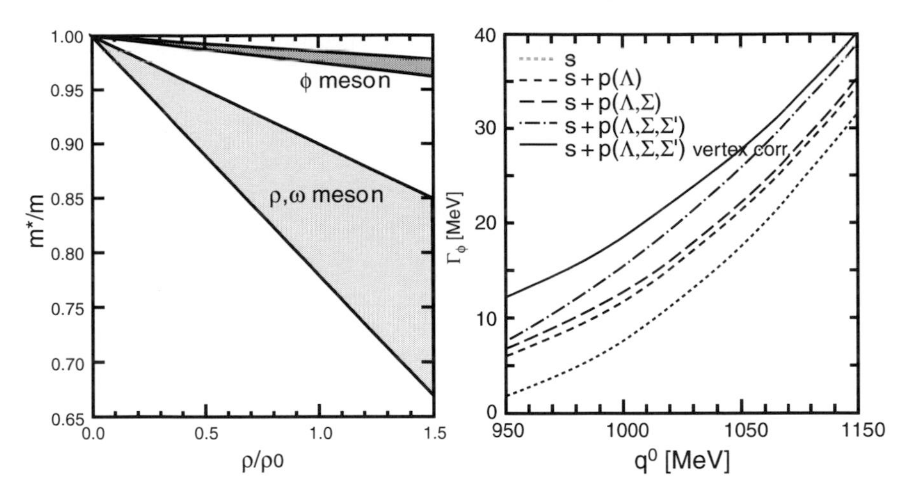

Fig. 16. Theoretical calculation of the in-medium mass shift of the ϕ by the QCD sum rule (*left*) [59], and the in-medium decay width by using the chiral unitary model (*right*) [64].

in quite good agreement with theoretical predictions, except for the strength of the attraction of the chiral unitary model.[2] Therefore, let us take the mass shift and width from the previous experiment reported in Ref. [49].

This means that a nucleus is nothing more than an energy pocket for the ϕ meson of about $\Delta m_\phi \sim 30\,\mathrm{MeV}$ at the relative distance r below the nuclear radius R_0 ($\propto A^{1/3}$). Another feature is that this spatial energy pocket (a nucleus) is absorptive, but the strength is quite weak. The situation is illustrated in Fig. 17. A weak absorptive interaction makes the ϕ a bit unstable, which broadens the width, but one can expect that the width would be less than

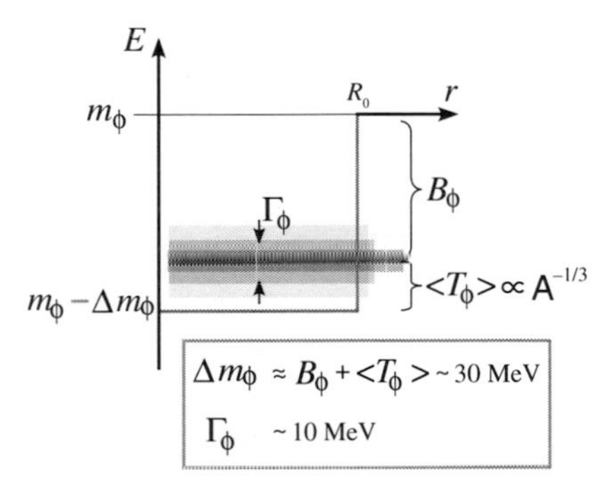

Fig. 17. A cartoon of the bound state formation trapped in an energy pocket caused by the in-medium mass shift of the ϕ.

---

[2] The interaction obtained by Oset–Ramos is attractive [64], but much weaker than what one can expect from Ref. [59].

16 MeV. Therefore, the ϕ will form a meta-stable, though still discrete, bound state in nuclei. Only the free decay channel remains outside the nucleus, while all decay channels realized in nuclear media are closed.

The situation is almost analogous to the case of Λ hypernuclei. The Δm_ϕ is almost the same as the potential depth of the Λ, whose mass, 1,115 MeV/c^2, is almost the same as that of the ϕ. Systematic study of Λ hypernuclei is giving us another remarkable insight for the study of the in-medium study of the ϕ meson. The study shows that the Λ behaves as if it is a free particle inside nuclei, because the Pauli principle does not contribute. Therefore, the Λ binding energy (B_Λ) has a clear A dependence, as shown in Fig. 18. For the Λ ground state,

$$B_\Lambda \approx V_\Lambda \left(1 - \left(\frac{A}{A^0_\Lambda}\right)^{-\frac{2}{3}}\right) \tag{5}$$

and the averaged kinetic energy term $\langle T_\Lambda \rangle$ is as follows:

$$\langle T_\Lambda \rangle \approx V_\Lambda \left(\frac{A}{A^0_\Lambda}\right)^{-\frac{2}{3}}, \tag{6}$$

where $\langle T_\Lambda \rangle$ is the mean Λ kinetic energy, V_Λ is the real part of the Λ potential strength, and A^0_Λ is the critical mass number for the Λ to bind ($A^0_\Lambda \sim 3$). The A dependence is quite easy to understand from the uncertainty principle $\Delta p \cdot \Delta x \sim \hbar$, and the Λ hypernuclei should have the size which are proportional to $A^{-\frac{1}{3}}$. As shown in Fig. 18, the simple $A^{-\frac{2}{3}}$ rule does not hold precisely, especially for large A.

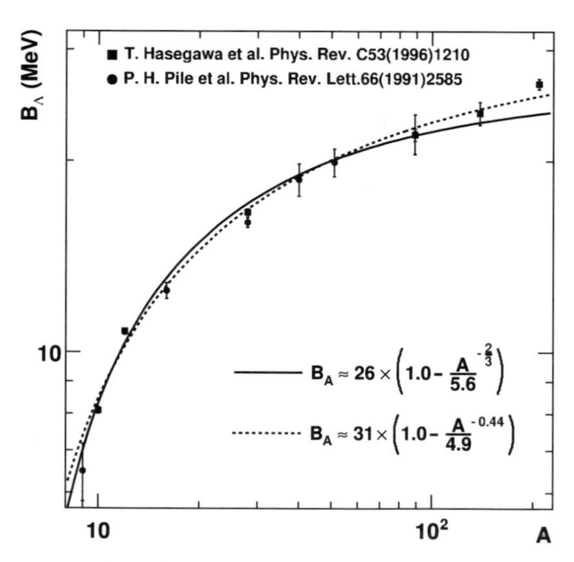

Fig. 18. The mass number dependence of the binding energy ground states of Λ hypernuclei. The simple $A^{-\frac{2}{3}}$ rule does not hold, especially for large A.

Analogous to the Λ state in hypernuclei, the ϕ binding energy (B_ϕ) of the ground states should have clear mass number dependence again, which can be given as

$$B_\phi \approx \Delta m_\phi \left(1 - \left(\frac{A}{A_\phi^0}\right)^{-\alpha_\phi}\right), \tag{7}$$

where α_ϕ represents the coefficient in the case of the ϕ-meson bound state and A_ϕ^0 is the critical mass number for the ϕ to be bound which should be around $A_\phi^0 \sim (A_\Lambda^0 - 1 \approx 4)$. The formation of the ϕ-meson bound state could be observed in the energy spectrum as a discrete peak for the case if $A > A_\phi^0$. More detailed theoretical study of the A dependence is required, which can also be experimentally deduced by a systematic study.

Note that the kinetic energy $\langle T_\phi \rangle$ is not a good quantum number having a finite width in a finite nucleus. Even if the total energy is defined, the ϕ momentum should have a distribution in a nucleus, which is simply a square of the wave function of ϕ in a momentum representation. Actually, this is one of the key reasons for the criticism given by Yamazaki–Akaishi [62]. The invariant mass is not a good quantum number, because it is a kind of operator to give residual energy (called rest mass) by subtracting the kinetic energy from its total energy $\langle T_\phi \rangle$, namely,

$$M_{inv}(\phi) = E_\phi - \langle T_\phi \rangle \,, \tag{8}$$

and the kinetic energy should have a size-dependent width in finite nucleus. Therefore, the invariant mass should distribute around $m_\phi - \Delta m_\phi$, even without considering its decay. When the ϕ is in a highly excited state, the central value and the width of the invariant mass could be moved and broadened by the core excitation of the residual nucleus, so instead they called the quantity the *quasi-invariant mass* in their paper.

It is clear that one can resolve the situation by specifically identifying the ϕ-meson nuclear bound state, which enables us to be free from the criticism described above.

5.3 Missing Mass Approach

Experimentally, the calculation of the missing mass is also quite simple. Let us discuss the reaction ${}^A_Z\mathrm{X}(B, C)$, namely

$$B + {}^{A'}_{Z'}\mathrm{X}' \rightarrow \{\phi\ {}^A_Z\mathrm{X}\} + C, \tag{9}$$

where B and C can be any particle and $\{\phi\ {}^A_Z\mathrm{X}\}$ represents the ϕ mesonic nuclei. If we measure the momenta of B and C, then the square of the missing mass for $\{\phi\ {}^A_Z\mathrm{X}\}$ can be given as

$$M_{missing} = \sqrt{(p_B + p_{^{A'}_{Z'}\mathrm{X'}} - p_C)^2}, \tag{10}$$

where p_B, $p_{^{A'}_{Z'}\mathrm{X'}}$, and p_C are the four momenta of each particle. The missing mass can also be given as

$$M_{missing} = M_{^A_Z\mathrm{X}} + m_\phi - B_\phi, \tag{11}$$

where $M_{^A_Z\mathrm{X}}$ is the mass of the residual (spectator) nuclei $^A_Z\mathrm{X}$. If $\{\phi\ ^A_Z\mathrm{X}\}$ is formed, the missing mass $M_{missing}$ should be located below the threshold energy $M_{^A_Z\mathrm{X}} + m_\phi$. Of course in a real experiment, the situation is not that simple, because C can be generated in other channels even without producing a ϕ in the reaction. The strong interaction cross section is in the order of 10 mb, while the ϕ production is in the order of 10 μb, so that the background is expected to be about 10^3 times higher.

However, if one can efficiently select reaction (9) exclusively and drastically improve the signal-to-noise ratio (S/N) at the same time, then the ϕ must decay inside the nuclei and the signal can be observed in the sub-threshold energy region. This is one of the remarkable points of missing mass spectroscopy.

5.4 Invariant Mass Approach

What happens if a ϕ is in a nucleus? One can naturally expect that six $\phi N \to KY$ channels will open. Table 2 shows a summary of possible decay/reaction channels. In the table, the binding energy of the proton ("p") and neutron ("n") in a nucleus is expressed as B_p and B_n, respectively. Because the partial decay widths to each KY channel are not given in Ref. [64], we computed the branching ratio from the given partial widths of $K\Lambda$ and $K\Sigma$ channels, by assuming the isospin relation 1:2 between $K^0\Sigma^0$ and $K^+\Sigma^-$ channels, and ϕp and ϕn have the same coupling strength.

The free-space decay channel is easy to analyze using the quasi-invariant mass, but it is difficult to observe. The di-lepton channel has a quite small branching ratio of $\sim 10^{-4}$. Because of the small Q-value, the $K\overline{K}$ channel is also difficult to observe if the ϕ is produced near or below the threshold. At the threshold, the $K\overline{K}$ pair produced at around 126 MeV/c can easily stop around the target.

Therefore, let us focus on the KY decay channel of the ϕ mesonic nuclei, ignoring the effect of the final state interaction (FSI) for simplicity. Especially, the $K^+ - \Lambda$ channel is quite promising, not only because of its large branching ratio but also because of the large $p\pi^-$ decay branch (as much as 60%) of the Λ. This means that a large fraction of the channel can be reconstructed by measuring only charged particles. Unfortunately, observation of $K^+\Sigma$ channel is difficult not only from its small branching ratio but also because (1) the $\Sigma\Lambda$ conversion is known to be quite strong and (2) most of the charged Σ decays emit neutral particles.

Table 2. Decay/reaction channels of the ϕ in nuclei.

Initial channel	Decay/KY channel	Free BR	Branching ratio (%)*	Q-value (MeV)	Suppression
ϕ	$e^+\ e^-$	3×10^{-4}	–	1018	EM
	$\mu^+\ \mu^-$	3×10^{-4}	–	808	EM
	$K^+\ K^-$	0.49	–	32	OZI
	$K^0\ \overline{K}^0$	0.33	–	32	OZI
ϕ "p"	$K^+\ \Lambda$	–	37	$348 - B_\phi - B_p$	None
	$K^+\ \Sigma^0$	–	1	$271 - B_\phi - B_p$	None
	$K^0\ \Sigma^+$	–	2	$275 - B_\phi - B_p$	None
	$K\ \Sigma^*$ etc.	–	10	–	–
ϕ "n"	$K^0\ \Lambda$	–	37	$346 - B_\phi - B_n$	None
	$K^0\ \Sigma^0$	–	1	$269 - B_\phi - B_n$	None
	$K^+\ \Sigma^-$	–	2	$268 - B_\phi - B_n$	None
	$K\ \Sigma^*$ etc.	–	10	–	–

* The decay branch is calculated from the theoretical partial widths contributed from $K\Lambda$ and $K\Sigma$ channels given in Ref. [64].

In a nucleus, the K^+–Λ channel can be observed as

$$\{\phi\ {}^A_Z\mathrm{X}\} \to K^+ + \Lambda + {}^{A-1}_{Z-1}\mathrm{X}'', \tag{12}$$

in which a ΛK^+ pair is produced by the simple quark re-configuration as

$$\phi(s\bar{s}) + \text{“}p\text{”}(uud) \to K^+(u\bar{s}) + \Lambda(uds). \tag{13}$$

If the ϕ is in a bound state or low-momentum region (roughly below proton Fermi motion), then the ΛK^+ pair should be produced in a *back-to-back* direction at the center of mass (CM) of the $\{\phi\ {}^A_Z X\}$ system, because the kinetic energy of the ϕ is well below the Q-value. If we neglect the kinetic energy carried out by residual nuclei, then the invariant mass should be given as

$$m_{inv}(\Lambda K^+) \approx \sqrt{(m_\phi - \Delta m_\phi)^2 + \mathbf{p}^2_{cm'}} + \sqrt{{m_p^{\mathrm{eff}}}^2 + \mathbf{p}^2_{cm'}}, \tag{14}$$

where m_ϕ and m_p are the mass of the ϕ and proton, respectively, $\mathbf{p}_{cm'}$ is the three momenta of the ϕ (or proton) in the center of mass of the ϕ–p subsystem in $\{\phi\ {}^A_Z\mathrm{X}\}$, $\Delta m(\phi)$ is the mass shift, and m_p^{eff} is the effective mass of the proton. If $p_{cm'}$ is small, it can be simplified as

$$m_{inv}(\Lambda K^+) \approx m_\phi + m_p - \Delta m_\phi - V_p + \left\langle \frac{m_\phi + m_p}{2 m_\phi m_p} \mathbf{p}^2_{cm'} \right\rangle, \tag{15}$$

where V_p is the real part of the optical potential of the proton.

Very unfortunately, there is no good way to deduce $\mathbf{p}_{cm'}$ experimentally. Thus, the invariant mass study of the ΛK^+ pair is even more difficult than that of the free decay mode of the ϕ. The invariant mass will be smeared out not only from the imaginary part but also by the internal kinetic term which is roughly 40 MeV, if we assume $|\mathbf{p}_{cm'}|$ is of the order of the Fermi motion ~270 MeV/c. However, this invariant mass is useful for defining an energy window to identify the ϕ production in the reaction using a rather wide window to cover the smearing effect.

5.5 Q-Value Approach

Is there any other way to be free from the hidden ambiguity in the invariant mass caused by other effects, such as finite size, core excitation? If one knows the rest frame of the ϕ mesonic nucleus, and if it decays to three bodies $\{\phi\ {}^A_Z\mathrm{X}\} \to \Lambda + K^+ + {}^{A-1}_{Z-1}\mathrm{X}''$, then the Q-value can be given as

$$\begin{aligned} Q &= (m_\phi - B_\phi) + (m_p - B_p) - (m_{K^+} + m_\Lambda) \\ &= T_\Lambda + T_{K^+} + T_{{}^{A-1}_{Z-1}\mathrm{X}''}, \end{aligned} \tag{16}$$

thus

$$\begin{aligned} Q + m_{K^+} + m_\Lambda &= E^{rf}_\Lambda + E^{rf}_{K^+} + \frac{\mathbf{p}^2_{\Lambda K^+}}{2M_{{}^{A-1}_{Z-1}\mathrm{X}''}} \\ &= m_\phi + m_p - B_\phi - B_p, \end{aligned} \tag{17}$$

where $\langle T_\phi \rangle$ and $\langle T_{{}^{A-1}_{Z-1}\mathrm{X}''} \rangle$ are the kinetic energies of the ϕ and the residual nucleus (with proton hole), respectively, $M_{{}^{A-1}_{Z-1}\mathrm{X}''}$ is the mass of the residual nucleus, E^{rf}_Λ and $E^{rf}_{K^+}$ are the total energies of Λ and K^+ in the rest frame of $\{\phi\ {}^A_Z\mathrm{X}\}$, respectively, and $\mathbf{p}_{\Lambda K^+}$ is the three momenta of the ΛK^+ pair in the rest frame. All the quantities used to compute E_{rest} can be observed experimentally, thus this is an extremely interesting quantity for deducing B_ϕ, if one can define the frame by knowing the production channel. Note that formula (17) is obtained simply from energy conservation. In contrast to the invariant mass formulation, there is no ambiguity in this representation except for the possible core excitation energy of the residual nucleus.

5.6 The Combined Approach

None of the approaches described above are simply enough to deduce the mass shift directly. However, the problem can be solved by selecting the simplest event by observing the formation and the decay reaction at the same time.

As discussed, both Λ and K^+ are tagged with strangeness, thus with a loose cut for ΛK^+-invariant mass and by requiring *back-to-back* Λ and K^+ emission, we can expect a clear missing mass spectrum. From the missing mass, one can also select events which correspond to a specific ϕ-bound state formation. It is also possible to extend the study further, using formulae (11) and (17) to check the consistency of the resultant binding energy. Namely, from a missing mass study of the ground state of the ϕ,

$$B_\phi = -\left(M_{missing} - M_{^A_Z\mathrm{X}} - m_\phi\right), \tag{18}$$

and from the energy conservation rule in the rest frame of $\{\phi\ ^A_Z\mathrm{X}\}$,

$$B_\phi = -\left(E^{rf}_\Lambda + E^{rf}_{K^+} + \frac{\mathbf{p}^2_{\Lambda K^+}}{2M_{^{A-1}_{Z-1}\mathrm{X}''}} - m_\phi - m_p + B_p\right), \tag{19}$$

where $^A_Z\mathrm{X}$ is the core nucleus of the ϕ mesonic nucleus $\{\phi\ ^A_Z\mathrm{X}\}$, and $^{A-1}_{Z-1}\mathrm{X}''$ is the residual nucleus of the decay. Note that the mass $M_{^A_Z\mathrm{X}}$ and the proton binding energy B_p are not unique, but can be assigned if the width broadening of the ϕ-bound state is relatively small. As described, one can deduce the mass shift of the ϕ meson in the nuclear medium by a systematic study of the ϕ-bound states as

$$\Delta m_\phi = \left(1 - \left(\frac{A}{A^0_\phi}\right)^{-\alpha_\phi}\right)^{-1} B_\phi. \tag{20}$$

There are several remarkable points, which can be summarized as follows:

- decay channel, Λ and K^+, open only when ϕ is in a nucleus
- both Λ and K^+ are labeled by strangeness coming from the $s\bar{s}$ pair of the ϕ
- all charged final states in $\Lambda \to p\pi^+$ and K^+ make for efficient detection
- FSI effect should be minimized for the *back-to-back* condition
- background-free missing mass spectrum is expected
- ϕ mass shift can be detected by two independent methods

Even if the branching ratio for the decay mode of $K^+\Sigma^0$ is small, there is a question whether the $K^+\Sigma^0$ channel could arise in the event if one could not detect the γ-ray emission of the Σ^0 decay $\Sigma^0 \to \gamma\Lambda$ ($BR{\sim}100\%$). It should be noted that $K^+\Lambda$-invariant mass spectrum will be shifted by 70 MeV for the events from $K^+\Sigma^0$ if experimental detector is not sensitive for the neutral particles, i.e., missing γ-ray from Σ^0 decay. However, if the overall energy resolution for the detector is better than 70 MeV, we will separate event of $K^+\Sigma^0$ channel from $K^+\Lambda$ direct decay channel.

5.7 Possible Production Channels of the ϕ-Bound State

To form a ϕ-bound state, the momentum transfer of the production reaction should be small. In reaction $B+^{A'}_{Z'}\mathrm{X}' \to \{\phi\ ^A_Z\mathrm{X}\}+C$, the minimum momentum transfer (recoilless condition) can be realized when the mass difference between

C and B is bigger than the mass of the ϕ. Unfortunately, there is no ideal reaction channel to fit this condition.

The pair annihilation is an alternative solution to achieve the minimum momentum transfer. In this sense, the most interesting formation channel is $\overline{p} + p \rightarrow \phi + \phi$. Among the ϕ production channels in which the $s\overline{s}$ pair can be produced in the residual nucleus, the threshold energy of this channel is practically the lowest, because the lower threshold channel $\overline{p} + p \rightarrow K + \overline{K} + \phi$ has a much smaller production cross section compared to the $\phi + \phi$ final state at around the threshold energy ($\overline{p}$ below 1.4 GeV/c). Therefore, $(\overline{p}, \phi)$ spectroscopy could be possible without any background process. The momentum transfer to the backward ϕ is much smaller compared to the Fermi motion, which is < 200 MeV/c.

Other promising formation channels close to the condition are $\pi + N' \rightarrow \phi + N$ reactions or $\gamma + N \rightarrow \phi + N$ reactions. In the case of (π, N) reactions, which can be realized at J-PARC, the experiment will be limited by the incident pion beam rate. In the case of (γ, N) reactions, which can be realized, e.g., at SPring-8, it will be limited by its small cross section and the small incident γ-ray yield. For both cases, one can expect similar or slightly better S/N ratio among the two different elementary reactions to produce the ϕ meson. It should be noted that, in both cases, the momentum transfers are almost the same and rather large ($300 \sim 400$ MeV/c) compared to the Fermi motion.

5.8 The ϕ-Meson Production via ($\overline{p}$, ϕ) Reactions

To search for gluonic matter and exotic quark-gluon formations, intensive studies have been performed at the CERN/LEAR facility for the "OZI-forbidden" formation reaction of the type $\overline{p}p \rightarrow M_1 M_2$, where M_1 and M_2 are vector mesons. One striking result is rather large $\phi\phi$ production cross section near the production threshold (~0.9 GeV/c), namely incident $\overline{p}$ momentum at around $1.3 \sim 1.4$ GeV/c. Figure 19 (left) shows the production cross section for double ϕ meson production as a function of incident momentum for $\overline{p}$ together with direct production of $\phi\ K^+K^-$ and $K^+K^-K^+K^-$ (non-resonant $K\overline{K}$ pairs).

Let us focus on the double-ϕ elementary reaction channel. In this reaction, one can use the backward ϕ as the source of quasi-recoilless ϕ meson production channel, and the forward ϕ as a spectroscopic analyzer of the missing mass of the backward ϕ-mesonic nuclei. The cross section of the double ϕ channel has a peak at around a $\overline{p}$ momentum of $1.3 \sim 1.4$ GeV/c. In this momentum region, other kaon-associated reactions, ϕK^+K^- and $K^+K^-K^+K^-$, have considerably lower cross sections (~10%) as shown in Fig. 19 (left), in spite of their lower Q-values. The incident ($\overline{p}$, ϕ) reaction can be written as follows:

$$\overline{p} + {}^{A+1}_{Z+1}\mathrm{X}' \rightarrow \{\phi\ {}^{A}_{Z}\mathrm{X}\} + \phi. \tag{21}$$

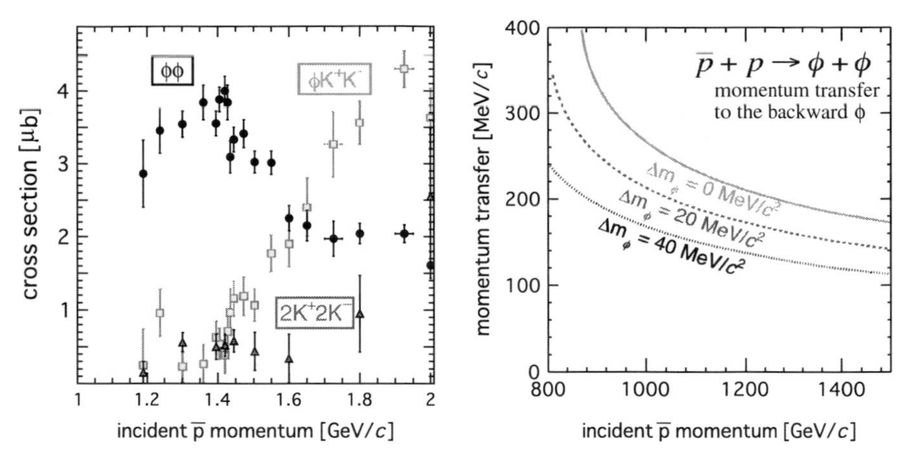

Fig. 19. *Left panel* shows the cross section for double ϕ meson production around the production threshold [65]. *Right panel* shows the momentum transfer for backward ϕ production as a function of the incident $\overline{p}$ momentum.

For these channels, the missing masses of the ϕ mesonic nuclei $\{\phi\ {}^A_Z X\}$ and their rest frames can be obtained by the forward ϕ meson momentum using $\phi \to K^+K^-$ decay.

The momentum transfer of the reaction is shown in Fig. 19 (right). As shown in the figure, the momentum transfer to the backward ϕ is below 200 MeV/c at around 1.3–1.4 GeV/c, where the double ϕ cross section is maximum. The momentum transfer is below the Fermi motion, so the ϕ-bound state formation rate would be more than for the case of (π, N) reaction channel, where the momentum transfer of the reaction is an order of 400 MeV.

Let us consider how to ensure that the backward ϕ is in a nucleus. As discussed, we can tag the events which have *back-to-back* ΛK^+ pair production at the momentum region around 200–600 MeV/c for both Λ and K^+, whose quasi-invariant mass is in the region of interest. By this tagging, we can select the cascade reactions $\overline{p}+{}^{A+1}_{Z+1}\mathrm{X}' \to \{\phi\ {}^A_Z\mathrm{X}\}+\phi$ and $\{\phi\ {}^A_Z\mathrm{X}\} \to {}^{A-1}_{Z}\mathrm{X}''+\Lambda+K^+$.

The most distinguishable feature of this reaction channel is its *fully background-free* nature. The yield of the kaon-associated ϕ production channel, ϕK^+K^- and $K^+K^-K^+K^-$, is much smaller than the double ϕ production channel for the incident $\overline{p}$ momentum below 1.4 GeV/c, and those events can be discriminated by the invariant mass analysis, so no background processes exist in the primary reaction. Another unique feature is that all the particles we shall observe, including forward $\phi \to K^+K^-$ decay, are labeled with strangeness so the discrimination from other process is quite clear, which ensures that it is free from any accidental background formation.

However, the hardware trigger for these events is difficult. Figure 20 shows cross sections of possible hardware trigger sources as a function of incident $\overline{p}$ momentum. For this experiment, the forward ϕ meson can be detected

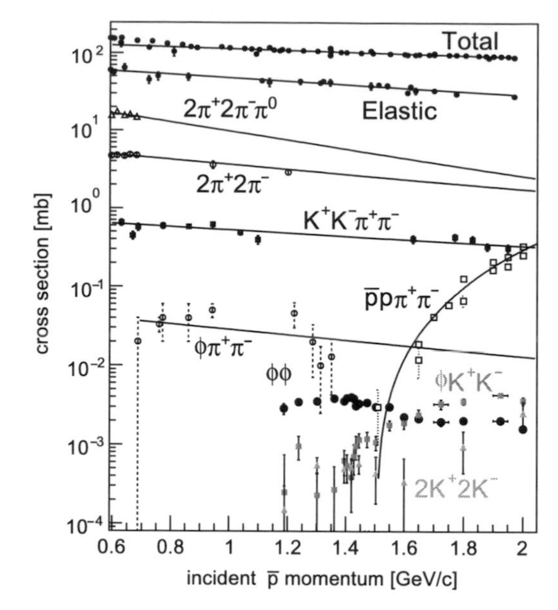

Fig. 20. Cross section of various $\overline{p}p$ reaction branches as a function of incident momentum of $\overline{p}$.

with a K^+K^- pair. Therefore the forward K^+K^- is a requirement of the experiment. As we can see in Fig. 20, it suppresses many of major hardware trigger sources like multi-pion production. Therefore, leftover hardware trigger sources are the $K^+K^-\pi^+\pi^-$ channels, if we simply require at least one charged particle around the target region in the hardware trigger level (note that the $\overline{p}p\pi^+\pi^-$ channel open only when the incident $\overline{p}$ momentum is above 1.5 GeV/c). Therefore, one needs a second-level trigger to identify a positive kaon track around the target region.[3]

In summary, the ideal completely background-free experiment can be done by $(\overline{p}, \phi)$ spectroscopy, in which all the particles to be measured are labeled with strangeness. The cross section of $p(\overline{p}, \phi)\phi$ is small but not in an unrealistic level. It would be possible to achieve an excellent experiment at J-PARC.

6 Summary

The experimental study of the meson bound state in nuclei was started only very recently. The difficulty of the study is the width of the state. We are presently accumulating techniques how to handle this difficulty and how to derive useful information from the experimental data. Thus, we need more

[3] If the event rate is acceptable for requiring single K^+ track in tracking detector around target, then one can extend the purpose of the experiment to include a di-lepton-invariant mass study.

time to judge how effective and efficient these probes are. Even with this difficulty, property of "real" mesons in nuclei relates many interesting subjects in physics today. Therefore, extensive study must be performed.

J-PARC is a unique facility which can provide intense kaon beam in DC mode. In the kaonic atom study, *J-PARC E17* will study kaonic helium-3 atomic transition from $3d$ to $2p$ using at-rest kaon as a "day-1" experiment. As a possible and interesting subject as an extension of the study in future, kaonic deuteron $2p$ to $1s$ transition to deduce isospin dependence of the $\overline{K}N$ interaction would be a good candidate. Another interesting subject is the kaonic helium-4 atomic transition from $2p$ to $1s$ to study the s-wave interaction precisely.

In the kaonic nuclear search, *J-PARC E15* will study "K^-pp" system and its decay using in-flight $^3\text{He}(K^-, n)$ at 1 GeV/c. As for the future extension, one needs to wait for the experimental result of the experiment, but it would be very important to revisit "K^-ppn" system studied at KEK and also carbon target examined by *KEK PS-E548* group. As another possible extension of the experimental apparatus of *J-PARC E15*, a letter of intent to search for η-mesonic nuclei was also submitted to J-PARC PAC [66].

This chapter also discussed about the possibility of the ϕ-meson bound state. The feasibility of the experiment also depends on the width of the state. However, there exists a unique and challenging gate to access this area using anti-proton beam in GeV/c region. J-PARC is an ideal facility for all of these experiments described in this chapter.

Acknowldgments

The author is grateful to all the collaborators for the useful discussion. The author thanks especially Dr. H. Ohnishi and Dr. F. Sakuma for their contribution to the idea of ϕ-bound state search, Dr. T. Suzuki and Dr. H. Outa for the idea of possible extension of the kaonic atom experiments at J-PARC, and finally Dr. H. Fujioka for the idea of the Λp tagging method applicability for K^-pp search experiment at J-PARC.

References

1. Yamazaki, T., et al.: Phys. Lett. B **418**, 246 (1998)
2. Geissel, H., et al.: Phys. Rev. Lett. **88**, 122301 (2002)
3. Hirenzaki, S., Toki, H.: Phys. Rev. C **55**, 2719 (1997)
4. Gell-Mann, M., Oakes, R.J., Renner, B.: Phys. Rev. **175**, 2195 (1968)
5. Suzuki, K., et al.: Phys. Rev. Lett. **92**, 072302 (2004)
6. Hatsuda, T., Kunihiro, T.: Phys. Rev. Lett. **55**, 158 (1985)
7. Hatsuda, T., Kunihiro, T.: Phys. Lett. B **185**, 304 (1987)
8. Kolomeitsev, E., Kaiser, N., Weise, W.: Phys. Rev. Lett. **90**, 092501 (2003)

9. Kolomeitsev, E., Kaiser, N., Weise, W.: Nucl. Phys. A **721**, 835 (2003)
10. Deser, S., et al.: Phys. Rev. **96**, 774 (1954)
11. Trueman, T.L.: Nucl. Phys. **26**, 57 (1961)
12. Davies, J.D., et al.: Phys. Lett. **83B**, 55 (1979)
13. Izycki, M., et al.: Z. Phys. A **297**, 11 (1980)
14. Bird, P.M., et al.: Nucl. Phys. A **404**, 482 (1983)
15. Iwasaki, M., et al.: Phys. Rev. Lett. **78**, 3067 (1997)
16. Beer, G., et al.: Phys. Rev. Lett. **94**, 212302 (2005)
17. Wiegand, C.E., Pehl, R.: Phys. Rev. Lett. **27**, 1410 (1971)
18. Batty, C.J., et al.: Nucl. Phys. A **326**, 455 (1979)
19. Baird, S., et al.: Nucl. Phys. A **392**, 297 (1983)
20. Hirenzaki, S., et al.: Phys. Rev. C **61**, 055205 (2000)
21. Wycech, S.: Nucl. Phys. A **450**, 399C–402C (1986)
22. Akaishi, Y.: Proceedings for international conference on exotic atoms – EXA05, p. 45, Austrian Academy of Sciences Press, Vienna, 2005
23. Okada, S., et al.: Phys. Lett. B **653**, 38 (2007)
24. Beer, G., et al.: J-PARC E17 proposal, http://j-parc.jp/NuclPart/pac_0606/pdf/p17-Hayano.pdf
25. Akaishi, Y., Yamazaki, T.: Phys. Rev. C **65**, 044005 (2002)
26. Shevchenko, N.V., Gal, A., Mares, J.: Phys. Rev. Lett. **98**, 082301 (2007)
27. Shevchenko, N.V., et al.: Phys. Rev. C **76**, 044004, (2007)
28. Hyodo, T., Weise, W.: Phys. Rev. C **77**, 035204, (2008)
29. Doté, A., Hyodo, T., Weise, W.: Nucl. Phys. **A804**, 197, (2008)
30. Doté, A., Hyodo, T., Weise, W.: Phys. Rev. **C79**, 014003 (2009)
31. Batty, C.J., Friedman, E., Gal, A.: Phys. Rep. **287**, 385 (1997)
32. Iwasaki, M., et al.: Nucl. Instr. Meth. A **473**, 286 (2001)
33. Suzuki, T., et al.: Phys. Lett. B **597**, 263 (2004)
34. Sato, M., et al.: Phys. Lett. B **659**, 107 (2008)
35. Iwasaki, M., et al.: Nucl. Phys. A **804**, 804 (2008)
36. Suzuki, T., et al.: Phys. Rev. C **76**, 068202 (2007)
37. Agnello, M., et al.: Phys. Lett. B **654**, 80 (2007)
38. Agnello, M., et al.: Phys. Rev. Lett. **94**, 212303 (2005)
39. Katz, P.A., et al.: Phys. Rev. D **1**, 1267 (1970)
40. Suzuki, T., et al.: arXiv:0711.4943 (2007)
41. Kishimoto, T.: Phys. Rev. Lett. **83**, 4701 (1999)
42. Kishimoto, T., et al.: Prog. Theo. Phys. Supp. **168**, 573 (2007)
43. Morimatsu, O., Yazaki, K.: Nucl. Phys. A **435**, 727 (1985)
44. Morimatsu, O., Yazaki, K.: Nucl. Phys. A **483**, 493 (1988)
45. Lio, M., et al.: J-PARC E15 proposal, http://j-parc.jp/NuclPart/pac_0606/pdf/p15-Iwasaki.pdf
46. Koike, T., Harada, T.: Phys. Lett. B **652**, 262 (2007)
47. Hatsuda, T., Lee, S.H.: Phys. Rev. C **46**, 34 (1992)
48. Muto, R., et al.: Nucl. Phys. A **774**, 723 (2006)
49. Muto, R., et al.: Phys. Rev. Lett. **98**, 042501 (2007)
50. Yokkaichi, S., et al.: J-PARC E16 proposal, http://j-parc.jp/NuclPart/pac_0606/pdf/p16-Yokkaichi_2.pdf
51. Iwasaki, M., et al.: J-PARC Letter of Intent, http://j-parc.jp/NuclPart/pac_0801/pdf/LoI-Iwasaki.pdf
52. Agakishiev, G., et al.: Phys. Rev. Lett. **75**, 1272 (1995)

53. Lolos, G.J., et al.: Phys. Rev. Lett. **80**, 241 (1998)
54. Ozawa, K., et al.: Phys. Rev. Lett. **86**, 5019 (2001)
55. Trnka, D., et al.: Phys. Rev. Lett. **94**, 192303 (2005)
56. Naruki, M., et al.: Phys. Rev. Lett. **96**, 09230 (2006)
57. Arnaldi, R., et al.: Phys. Rev. Lett. **96**, 162302 (2006)
58. Brown, G.E., Rho, M.: Phys. Rev. Lett. **66**, 2720 (1991)
59. Hatsuda, T., Shiomi, H., Kuwabara, H.: Prog. Theor. Phys. **95**, 1009 (1996)
60. Klingl, F., Waas, T., Weise, W.: Phys. Lett. B **431**, 254 (1998)
61. Oset, E., Ramos, A.: Nucl. Phys. A **679**, 616 (2001)
62. Yamazaki, T., Akaishi, Y.: Phys. Lett. B **453**, 1 (2000)
63. Klingl, F., Waas, T., Weise, W.: Phys. Lett. B **431**, 254 (1998)
64. Oset, E., et al.: Acta Phys. Hung. A **27**, 115 (2006)
65. Evangelista, C., et al.: Phys. Rev. D **57**, 5370 (1996)
66. Itahashi, K., et al.: J-PARC Letter of Intent, http://j-parc.jp/NuclPart/pac_0707/pdf/LoI-Itahashi.pdf

Muon Particle and Nuclear Physics Programs at J-PARC

Yoshitaka Kuno

Osaka University, Toyonaka, Osaka, Japan 305-0042
kuno@phys.sci.osaka-u.ac.jp

1 Overview

An overview (but not comprehensive) list of intense slow muon physics programs at J-PARC, ranging from particle physics to nuclear physics, is given in Table 1. Potential research projects with muons at J-PARC are mostly focused on high-precision measurements and/or searches for unseen rare processes with muons. The reason why muons are selected is that technical improvements facilitating production of large numbers of muons, predicted to be about 10^{10}–10^{12} muons/s in the future, are expected. This is in contrast to the number of taus capable of being produced, 10^8–10^9/year.

Among the particle physics programs with muons [1], very important are (a) precision measurement of the muon $g-2$ anomalous magnetic moment, (b) search for T-odd muon electric dipole moment (EDM), and (c) search for lepton flavor violation of charged leptons (LFV).[1] These three are often called "the muon trio." They are significantly important to search for new physics and also closely related to one another in some cases of new physics, such as supersymmetry (SUSY).

Table 1. Overview list of slow muon physics programs.

Categories	Topics	Comments	Beam
Precision	Muon lifetime	G_F determination	Pulsed
measurements	Muon capture rates	Nuclear physics	Pulsed or DC
	Muonic X-ray	Nuclear physics	Pulsed or DC
	Muon $g-2$	SM allowed, new physics	Pulsed
Rare processes	**Muon EDM**	SM suppressed, new physics	Pulsed
	Muon LFV	SM forbidden, new physics	Pulsed or DC

[1] It is lepton flavor violation for charged leptons.

Kuno, Y.: *Muon Particle and Nuclear Physics Programs at J-PARC.*
Lect. Notes Phys. **781**, 231–262 (2009)
DOI 10.1007/978-3-642-00961-7_9

The goals of next-generation experiments on the muon trio might be given as follows:

Muon $g-2$	:	0.5 ppm	$\rightarrow$	0.05 ppm
Muon EDM	:	$10^{-19}e{\cdot}$cm	$\rightarrow$	$10^{-24}e{\cdot}$cm
Muon LFV B($\mu^+ \rightarrow e^+\gamma$)	:	10^{-11}	$\rightarrow$	10^{-13}
Muon LFV B($\mu^- N \rightarrow e^- N$)	:	10^{-12}	$\rightarrow$	10^{-18}

The physics motivation of the muon trio should be robust for the next decade. They are briefly shown in the following.

1.1 Lepton Flavor Violation with Muons

Among the research topics, we will be particularly focusing on the search for LFV of charged leptons. This is strictly forbidden in the Standard Model with massless neutrinos. Even with massive neutrinos, LFV of charged leptons is highly suppressed by the GIM mechanism and the small neutrino masses. However, new physics beyond the Standard Model predicts sizable effects for the LFV processes [1]. The muon system is considered to be the best system in which to study LFV of charged leptons since the number of muons available is the highest. In consideration of future LFV experiments with muons, the search for muon to electron conversion ($\mu-e$ conversion, $\mu^- + N \rightarrow e^- + N$) is chosen.

1.2 The Anomalous Magnetic Moment of the Muon

The E821 at Brookhaven National Laboratory (BNL) carried out precision measurements of the anomalous $g-2$ magnetic moments of μ^+ and μ^-. Their combined result is [2]

$$a_\mu = \frac{g-2}{2} = 11659208(6) \times 10^{-10} \quad (0.5\,\text{ppm}). \tag{1}$$

There is a difficulty in estimation of the Standard Model (SM) value of the muon $g-2$, since the hadronic contribution has a large uncertainty, depending on which experimental data are used. If the data from e^+e^- collision experiments are used, a deviation from the SM prediction is $\Delta a_\mu = (23.9 \pm 9.9) \times 10^{-10}$ yielding 2.6 σ level, whereas when the data from τ decays are used, $\Delta a_\mu = (7.6 \pm 8.9) \times 10^{-10}$, yielding 0.9 σ level.

On the other hand, the SUSY contribution to the muon $g-2$ is given as follows:

$$a_\mu^{\text{SUSY}} \sim 13 \times 10^{-10} \cdot \tan\beta \cdot \left(\frac{100\text{GeV}}{\tilde{m}}\right)^2. \tag{2}$$

From Eq. (2), the possible deviation can be explained by SUSY contribution with its right magnitude, if $\tan\beta$ and a typical SUSY mass $\tilde{m}$ are in the LHC range.

In the near future, a new experiment BNL-E969 to improve the muon $g-2$ value down to 0.2 ppm was scientifically approved and is awaiting funding. In the far future, a letter of intent should be submitted to J-PARC, Japan, to improve down to 0.05 ppm with higher muon beam intensity and the use of backward-decay muons.

1.3 Electric Dipole Moment of the Muon

The electric dipole moment (EDM) of an elementary particle is known to be a P-odd and T-odd observable. If CPT invariance is assumed, a non-vanishing EDM would imply CP violation. Since the contribution from the source of CP violation in the Standard Model is known to be very small, the observable effect should come from new physics beyond the Standard Model. Searches for EDM for various elementary particles, such as the electron, the neutron, and various atoms, have been extensively made in the past. Studying the EDM of the muon would greatly facilitate EDM searches, because elementary particles in the second generation such as the muon might receive a larger contribution from new physics than lighter particles, and less ambiguous results may be obtained than in searches involving strong interactions.

The SUSY contributions to the muon EDM and the muon $g-2$ are closely related, since they are imaginary and real parts of the same diagrams of new physics. A naive estimation of new physics contribution to the muon EDM (d_μ^{NP}) is made as follows [3, 4]:

$$d_\mu^{NP} \sim 3 \times 10^{-22} \left(\frac{\Delta a_\mu^{NP}}{3 \times 10^{-9}} \right) \tan \phi_{CP}, \tag{3}$$

where ϕ_{CP} is an unknown CP imaginary phase and Δa_μ^{NP} is new physics contribution to the muon $g-2$. From Eq. (3), the muon EDM of an order of 10^{-24} $e{\cdot}$cm at the maximum can be possible for a large ϕ_{CP}. In minimal SUSY-seesaw models consistent with neutrino oscillation data, d_μ has also been estimated, assuming that CP violation in soft SUSY breaking parameters induced neutrino Yukawa couplings. The prediction ranges from 10^{-26} to 10^{-28} $e{\cdot}$cm, but it could become larger for some parameters. It is also noticed that in SUSY models, a ratio of the muon EDM to the electron EDM, d_μ/d_e, may not be proportional to their mass ratio, m_μ/m_e, and could be enhanced more for some parameter cases.

The EDM of the muon can be searched for by storing muons in a storage ring. The muons would be subjected to a strong electric field produced by a relativistic effect resulting from an applied magnetic field in the storage ring. A letter of intent (LOI) for a search for the EDM of the muon at a sensitivity of 10^{-24} $e{\cdot}$cm was submitted to J-PARC, Japan, in 2003 [5].

1.4 R&D of Neutrino Factories and Muon Colliders

As future accelerator projects after the LHC and the ILC, neutrino factories and muon colliders are receiving much attention from experimentalists and theorists. A neutrino factory is a next-generation accelerator project for producing a highly intense neutrino beam from decays of muons accelerated to high energy (such as 50 GeV). The physics research topics of a neutrino factory would be (1) search for CP violation in the neutrino sector, (2) search for new physical interactions accompanying the neutrinos, and (3) precision testing of the unitarity of the Maki–Nakagawa–Sakata (MNS) neutrino-mixing matrix. A muon collider is an ultrahigh energy $\mu^+\mu^-$ collider of multiple TeVs in energy, where ultrahigh energies can be achieved using a circular accelerator thanks to the heavy muon mass, in contrast to the ILC. The critical R&D issues are related to the technology used to accelerate the muons. Establishing a cooling technology applicable to the muon is regarded as being of the highest priority. The R&D for neutrino factories and muon colliders would promote the construction of highly intense muon sources.

In the following sections, more detailed descriptions of muon LFV are given. The staging approach of our research scenario is also given. In the subsequent sections, each of the staging experiments of searching for muon LFV is presented briefly.

2 Physics Motivation of Lepton Flavor Violation of Charged Leptons

Recently, lepton flavor violation (LFV) of charged leptons[2] has attracted much theoretical and experimental interest, since it has a growing potential to help in finding important clues for new physics beyond the Standard Model [1]. Some of the notable features of LFV studies are that (1) LFV of charged leptons might have sizable contributions from new physics, which could be observed in future experiments, and (2) LFV of charged leptons does not have any sizable contributions from the Standard Model (such as from neutrino mixing), which could cause serious background signals otherwise.

Since the first search by Hincks and Pontecorvo in 1947 [6], experimental searches for LFV have been continuously carried out with various elementary particles, such as muons, kaons, and others. The upper limits have been improved at a rate of 2 orders of magnitude per decade, as can be seen in Fig. 1 (left). The present upper limits of various LFV decays are listed in Fig. 1 (right), where it can be seen that the sensitivity of the muon system to LFV is very high. This is mostly because of the large number of muons available for experimental searches nowadays (about 10^{14}–10^{15} muons/year). Moreover, an

[2] Hereafter, the term “LFV” is used for lepton flavor violation of charged leptons, even when charged leptons are not explicitly specified.

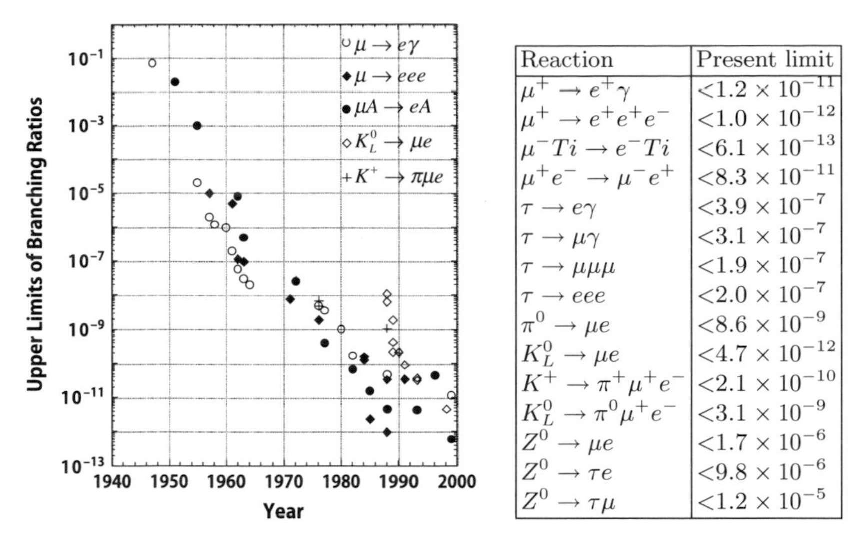

Reaction	Present limit
$\mu^+ \to e^+\gamma$	$<1.2 \times 10^{-11}$
$\mu^+ \to e^+e^+e^-$	$<1.0 \times 10^{-12}$
$\mu^- Ti \to e^- Ti$	$<6.1 \times 10^{-13}$
$\mu^+e^- \to \mu^-e^+$	$<8.3 \times 10^{-11}$
$\tau \to e\gamma$	$<3.9 \times 10^{-7}$
$\tau \to \mu\gamma$	$<3.1 \times 10^{-7}$
$\tau \to \mu\mu\mu$	$<1.9 \times 10^{-7}$
$\tau \to eee$	$<2.0 \times 10^{-7}$
$\pi^0 \to \mu e$	$<8.6 \times 10^{-9}$
$K_L^0 \to \mu e$	$<4.7 \times 10^{-12}$
$K^+ \to \pi^+\mu^+e^-$	$<2.1 \times 10^{-10}$
$K_L^0 \to \pi^0\mu^+e^-$	$<3.1 \times 10^{-9}$
$Z^0 \to \mu e$	$<1.7 \times 10^{-6}$
$Z^0 \to \tau e$	$<9.8 \times 10^{-6}$
$Z^0 \to \tau\mu$	$<1.2 \times 10^{-5}$

Fig. 1. (*Left*) History of searches for LFV. (*Right*) Present limits of LFV for various particle decay modes.

even greater number of muons (about 10^{19}–10^{20} muons/year) will be available in the future, if new highly intense muon sources are realized.

In the minimal Standard Model, lepton flavor conservation is built by assuming vanishingly small neutrino masses. However, neutrino mixing has been experimentally confirmed by the discovery of neutrino oscillations, and lepton flavor conservation is known to be violated. However, LFV of charged leptons has yet to be observed experimentally. It is known that the contribution of neutrino mixing to LFV is extremely small, since it is proportional to $(m_\nu/m_W)^4$, yielding the order of 10^{-52} in branching ratios. Therefore, discovery of LFV would imply new physics beyond "neutrino oscillations." As a matter of fact, any new physics or interactions beyond the Standard Model would predict LFV at some level. The motivation for studying the physics of LFV throughout the next decade is very robust. To illustrate that, let us consider, as just an example, the case of supersymmetry (SUSY).

2.1 Supersymmetry – Supergravity Models

It is known that LFV has significant contributions from SUSY, if SUSY particles exist in the LHC energy range. In SUSY models, LFV of charged leptons would occur through the mixing of their SUSY partners, sleptons $\tilde{l}$. Figure 2 shows one of the diagrams of SUSY contributing to a muon to electron transition, where the mixing of a smuon ($\tilde{\mu}$) and a selectron ($\tilde{e}$) is given by the off-diagonal slepton mass matrix element $m^2_{\tilde{\mu}\tilde{e}}$, where the slepton mass matrix ($m^2_{\tilde{l}}$) is given by the following equation:

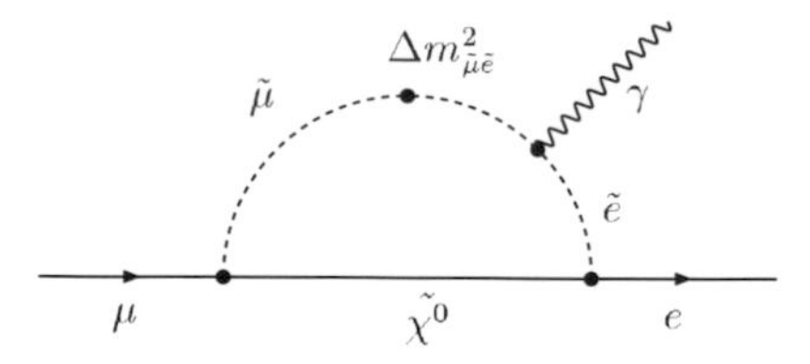

Fig. 2. One of the diagrams of SUSY contributions to a μ to e transition. $\Delta m^2_{\tilde{\mu}\tilde{e}}$ indicates the magnitude of the slepton mixing.

$$m^2_{\tilde{l}} = \begin{pmatrix} m^2_{\tilde{e}\tilde{e}} & \Delta m^2_{\tilde{e}\tilde{\mu}} & \Delta m^2_{\tilde{e}\tilde{\tau}} \\ \Delta m^2_{\tilde{\mu}\tilde{e}} & m^2_{\tilde{\mu}\tilde{\mu}} & \Delta m^2_{\tilde{\mu}\tilde{\tau}} \\ \Delta m^2_{\tilde{\tau}\tilde{e}} & \Delta m^2_{\tilde{\tau}\tilde{\mu}} & m^2_{\tilde{\tau}\tilde{\tau}} \end{pmatrix} . \quad (4)$$

In one type of SUSY model termed supergravity model, the slepton mass matrix is assumed to be diagonal at the Planck mass scale (10^{19} GeV), and no off-diagonal matrix elements exist ($\Delta m^2_{\tilde{\mu}\tilde{e}} = 0$). However, non-zero off-diagonal matrix elements can be induced by radiative corrections from the Planck scale to the weak scale ($\sim 10^2$ GeV) when new physics exists in between [7]. This could deduce grand unification theories (GUT), where Yukawa interactions at GUT create non-zero off-diagonal elements. This scenario is called a *SUSY-GUT* model [8]. In contrast, the new physics could be constituted by the neutrino seesaw mechanism, where the neutrino Yukawa interaction has the same effect. This scenario is called a *SUSY-seesaw* model [9–13]. These two scenarios are illustrated in Fig. 3. Both models predict large branching ratios for LFV, which are just a few orders of magnitude below the current experimental upper limits. Figures 4 and 5 show the predictions of the SUSY-GUT and SUSY-seesaw models, respectively. If we could improve experimental sensitivity by a few orders of magnitude, this would provide a great potential for new discoveries. If the LHC does not find evidence for SUSY, two cases can be considered: either SUSY does not exist at all or SUSY only exists for heavier masses on a multiple TeV scale. High-precision measurements

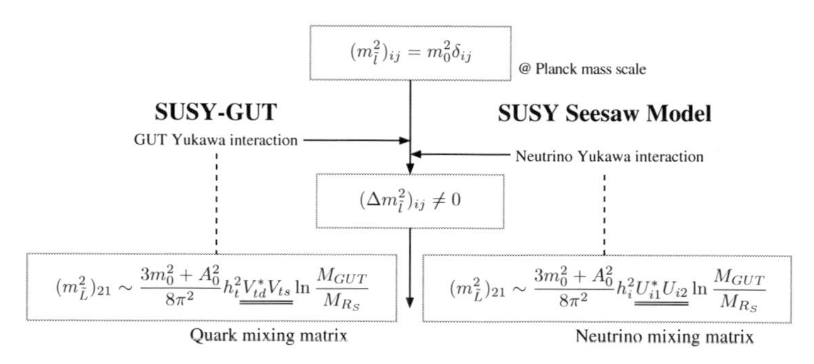

Fig. 3. Two physics mechanisms (SUSY-GUT and SUSY-seesaw) introducing slepton mixing into MSSM.

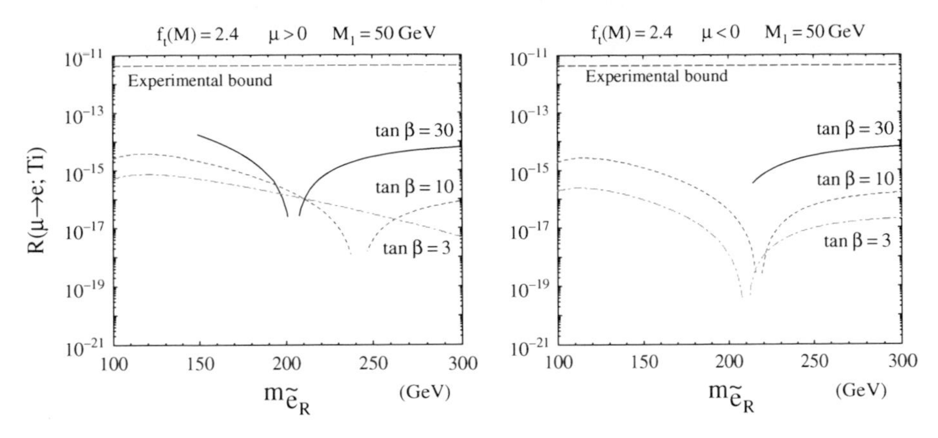

Fig. 4. Predicted branching ratios for $\mu^- - e^-$ conversion in SUSY-GUT. The present best published limit is 4.3×10^{-12}.

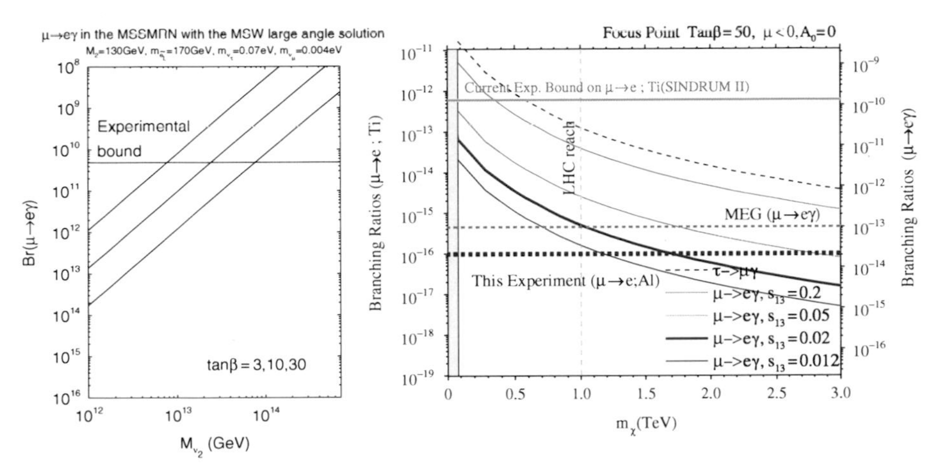

Fig. 5. (*Left*) Predictions of $\mu^+ \to e^+\gamma$ branching ratio in SUSY-seesaw models. The *three lines* correspond to the cases of $\tan\beta = 30, 10, 3$ from top to bottom, respectively. (*Right*) Prediction of the branching ratio of $\mu - e$ conversion in Ti in SUSY-seesaw models as a function of SUSY mass scale (neutralino). The aimed sensitivities of MEG and PRISM (hereafter referred to as "this experiment") are also shown.

with intense slow muons become very important, since such measurements are sensitive to a heavier mass scale than what can be reached by high-energy accelerators. For heavier SUSY, if the LFV search has sufficient experimental sensitivity (such as 10^{-18} for $\mu^- - e^-$ conversion), it could be sensitive to the SUSY mass scale up to several TeV, as shown in Fig. 5 (right). Therefore, the search for LFV would be worth carrying out even if the LHC does not find evidence for SUSY below the TeV energy scale. It should be noted that besides SUSY, there are other models that predict sizable effects of LFV.

These include heavy neutrino models, leptoquark models, composite models, two Higgs doublet models, second Z' models, and anomalous Z coupling.

3 Phenomenology of $\mu^- - e^-$ Conversion

3.1 What is a $\mu^- - e^-$ Conversion Process?

One of the most prominent muon LFV processes is coherent neutrino-less conversion of muons to electrons ($\mu^- - e^-$ conversion), $\mu^- + N(A, Z) \rightarrow e^- + N(A, Z)$. When a negative muon is stopped by some material, it is trapped by an atom and a muonic atom is formed. After it cascades down energy levels in the muonic atom, the muon is bound in its $1s$ ground state. The fate of the muon is then to either decay in orbit ($\mu^- \rightarrow e^- \nu_\mu \overline{\nu}_e$) or be captured by a nucleus of mass number A and atomic number Z, namely $\mu^- + (A, Z) \rightarrow \nu_\mu + (A, Z-1)$. However, in the context of physics beyond the Standard Model, the exotic process of neutrino-less muon capture, such as

$$\mu^- + (A, Z) \rightarrow e^- + (A, Z) \quad (5)$$

is also expected. This process is called $\mu^- - e^-$ conversion in a muonic atom. This process violates the conservation of lepton flavor numbers, L_e and L_μ, by one unit, but the total lepton number, L, is conserved. The final state of the nucleus (A, Z) could be either the ground state or one of the excited states. In general, the transition to the ground state, which is called coherent capture, is dominant. The rate of the coherent capture over non-coherent capture is enhanced by a factor approximately equal to the number of nucleons in the nucleus, since all the nucleons participate in the process.

3.2 Signal and Background Events

The event signature of coherent $\mu^- - e^-$ conversion in a muonic atom is a mono-energetic single electron emitted from the conversion with an energy of $E_e \sim m_\mu - B_\mu$, where m_μ is the muon mass and B_μ is the binding energy of the $1s$ muonic atom.

From an experimental point of view, $\mu^- - e^-$ conversion is a very attractive process. First, the e^- energy of about 105 MeV is far above the endpoint energy of the muon decay spectrum (~52.8 MeV). Second, since the event signature is a mono-energetic electron, no coincidence measurement is required. The search for this process has the potential to improve sensitivity by using a high muon rate without suffering from accidental background events, which would be serious for other processes, such as $\mu^+ \rightarrow e^+ \gamma$ and $\mu^+ \rightarrow e^+ e^+ e^-$ decays.

The electron is emitted with an energy $E_e \approx m_\mu$, which coincides with the end point of muon decay in orbit (DIO), which is the only relevant intrinsic physics background event. Since the energy distribution of DIO falls steeply above $m_\mu/2$, the experimental setup may have a large signal acceptance and the detectors can still be protected against the vast majority of decay and capture background events. Energy distributions for DIO electrons have been calculated for a number of muonic atoms [14–16] and energy resolutions of the order of 0.1% are sufficient to keep this background below 10^{-18}.

There are several other potential sources of electron background events in the energy region around 100 MeV, involving either beam particles or cosmic rays. Beam-related background events may originate from muons, pions, or electrons in the beam. Apart from DIO, muons may produce background events by muon decay in flight or radiative muon capture (RMC). Pions may produce background events by radiative pion capture (RPC). Gamma rays from RMC and RPC produce electrons mostly through e^+e^- pair production inside the target.

Beam-related background events can be suppressed by various methods:

- *Beam pulsing*: Since muonic atoms have lifetimes of the order of 100 ns, a pulsed beam with buckets that are short compared with this lifetime would allow one to remove prompt background events by performing measurements in a delayed time window. As will be discussed below there are stringent requirements on beam extinction during the measuring interval. Two approaches to achieving beam extinction at the required levels are considered. One is extinction for protons and the other is extinction for muons. For the latter, a kicker magnet for injection/extraction of the muon storage ring can be considered as an additional extinction device.
- *Beam purity*: A low-momentum ($<$70 MeV/c) μ^- beam with no pion contamination ($<10^{-20}$) would keep prompt background events at a negligible level. This could be achieved by adopting a muon storage ring where pions decay with a flight length of the order of a few hundreds of meters. A major advantage of the method is that heavy targets such as gold, with which muon lifetimes become around 70 ns, can be studied. This scheme will be applied by PRISM, as will be described later.

3.3 $\mu^-\!-e^-$ Conversion vs. $\mu^+ \to e^+\gamma$

Two possible contributions in the $\mu^-\!-e^-$ transition diagrams are to be considered. One is a photonic contribution and the other is a non-photonic contribution. For the photonic contribution, there is a definite relation between the $\mu^-\!-e^-$ conversion process and the $\mu^+ \to e^+\gamma$ decay. Suppose the photonic contribution is dominant, the branching ratio of the $\mu^- - e^-$ conversion process is expected to be smaller than that of $\mu^-\!-e^-$ decay by a factor of about a few hundred. This implies that the search for $\mu^-\!-e^-$ conversion at the level of 10^{-16} is comparable to that for $\mu^+ \to e^+\gamma$ at the level of 10^{-14}.

If the non-photonic contribution dominates, the $\mu^+ \to e^+\gamma$ decay would be small whereas the $\mu^-\!-e^-$ conversion could be sufficiently large to be observed. It is worth noting the following. If a $\mu^+ \to e^+\gamma$ signal is found, the

$\mu^-{-}e^-$ conversion signal should also be found. If no $\mu^+ \to e^+\gamma$ signal is found, there will still be an opportunity to find a $\mu^-{-}e^-$ conversion signal because of the potential existence of non-photonic contributions.

3.4 Present Experimental Status

The experimental status of searches for $\mu^-{-}e^-$ conversion and $\mu^+ \to e^+\gamma$ decay is presented. Table 2 summarizes the history of searches for $\mu^-{-}e^-$ conversion. The latest search for $\mu{-}e$ conversion was performed by the SINDRUM II collaboration at PSI. Figure 6 shows their results. The main spectrum, taken at 53 MeV/c, shows the steeply falling distribution expected from muon DIO. Two events were found at higher momenta, but just outside the region of interest. The agreement between measured and simulated positron distributions from μ^+ decay means that confidence can be had in the accuracy of the momentum calibration. At present there are no hints concerning the nature of the two high-momentum events: they might have been induced by cosmic rays or RPC, for example.

There was another proposal at BNL, the MECO experiment [25], aiming to search with a sensitivity of 10^{-16}. This project was planned to combat beam-related background events with the help of a pulsed 8 GeV/c proton beam. Figure 7 (left) shows the proposed layout. Pions are produced by 8 GeV/c protons crossing a 16-cm-long tungsten target, and muons from the decays of the pions are collected efficiently with the help of a graded magnetic field. Negatively charged particles with 60–120 MeV/c momenta are transported by a curved solenoid to the experimental target. In the spectrometer magnet, a graded field is also applied. A major challenge is the requirement for proton extinction in between the proton bursts. In order to maintain the pion stop rate in the "silent" interval, a beam extinction factor better than 10^{-8}–10^{-9} is required. Unfortunately, the MECO experiment was cancelled in 2005 due to funding problems. However, the revival of the MECO experiment (the mu2e experiment) is now being seriously considered at Fermi National Laboratory.

Table 2. Past experiments on $\mu^-{-}e^-$ conversion.

Process	Upper limit	Place	Year	Reference
$\mu^- + \mathrm{Cu} \to e^- + \mathrm{Cu}$	$< 1.6 \times 10^{-8}$	SREL	1972	[17]
$\mu^- + {}^{32}\mathrm{S} \to e^- + {}^{32}\mathrm{S}$	$< 7 \times 10^{-11}$	SIN	1982	[18]
$\mu^- + \mathrm{Ti} \to e^- + \mathrm{Ti}$	$< 1.6 \times 10^{-11}$	TRIUMF	1985	[19]
$\mu^- + \mathrm{Ti} \to e^- + \mathrm{Ti}$	$< 4.6 \times 10^{-12}$	TRIUMF	1988	[20]
$\mu^- + \mathrm{Pb} \to e^- + \mathrm{Pb}$	$< 4.9 \times 10^{-10}$	TRIUMF	1988	[20]
$\mu^- + \mathrm{Ti} \to e^- + \mathrm{Ti}$	$< 4.3 \times 10^{-12}$	PSI	1993	[21]
$\mu^- + \mathrm{Pb} \to e^- + \mathrm{Pb}$	$< 4.6 \times 10^{-11}$	PSI	1996	[22]
$\mu^- + \mathrm{Ti} \to e^- + \mathrm{Ti}$	$< 6.1 \times 10^{-13}$	PSI	1998*	[23]
$\mu^- + \mathrm{Au} \to e^- + \mathrm{Au}$	$< 7 \times 10^{-13}$	PSI	2001*	[24]

* Reported only in Conference Proceedings.

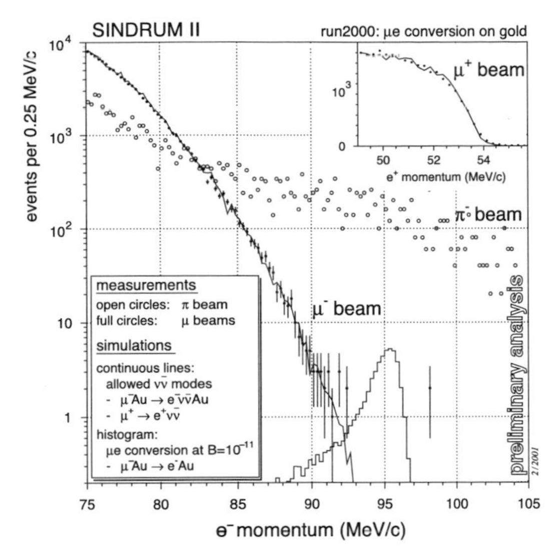

Fig. 6. Recent results by SINDRUM II. Momentum distributions of emitted electrons/positrons for three beams with different momenta and polarities: (i) 53 MeV/c negative, optimized for μ^- stops, (ii) 63 MeV/c negative, optimized for π^- stops, and (iii) 48 MeV/c positive, optimized for μ^+ stops. The 63 MeV/c data were scaled to the different measuring times. The μ^+ data were taken using a reduced spectrometer field.

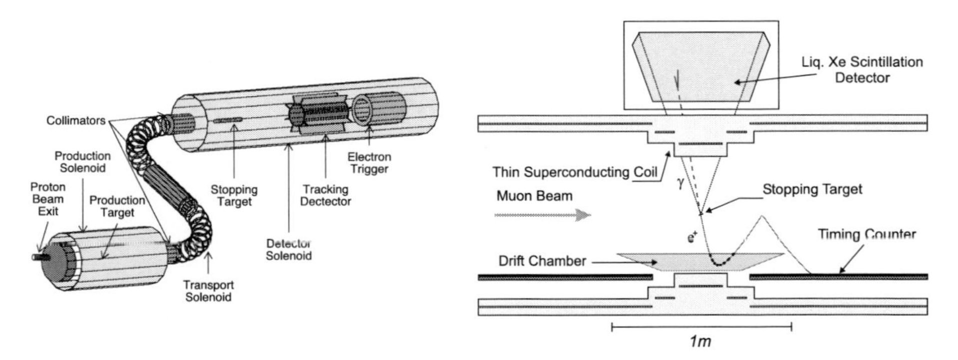

Fig. 7. (*Left*) Setup of the MECO experiment. (*Right*) Setup of the MEG experiment.

The present experimental limit for $\mu^+ \to e^+\gamma$ is 1.2×10^{-11}, which was obtained by the MEGA experiment at LANL in the United States [26]. A new experiment called MEG at PSI [27], which aims to achieve a sensitivity of 10^{-13} in the $\mu^+ \to e^+\gamma$ branching ratio, is under construction. A schematic view of the detector is shown in Fig. 7 (right). The improved experiment will be expected to utilize a continuous DC muon beam of 100% duty factor at PSI. Utilizing the same instantaneous beam intensity as MEGA, the total number of muons available can be increased by a factor of 16. A further improvement is the use of a novel liquid xenon scintillation detector which is a 0.8 m^3 volume of liquid xenon observed by an array of 800 photomultipliers from all sides (so-called "mini-Kamiokande" type). For e^+ detection, a solenoidal magnetic spectrometer with a graded magnetic field is to be adopted. Physics data taking are expected to start in the year 2007 or later.

3.5 Why is $\mu^-{-}e^-$ Conversion the Next Step?

Considering its marked importance to physics, it is highly desirable to consider a next-generation experiment to search for LFV. There are three processes to be considered, namely $\mu^+ \to e^+\gamma$, $\mu^+ \to e^+e^+e^-$, and $\mu^-{-}e^-$ conversion.

The three processes have different experimental issues that need to be solved to realize improved experimental sensitivities. They are summarized in Table 3. The processes of $\mu^+ \to e^+\gamma$ and $\mu^+ \to e^+e^+e^-$ are detector limited. To consider and go beyond the present sensitivities, the resolutions of detection have to be improved, which is, in general, very hard. In particular, improving the photon energy resolution is difficult. On the other hand, for $\mu^-{-}e^-$ conversion, there are no accidental background events, and an experiment with higher rates can be performed. If a new muon source with a higher beam intensity and better beam quality for suppressing beam-associated background events can be constructed, measurements of higher sensitivity can be performed.

Table 3. LFV processes and issues.

Process	Major background events	Beam requirements	Sensitivity issues
$\mu^+ \to e^+\gamma$	Accidental events	DC beam	Detector resolution
$\mu^+ \to e^+e^+e^-$	Accidental events	DC beam	Detector resolution
$\mu^-{-}e^-$ conversion	Beam associated	pulsed beam	Beam qualities

Furthermore, it is known that in comparison with $\mu^+ \to e^+\gamma$, there are more physical processes that $\mu^-{-}e^-$ conversion and $\mu^+ \to e^+e^+e^-$ could contribute to. Namely, in SUSY models, photon-mediated diagrams can contribute to all the three processes, but the Higgs-mediated diagrams can contribute to only $\mu^-{-}e^-$ conversion and $\mu^+ \to e^+e^+e^-$. In summary, with all the above considerations, a search for $\mu^-{-}e^-$ conversion would be the natural next step to take to realize an improved experiment.

4 Overview of LFV Searches at J-PARC

We intend to perform the searches for $\mu^-{-}e^-$ conversion in a staging approach as follows:

- Phase 1 **COMET**: $B(\mu^- + \mathrm{Al} \to e^- + \mathrm{Al}) < 10^{-16}$ with a highly intense muon source (Fig. 8, left), and

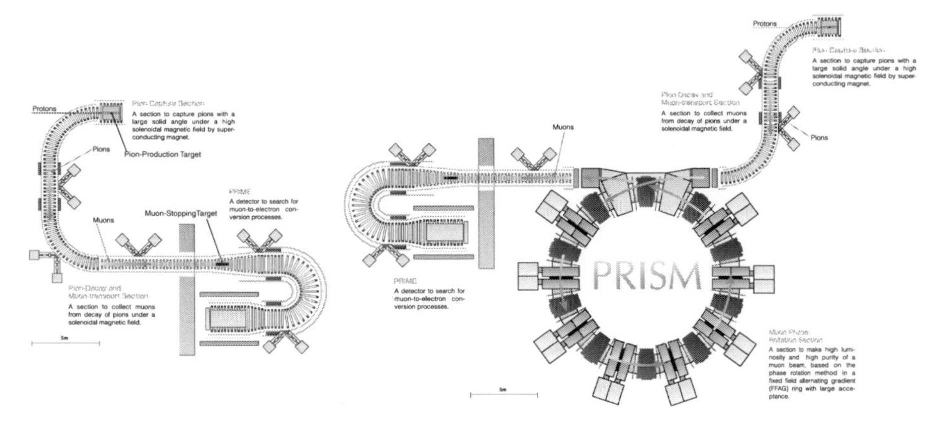

Fig. 8. Phase 1 layout (*left*) and Phase 2 layout (*right*).

- Phase 2 **PRISM**: $B(\mu^- + \mathrm{Ti} \rightarrow e^- + \mathrm{Ti}) < 10^{-18}$ with an additional muon storage ring (Fig. 8, right).

The target sensitivity for Phase 1 would be a factor of 10,000 better than the present published limit of $B(\mu + \mathrm{Ti} \rightarrow e + \mathrm{Ti}) = 4.3 \times 10^{-12}$ and that at Phase 2 would be a factor of 1,000,000 better. This would allow a very large window in which to search for new physics beyond the Standard Model.

The reason why the staging approach is taken is to realize an early start. The Phase 2 experiment, which is called "PRISM", has several issues to overcome before its realization. First, it needs a pulsed proton beam with fast extraction, whose timing characteristics should match those of the muon storage ring (PRISM-FFAG ring). For instance, in the example case of J-PARC, to obtain such a proton beam, a new proton beamline and a new experimental hall must be built, but the required funding has not yet been approved.

Second, R&D on the PRISM-FFAG ring needs time to develop. On the other hand, Phase 1 could use a pulsed beam with slow extraction and can be accommodated in the hadron hall (NP hall) currently under construction. Therefore, an earlier start can be realized. For the other locations, the staging approach would benefit as well.

5 Phase 1: COMET: $\mu - e$ Conversion at 10^{-16}

The Phase 1 experiment [28] is focused on $B(\mu^- + \mathrm{Al} \rightarrow e^- + \mathrm{Al}) < 10^{-16}$. To improve the sensitivity by a factor of 10,000 over the current limit, several important features have been considered, as highlighted below.

- **Highly Intense Muon Source**: To achieve an experimental sensitivity of 10^{-16}, the total number of muons needed is of the order of 10^{18}, and, therefore, a highly intense muon beamline has to be constructed. To increase the muon beam intensity, two methods are adopted. One is to use a proton beam of high beam power. The other is to use a system of collecting pions, which are parents of muons, with high efficiency. In the muon collider and neutrino factory R&D, *superconducting solenoid magnets* producing a high magnetic field surrounding the pion-production target have been proposed and studied for pion capture over a large solid angle. With the pion capture solenoid system, about 8×10^{20} protons of 8 GeV are needed to achieve the number of muons of the order of 10^{18}.

- **Pulsed Proton Beam**: There are several potential sources of electron background events in the energy region around 100 MeV, where the $\mu^- - e^-$ conversion signal is expected. One of them is beam-related background events. To suppress the occurrence of beam-related background events, a pulsed proton beam utilizing *beam pulsing* is proposed. Since muons in muonic atoms have lifetimes of the order of 100 ns, a pulsed beam with beam buckets that are short compared with these lifetimes would allow removal of prompt beam background events by allowing measurements to be performed in a delayed time window. As will be discussed below, there are stringent requirements on the *beam extinction* during the measuring interval. Tuning of a proton beam in the accelerator ring as well as extra extinction devices should be installed to achieve the required level of beam extinction.
- **Muon Transport System with Curved Solenoids**: The captured pions decay to muons, which are transported with high efficiency through a superconducting solenoid magnet system. Beam particles with high momenta would produce electron background events in the energy region of 100 MeV, and therefore, they must be eliminated with the use of curved solenoids where the centers of the helical motion of the electrons drift perpendicular to the plane in which their paths are curved, and the magnitude of the drift is proportional to their momenta. By using this effect and by placing suitable collimators at appropriate locations, beam particles of high momenta can be eliminated.
- **Spectrometer with Curved Solenoids**: To reject electron background events and reduce the probability of false tracking owing to high counting rates, a curved solenoid spectrometer is considered to allow selection of electrons on the basis of their momenta. The principle of momentum selection is the same as that used in the transport system, but, in the spectrometer, electrons of low momenta which mostly come from muon decay in orbit (DIO) are removed. The detection rate of DIO electrons would be about 1,000 tracks per second (1,000 Hz), whereas the MECO experiment expected hit rates of about 500 kHz per single wire of the tracking device.

Among several candidate locations, one candidate experimental location would be J-PARC, which is currently under construction in Japan. It will have a slow extraction beamline and an experimental hall called the hadron hall. It will have a slow extraction proton beam with an energy of 30 GeV (maximum)

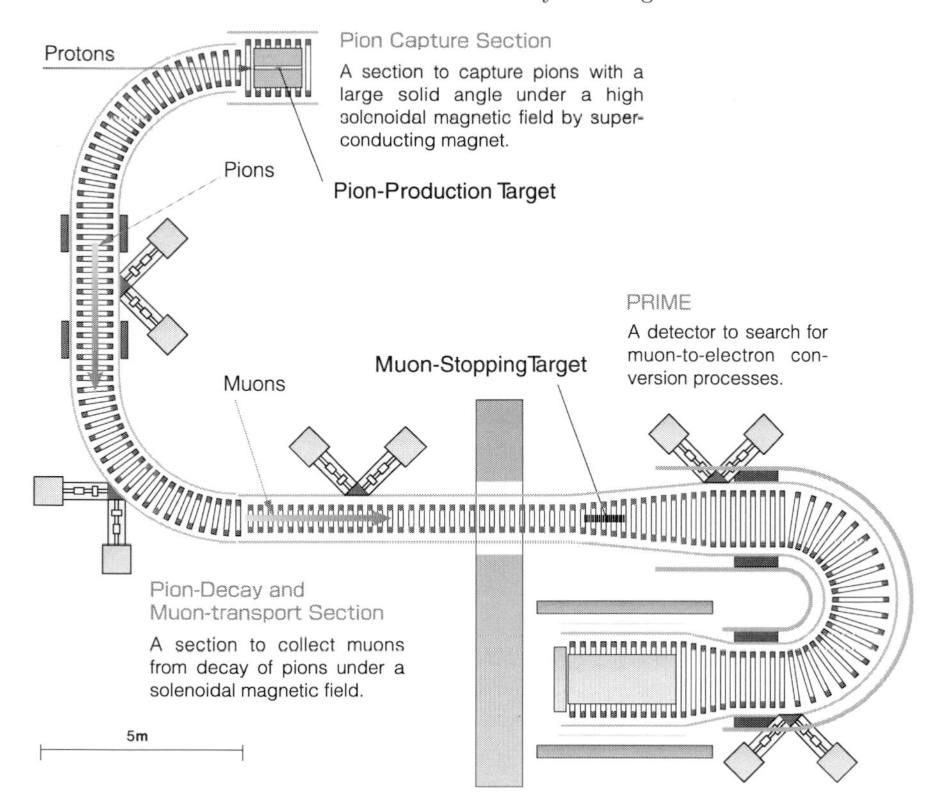

Fig. 9. Schematic layout of muon beamline and detector for $\mu^- - e^-$ conversion search.

and a beam current of 15 μA (maximum).[3] The proton beam power will be about 450 kW, which is the strongest worldwide among GeV proton machines. In the following sections, the requirements for the proton beam are described (Fig. 9).

5.1 The Muon Source

Pulsed Proton Beam

To realize the target sensitivity of 10^{-16} in the branching ratio of $\mu^- - e^-$ conversion, about 10^{18} muons in total are required. On the basis of the current design of the muon beamline, about 8×10^{20} protons with an energy of 8 GeV are needed. For a 2×10^7 s running time, a beam intensity of 4×10^{13} protons per second (which corresponds to a beam current of about 7 μA) is needed. This requirement of beam power is about the same as that in the MECO

[3] In the initial stage, because the proton linear accelerator (LINAC) beam energy is lower, the beam current is expected to be about 9 μA.

experiment at BNL-AGS (MECO Technical Proposal, August 1, 2001, unpublished). It is noted that the reason why a low-proton beam energy like 8 GeV is considered is twofold. One is to suppress production of antiprotons, as will be discussed in Sect. 5.5, and the other is related to beam extinction, where a lower beam energy is easier to kick off.

To suppress beam-related background events for $\mu^{-}-e^{-}$ conversion, a proton beam is required to be supplied in bunches with a time separation of about μs. Detection is carried out between beam bunches. The time separation of μs corresponds to a negative muon lifetime, for instance, in aluminum. For instance, this can be performed at the J-PARC MR ring by filling only every other (or every third) beam bunch in the ring. The cycle time structure of the J-PARC MR ring is illustrated in Fig. 10 (left). The time period of every bunch is about 598 ns (1.67 MHz). For instance, the operation of every other mode can be performed by filling one of the two bunches in the 3-GeV RCS ring. Then, proton bunches in the ring can be extracted in a slow extraction mode. A proton bunch train in a slow extraction mode is shown in Fig. 10 (right).

The beam extinction between beam bunches is of critical importance. For the MECO experiment (MECO Technical Proposal, August 1, 2001, unpublished), some tests to measure proton extinction were performed at BNL-AGS. Figure 11 shows the relative intensity as a function of time with respect to the filled bucket. The proton extinction between buckets is below 10^{-6} and in unfilled buckets is of the order of 10^{-3}. At BNL-AGS, a second test using the E787 detector was performed to measure proton beam extinction, and an extinction of 10^{-7} was measured. It was concluded that the proton extinction of 10^{-9} is unlikely to be achieved by just tuning the AGS, and two possible ways

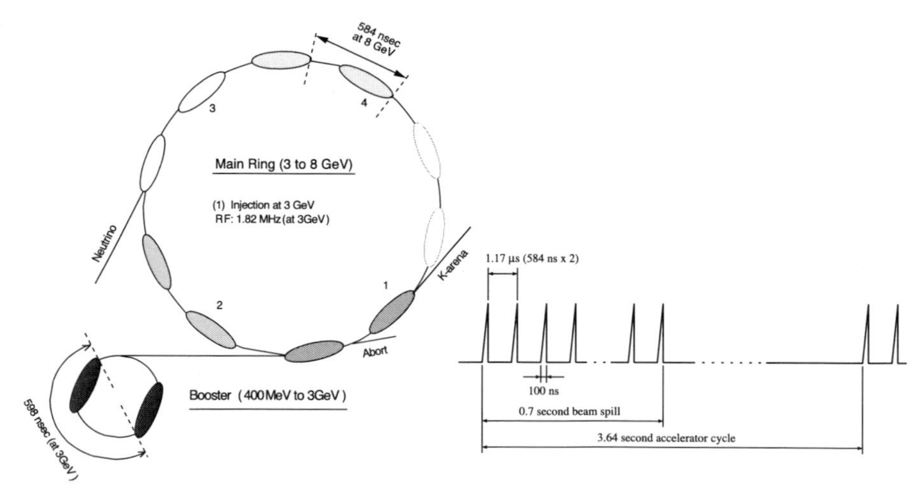

Fig. 10. (*Left*) Typical J-PARC main-ring cycle structure. (*Right*) Proposed bunched proton beams in a slow extraction mode.

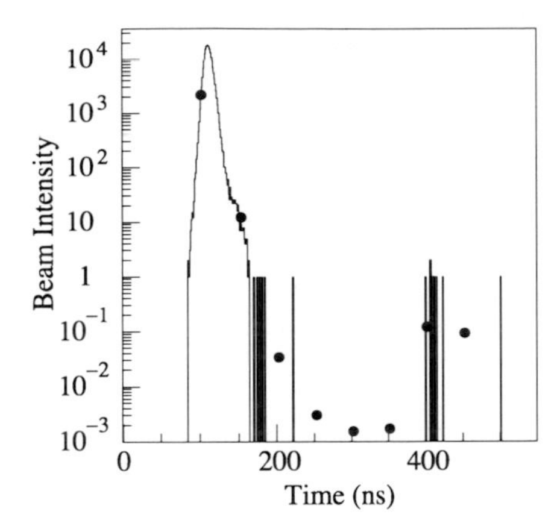

Fig. 11. Results of the proton beam extinction measured at BNL-AGS. The beam intensity is shown as a function of time with respect to pulses in the bunched beam extracted from the AGS. The solid histogram and dots are the results from the measurements with a QVT and scalers, respectively.

to improve the extinction were proposed. One is to install a system of kickers in the ring. The second method is to install a pulsed electric or magnetic kicker in the proton transport line. The additional beam extinction devices are being considered for implementation both in the ring and in the proton transport line.

Pion-Production Target

Experiments on searching for $\mu^- - e^-$ conversion would use low-energy muons which can be stopped in a target. Therefore, capture of low-energy pions is of interest. At the same time, high-energy pions, which could potentially cause background events, should be eliminated. For those reasons, it has been decided to collect pions emitted backward with respect to the proton beam direction. To study pion capture, a simulation was performed using both MARS and GEANT3 with FLUKA.

Figure 12 shows the momentum spectra of π^- produced from a graphite target. It can be seen that the maximum of transverse momentum (p_T) is around 100 MeV/c for a longitudinal momentum (p_L) of $0 < p_L < 200$ MeV/c for both the forward- and backward-scattered pions. The maximum of the total momentum for the backward-scattered pions is about 120 MeV/c, whereas that for the forward-scattered pions is about 200–400 MeV/c. It can also be seen that high energy pions are suppressed in the backward direction.

Pion Capture

To collect as many pions (and cloud muons) of low energy as possible, the pions are captured using a high solenoidal magnetic field. In this case, pions

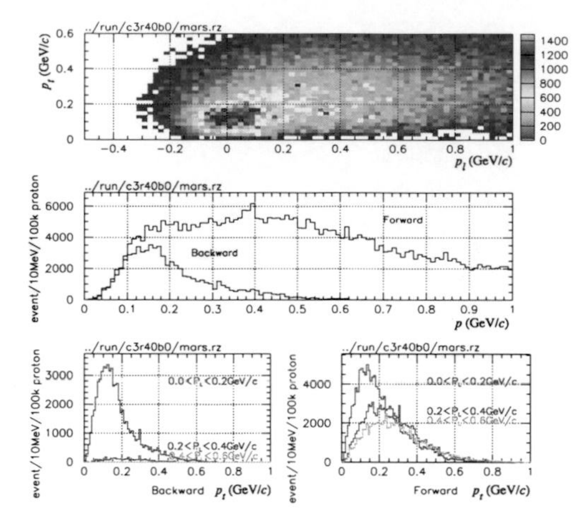

Fig. 12. Pion production in a graphite target. (*Top*) Correlation between p_L and p_T. (*Middle*) Total momentum distributions for forward and backward π^-s. (*Bottom*) p_T distributions for $0 < p_L < 0.2$ GeV/c, $0.2 < p_L < 0.4$ GeV/c, and $0.4 < p_L < 0.6$ GeV/c.

emitted into a half hemisphere can be captured within the transverse momentum threshold ($p_t^{\max}$). $p_t^{\max}$ is given by the magnetic field strength (B) and the radius of the inner bore of solenoid magnet (R) as

$$p_T^{\max}(\mathrm{GeV}/c) = 0.3 \times B(\mathrm{T}) \times R(\mathrm{m})/2. \tag{6}$$

In the current design, we employ conservative design values, namely $B = 5$ T, $R = 15$ cm, and length of 1.4 m.

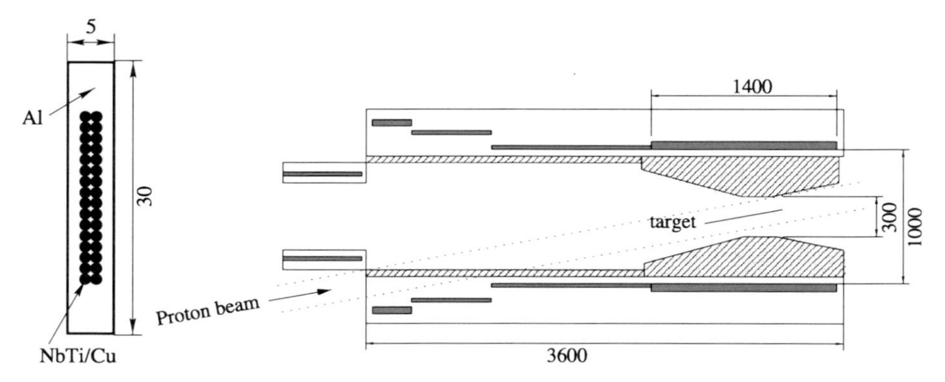

Fig. 13. (*Left*) Cross section of superconducting coil for capture solenoid. (*Right*) Schematic layout of the capture solenoid system. *Shaded areas* represent radiation shields made of tungsten. *Gray regions* represent superconducting coils. A proton beam is injected from the lower-left region of the figure, and captured pions are transported toward left. Dimensions are in mm.

Figure 13 shows a schematic view of the system of pion production and capture. It consists of a pion-production target, a surrounding radiation shield, a superconducting solenoid magnet for pion capture with a 5 T magnetic field, and a matching section connected to the transport solenoid system with a 2 T field. The 30-cm-thick radiation shield, which is necessary to achieve a heat load below 100 W for the superconducting magnets, is inserted between the pion-production target and the superconducting coils. The inner bore of the shield is tapered to keep it away from beam protons and high-energy pions, which are scattered forward. To collect backward-scattered pions, the proton beam should be injected through the barrel of the solenoid and should be tilted with respect to the solenoid axis by about 10°.

5.2 Muon Beamline

Pions and muons are transported to a muon-stopping target through the muon beamline, which consists of curved and straight solenoids. The key requirement for the muon beamline is that it should be possible to select the electric charge and momentum of beam particles. In addition, it is required to provide a high-efficiency transportation of muons having a momentum of around 40 MeV/c. At the same time, it is necessary to eliminate energetic muons having a momentum larger than 75 MeV/c, since their decays in flight would produce spurious signals of ~105 MeV electrons.

A schematic layout of the muon beamline including the capture and detector sections is shown in Fig. 14. Tracking simulation studies were performed using a single-particle tracking code based on GEANT. The magnetic field of the solenoids can be computed using a realistic configuration of coils and their current settings.

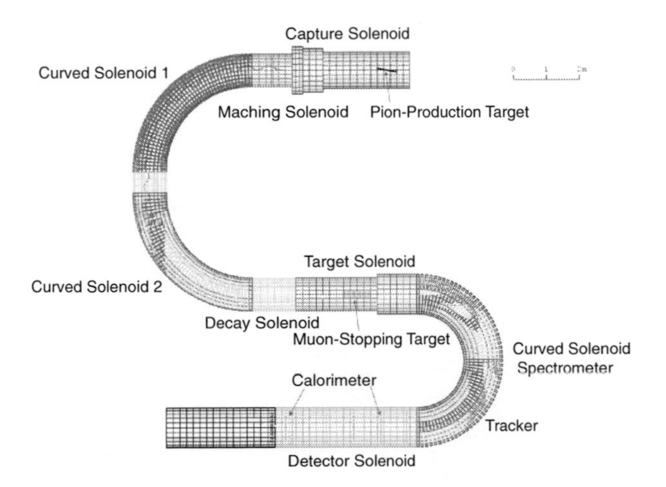

Fig. 14. Present design of solenoid channel used in tracking studies.

The selection of the electric charge and the momentum of beam particles can be performed by using curved solenoids. It is known that, in a curved solenoid, the center of the helical trajectory of a charged particle is shifted, and the drift (D[m]) is given by

$$D = \frac{q}{0.3 \times B} \times \frac{s}{R} \times \frac{p_l^2 + \frac{1}{2}p_t^2}{p_l}, \tag{7}$$

where q is the electric charge of the particle (with its sign), B[T] is the magnetic field at the axis, and s[m] and R[m] are the path length and the radius of curvature of the curved solenoid, respectively. Unless two curved solenoids bent in opposite directions are installed, a dipole magnetic field to compensate for the drift of particles having the momentum of interest might be needed.

The present design utilizes two curved solenoids with a bend angle of 90° in the same bend direction. Each has a magnetic field of 2 T and a radius of curvature of 3 m. Adjustment of the inner radius of the solenoid has a similar effect to that of a collimator. A compensation field of 0.038 T for the first 90° bend and of 0.052 T for the second 90° bend are applied. A pion yield of about $1{\times}10^{-5}\pi$/proton is obtained.

5.3 The Detector

The sole role of the detector is to identify genuine $\mu^-{-}e^-$ conversion events from the huge number of background events. The signature of a $\mu^-{-}e^-$ conversion event is, as mentioned in Sect. 3.1, a mono-energetic (∼105 MeV) electron scattered from a muon stopped in the target. In contrast, background events have various origins. They can be rejected using various combinations of different methods associated with the muon beamline and the detector. The background event rejection will be explained in detail in Sect. 5.5. The parameters that can be measured from the signals are momentum, energy, and timing only. Therefore, to enable signals of genuine events to be distinguished from those of background events, the events should be measured as

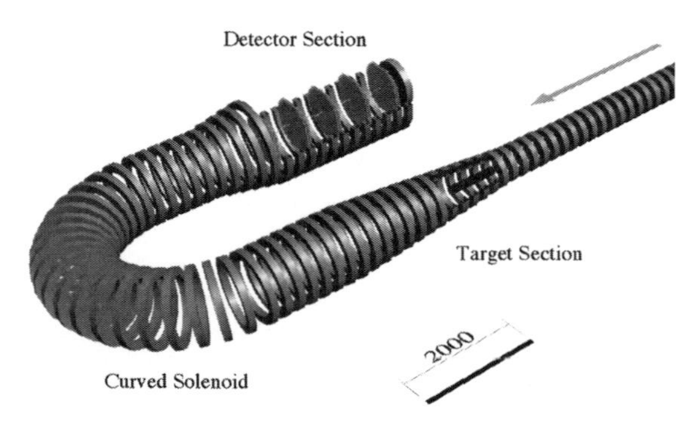

Fig. 15. Setup of proposed spectrometer.

precisely as possible. The detector being considered here is quite different from that planned in the MECO experiment (MECO Technical Proposal, August 1, 2001, unpublished). The detector consists of three sections. The first is a section where a muon-stopping target is placed in a graded magnetic field. The second is a section in which electron transport with curved solenoids is performed, and electrons and other particles of low energy are eliminated to reduce the occurrence of background events as well as the counting rate of the detector. The third is a section where the momentum and energy of remaining electrons are measured in a uniform solenoidal magnetic field (Fig. 15).

Muon-Stopping Target

In this Phase 1 experiment, to eliminate background events arising from both prompt and late-arriving beam particles, a detection window opens about 700 ns after the prompt. Therefore, it is not suitable to use heavy materials for which the lifetime of muonic atoms is short. It was determined to use aluminum for a muon-stopping target. It should be noted that, for the photonic diagrams, the branching ratio for aluminum ($Z = 13$) is smaller than titanium ($Z = 22$) by only a factor of 1.7. In the preliminary target configuration, the target is composed of 17 aluminum disks, each being 100 mm in radius and 200 μm in thickness, which are arranged with a disk spacing of 50 mm. A graded magnetic field is applied at the target location. The graded magnetic field would reflect electrons emitting backward toward the forward direction due to mirroring effects and at the same time make the directions in which electrons are emitted more parallel to the axis, by reducing their polar angles. Figure 16 (left) shows the baseline configuration of a graded magnetic field in the stopping target region. Monte Carlo simulations were carried out to study the stopping efficiency of muons for this configuration. Figure 16 (right) shows the momentum distributions of the muons approaching the target (open histogram) and those stopped by the target disks (shaded histogram). A muon-stopping efficiency of 0.29 was obtained.

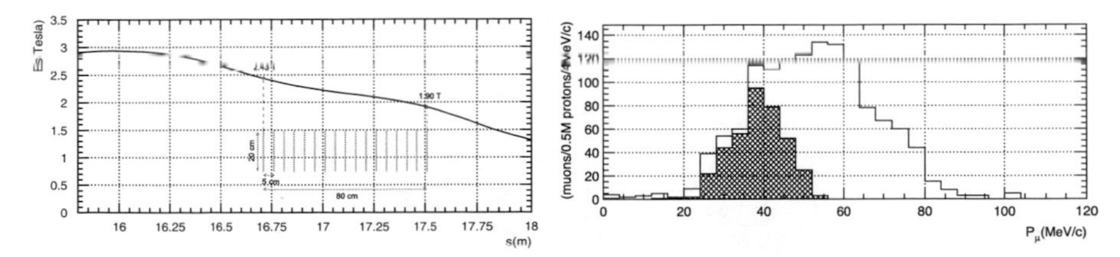

Fig. 16. (*Left*) Distribution of a graded magnetic field over the target region. (*Right*) Momentum distributions of muons approaching the target (an *open histogram*) and those stopped by the muon-stopping target (a *shaded histogram*).

Electron Transport Through Curved Solenoids

As in the muon beamline, the electron transport system adopts curved solenoids to remove charged particles of low momentum, where the single counting rate is mostly dominated by DIO (muon decay in orbit) electrons. It is noted that a counting rate of about 500 kHz per single wire in the straw gas chambers in the MECO experiment was estimated.

The transmission efficiency is estimated by GEANT. The present design of the curved solenoid spectrometer uses a bend angle of 180° and an applied magnetic field of 1 T. The case of application of a field gradient from 3 to 1 T in the target region is considered. This would give a transport efficiency of the signal (∼100 MeV/c) of 30–40% and survival rates for DIO electrons of 10^{-7}–10^{-8}. The detection rate is estimated to be of the order of 1 kHz for 10^{11} muons per second in the muon-stopping target. Since the geometrical acceptance of signal events in a graded magnetic field in the target region is about 0.73 and the transmission efficiency of signal events is about 0.44, the acceptance of signals of $\mu^- - e^-$ conversion signals is about 0.32.

Detection of Electrons

The main purpose of the electron detector is to distinguish electrons from other particles and to measure their energies, momenta, and timing. The electron detector consists of an electron tracking detector with straw-tube gas chambers for measuring momenta of electrons, an electromagnetic calorimeter for measuring their energies, and fast trigger counters. The detector is placed under a uniform solenoidal magnetic field for momentum tracking. Furthermore, to reduce multiple scattering in momentum measurements, the entire system is placed under vacuum. A candidate layout of the electron detector is shown in Fig. 17.

The required momentum resolution is better than 350 keV/c for a sensitivity of 10^{-16}. The electron tracking detector consists of five stations of straw-tube gas chambers, where each station is composed of two views (x and y), and one view has two staggered layers of straw tubes. Each of the straw tubes

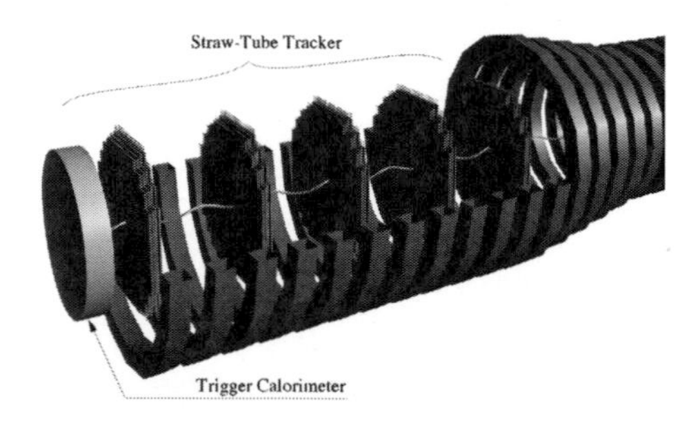

Fig. 17. Layout of an electron detector. It consists of five stations of straw-tube gas chambers, followed by an electron calorimetric detector.

is 5 mm in diameter and 25 μm in thickness. From a GEANT Monte Carlo simulation with 250 μm position resolution, a momentum resolution of 230 keV/c is obtained. The contamination of DIO background events in the signal region is estimated by using the events with χ^2 cuts. The contamination is estimated as 0.05 events for the entire measurement period.

The electron calorimeter, which is located downstream from the tracking detector, would serve three purposes. One is to measure the energy of electrons. High-energy resolution is required. The second is to provide a timing signal for the electron events and at the same time give a trigger signal which could be used to select events to be recorded for further analysis. In this regard, fast response and high efficiency are needed. The third is to provide additional data on hit positions of the electron tracks at the calorimeter location. This would be useful in eliminating false tracking. Candidate inorganic crystals, such as cerium-doped Gd_2SiO_5 (GSO) crystals, have been considered.

Cosmic-ray-induced electrons (or other particles misidentified as electrons) may cause background events. Therefore, passive and active shielding against cosmic rays covering the entirety of the detector is considered.

5.4 Acceptance

The acceptance is determined by the geometrical acceptance, which has been discussed before, and the analysis acceptance given by the analysis cuts. They are discussed in the following.

- *Energy*: To determine the energy region for the $\mu^- - e^-$ conversion signals, the $\mu^- - e^-$ conversion electrons were generated inside the muon-stopping target and reconstructed using a tracking program. Figure 18 shows the distribution of the reconstructed momentum (without correction for energy loss in the target), where a momentum spread of about 350 keV/c is seen. The signal region is determined to be from 104.0 to 105.2 MeV/c, which corresponds to one 1.7 sigma width of momentum spread. In this signal region, about 68% of total signal events is contained.
- *Transverse Momentum*: To eliminate background events, such as those from beam electrons and muon decay in flight, a transverse momentum of electrons greater than 52 MeV/c ($p_t > 52$ MeV/c) at the detector position is desired.
- *Timing*: Measurement starts about 700 ns after the prompt to avoid beam-related prompt background events. A schematic timing chart is shown in Fig. 18. The acceptance in the detection window is about 0.38 for aluminum.

Table 4 summarizes the acceptances. The total signal acceptance for spectrometer and detector is 0.07.

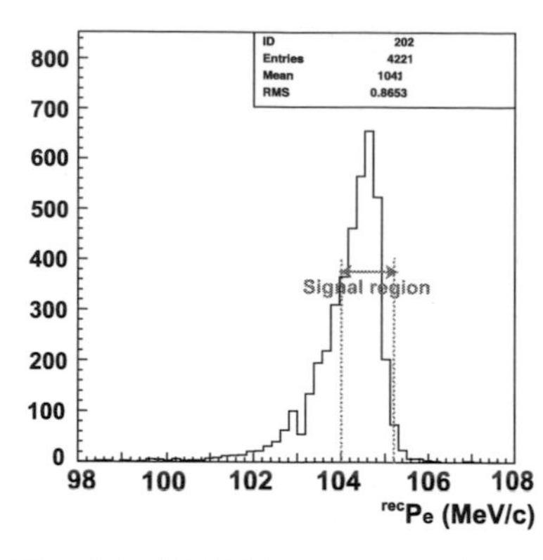

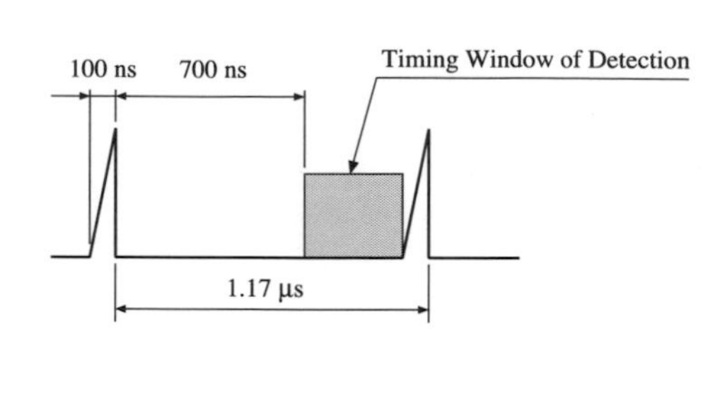

Fig. 18. (*Left*) Reconstructed momentum distribution of 105 MeV electrons. This is not corrected for average energy loss of electrons (of about 0.4 MeV/c). The energy region for the signal is set to that from 104.0 to 105.2 MeV/c for an uncorrected energy scale. (*Right*) Timing window of detection.

Table 4. Summary of signal acceptance.

	Acceptance
Geometrical acceptance	0.73
Electron transport efficiency	0.44
Energy selection	0.68
Transverse momentum ($p_t > 52$ MeV/c)	0.82
Timing window of detection	0.38
Total	0.07

5.5 Sensitivity and Background

Signal Sensitivity

We estimate the signal sensitivity of our search for $\mu^- - e^-$ conversion. The single-event sensitivity is defined by the number of muons stopping in the muon target (N_μ), the fraction of captured muons (f_{cap}), and the detector acceptance (A_e) as follows:

$$B(\mu^- + \mathrm{Al} \to e^- + \mathrm{Al}) \sim \frac{1}{N_\mu \cdot f_{cap} \cdot A_e}. \tag{8}$$

The total number of muons which are stopped in the muon-stopping target (N_μ) of about 5.6×10^{17} for 2×10^7 s is estimated as summarized in Table 5. For aluminum, the fraction of muons captured is about $f_{cap} = 0.6$. The acceptance A_e is summarized as shown in Table 4. The total acceptance for the signal is 0.07. By using N_μ, f_{cap}, and A_e, the single-event sensitivity is obtained as

$$B(\mu^- + \mathrm{Al} \to e^- + \mathrm{Al}) = \frac{1}{6 \times 10^{17} \times 0.6 \times 0.07} = 4 \times 10^{-17}. \tag{9}$$

Table 5. Total number of muons delivered to the muon-stopping target.

Proton intensity	4×10^{13} protons/s
Running time	2×10^7 s
Rate of muons per proton transported to the target	0.0024
Muon-stopping efficiency	0.29
Total	5.6×10^{17} stopped muons

Since a 90% confidence level (CL) upper limit is given by $2.3/(N_\mu \cdot f_{cap} \cdot A_e)$, the upper limit is obtained as

$$B(\mu^- + \mathrm{Al} \rightarrow e^- + \mathrm{Al}) < 10^{-16} \quad (90\% \text{ C.L.}), \tag{10}$$

which is about 10,000 times better than the current published limit obtained by SINDRUM II at PSI of $< 4.3 \times 10^{-12}$ (90% CL) [21].

Background Events and Their Rejection

Potential sources of background events for $\mu^- - e^-$ conversion are categorized into three different types. They are as follows:

1. Intrinsic physics background events: Intrinsic physics background events originate mostly from muons stopping in the muon-stopping target. They arise from muon decays in orbit, radiative muon capture, and particle emission after muon capture.

Table 6. Summary of background event rates at a sensitivity of 10^{-16}. Background events marked with an asterisk are proportional to the beam extinction, and the rates in the table assume 10^{-9} for the beam extinction.

Background	Events	Comments
Muon decay in orbit	0.05	230 keV (sigma) assumed
Pattern recognition errors	<0.001	
Radiative muon capture	<0.001	
Muon capture with neutron emission	<0.001	
Muon capture with charged particle emission	<0.001	
Radiative pion capture*	0.12	Prompt pions
Radiative pion capture	0.002	Due to late-arriving pions
Muon decay in flight*	<0.02	
Pion decay in flight*	<0.001	
Beam electrons*	0.08	
Neutron induced*	0.024	For high-energy neutrons
Antiproton induced	0.007	For 8 GeV protons
Cosmic rays induced	0.04	With 10^{-4} veto inefficiency
Total	0.34	

2. Beam-related background events: This type of background event is caused by particles in a beam, such as electrons, pions, muons, and antiprotons. There are two different types, one is a prompt background event and the other is a late-arriving background event. For the former, beam pulsing with a high beam extinction is a very effective way of rejecting the background events.
3. Cosmic-ray background events: The expected background rates at a sensitivity of 10^{-16} are summarized in Table 6.

6 Phase 2: PRISM: $\mu - e$ Conversion at 10^{-18}

To achieve the ultimate experimental sensitivity of $B(\mu^- + Ti \to e^- + Ti) < 10^{-18}$ at Phase 2, not only the quantity (high intensity) of the muon beam but also its qualities, such as monochromaticity and purity, become critical. We are proposing a new scheme for a next-generation muon beam source called PRISM [29–31]. In this chapter, a brief overview of PRISM is presented.

6.1 What is PRISM?

PRISM is a new-generation muon source that will provide a muon beam of high intensity with narrow energy spread and small beam contamination. PRISM stands for "*phase rotated intense slow muon source*". The target beam intensity is 10^{11}–$10^{12}\mu^{\pm}$/s, 4 orders of magnitude higher than that available at present. It is achieved by large solid-angle pion capture with a high solenoid magnetic field. The narrow energy spread can be achieved by phase rotation, which accelerates slow muons and decelerates fast muons by means of a radio-frequency (RF) field. The pion contamination in the muon beam can be removed by providing a long flight path for beam particles in PRISM so that most pions decay out.

PRISM consists of

- a pulsed proton beam (to produce a short pulsed pion beam);
- a pion capture system (with a large solid angle provided by a high solenoidal magnetic field);
- a pion decay and muon transport system (in a long solenoid magnet of about 10 m length); and
- a muon storage ring (in which phase rotation is performed to accelerate slow muons and decelerate fast muons by means of an RF field).

A schematic layout of PRISM is shown in Fig. 19. A major difference between the Phase 1 experiment and PRISM is the muon storage ring in which the energies of muons are equalized by the phase rotation technique. For the muon storage ring, a fixed-field alternating gradient synchrotron (FFAG) was adopted. A FFAG ring has several advantages, such as large momentum acceptance. Some of the key components are explained in detail in the following

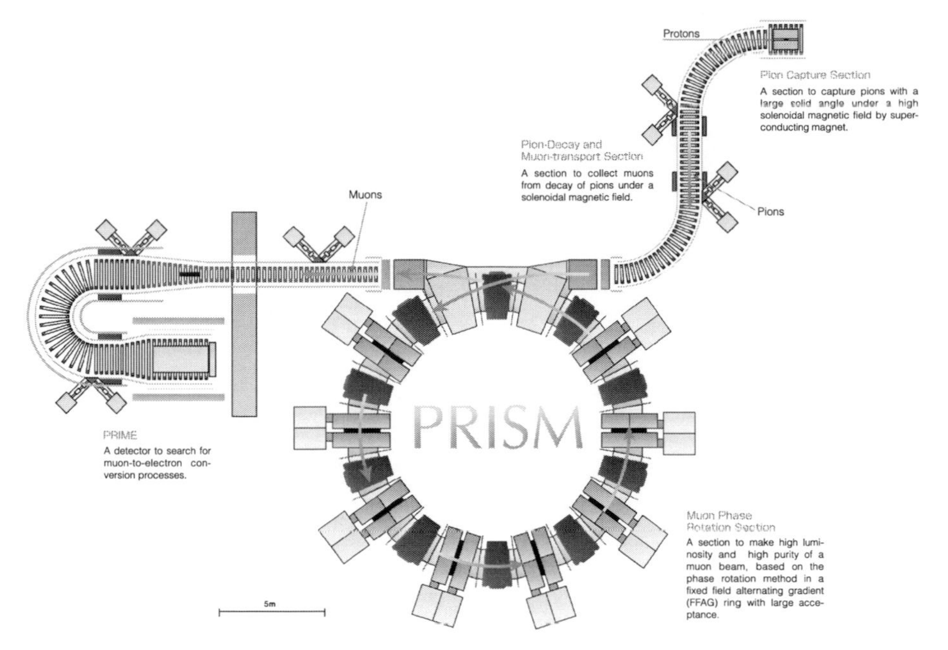

Fig. 19. Schematic layout of Phase 2 experiment; PRISM/PRIME.

Table 7. Anticipated design characteristics of PRISM beam

Parameters	Design goal	Comments
Beam intensity	10^{11}–$10^{12}\mu^{\pm}$/s	10^{14} protons/s is assumed
Muon kinetic energy	20 MeV	$P_{\mu} = 68$ MeV/c
Kinetic energy spread	±(0.5–1.0) MeV	
Beam repetition	100–1000 Hz	

sections. The PRISM beam characteristics are summarized in Table 7. The upgrade from the Phase 1 experiment to PRISM could be made by just adding the muon storage ring. The potential for future extension is well maintained.

6.2 Phase Rotation

Phase rotation is a method of achieving a beam of narrow energy spread. The principle of phase rotation is used to accelerate slow muons and decelerate fast muons by means of a strong radio-frequency (RF) electric field, in order to yield a narrow longitudinal momentum spread. This corresponds to a 90° rotation of the phase volume occupied by muons in a beam in the energy–time phase space, as schematically shown in Fig. 20. After phase rotation, the projection of the phase volume onto the energy axis becomes narrower and sharper. As a result of phase rotation, the initial time spread is converted

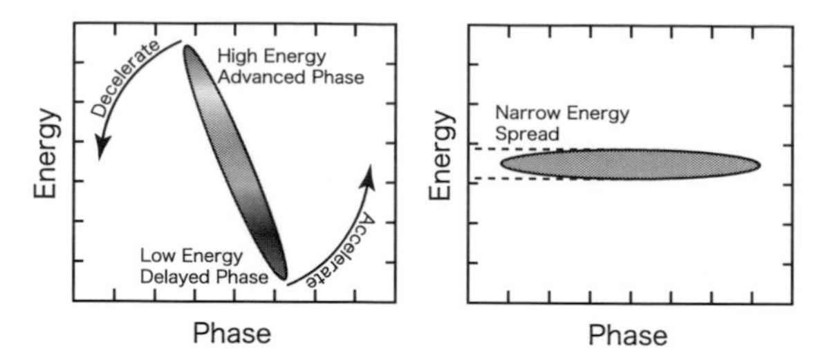

Fig. 20. Principle of phase rotation shown in energy–time phase space. The initial narrow time spread is converted into the final narrow energy spread.

into the final energy spread. Having a pulsed proton beam of narrow width is critical to the net performance of phase rotation.

6.3 PRISM Features and Background Rejection

PRISM would have several outstanding characteristics that would facilitate background rejection. They are listed as follows:

(1) *Long flight length*: The total flight length of particles in the muon storage ring having a circumference of about 40 m would allow most of the pions contaminating the beam to decay out. A pion survival rate of less than 10^{-20} can be obtained.
(2) *Narrow beam energy spread*: A narrow beam energy spread can be achieved by using phase rotation in the PRISM-FFAG ring. The goal is $\pm 3\%$ for the central momentum of 68 MeV/c. This narrow energy spread allows a thinner muon-stopping target to be used, which would yield better momentum resolution.
(3) *Muon beam energy selection*: A momentum slit allowing selection of the momentum of muons (such as $\pm 3\%$ of 68 MeV/c) after the PRISM-FFAG could be disposed. As a result no particles of $\sim$100 MeV/c would be able to pass the slit.
(4) *Beam extinction for muons*: A kicker magnet for extraction should serve as an additional beam extinction device. It would be more effective than proton beam extinction. This being the case, proton beam extinction might not be needed.
(5) *Low duty factor for detection*: The PRISM-FFAG ring runs with a low repetition rate of 100–1,000 Hz. With a 1 μs detection window, the duty factor of running is 10^{-4}.

These features would be tremendously advantageous in the removal of further background events in the search for $\mu^- - e^-$ conversion. The features are summarized in Table 8, where the numbers in the "How to eliminate" column correspond to the numbers in the list of features described above.

Table 8. Summary of methods of background rejection by the PRISM.

Background sources	How to eliminate	Comments
Muon decay in orbit (DIO)	(2)	1/10 of Phase 1 target thickness
Radiative pion capture	(1)	No pion contamination in beam
Beam electrons	(3)	No electrons of ∼100 MeV
Muon decay in flight	(3)	P_μ should be <77 MeV/c
Long transit background	(4)	Better beam extinction
Cosmic rays	(5)	No active cosmic-ray shielding needed

6.4 Muon Storage Ring: PRISM-FFAG

There are several advantages to using the muon storage ring for phase rotation instead of a linear phase rotating system. The ring system could be made smaller and more compact. Furthermore, the number of RF cavities and the required electric power are greatly reduced. Therefore, the total cost can be significantly reduced [32]. Among the many ring options, a fixed-field alternating gradient synchrotron (FFAG) has been selected because of (1) its capability of rapid acceleration and (2) its large longitudinal and transverse acceptance. It is referred to as the PRISM-FFAG ring hereafter.

In order to achieve a high-intensity muon beam, it is necessary for the PRISM-FFAG to have both large transverse acceptance and large momentum acceptance. Furthermore, long straight sections in which to install RF cavities are required to obtain a high surviving ratio of muons. The parameters of the beam optics of the PRISM-FFAG ring and its layout are shown in Fig. 21.

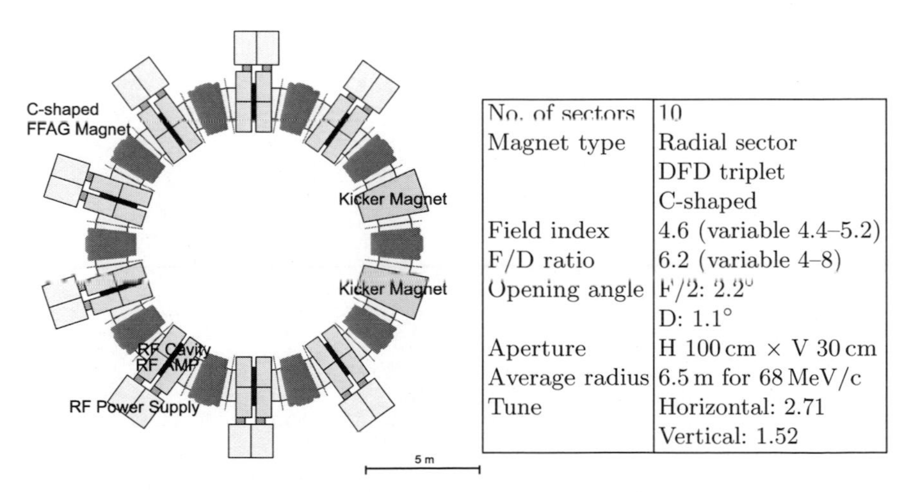

No. of sectors	10
Magnet type	Radial sector DFD triplet C-shaped
Field index	4.6 (variable 4.4–5.2)
F/D ratio	6.2 (variable 4–8)
Opening angle	F/2: 2.2° D: 1.1°
Aperture	H 100 cm × V 30 cm
Average radius	6.5 m for 68 MeV/c
Tune	Horizontal: 2.71 Vertical: 1.52

Fig. 21. Schematic layout of PRISM-FFAG (*left*). FFAG parameters (*right*).

6.5 PRISM-FFAG R&D

A large portion of the PRISM-FFAG ring is now under construction at Osaka University. The first three PRISM-FFAG magnets have been constructed, as shown in Fig. 22. The magnet has a large aperture of 30 cm vertically and 100 cm horizontally. According to the tracking simulations using a magnetic field map calculated by TOSCA, the PRISM-FFAG ring has a horizontal acceptance of more than 40,000 πmm·mrad and a vertical acceptance of about 6,500 πmm·mrad.

Since the muon is an unstable particle with a lifetime of 2.2 μs, it is crucial to complete the phase rotation as quickly as possible. In this regard, the requirement of the PRISM RF system is an RF field gradient of $\sim$170 kV/m at a low frequency of 4–5 MHz, which is markedly high in comparison with other ordinary RF cavities. We have already developed an RF system in which an RF voltage of $\pm$43 kV/gap is achieved. Figure 23 shows an RF core of magnetic alloy and the PRISM-FFAG RF system, which consists of an amplifier, an anode power supply, and an auxiliary power supply.

Fig. 22. Photograph of PRISM-FFAG magnet (*left*).

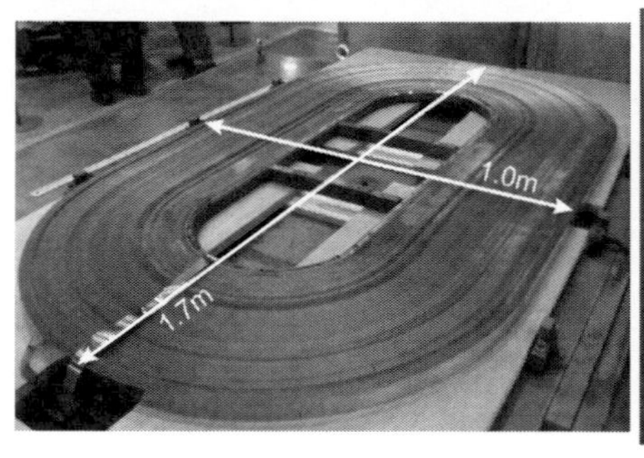

Fig. 23. RF core made of metal alloy (*left*). Power supply and amplifier for PRISM-RF (*right*).

7 Summary

The physics significance of LFV of charged leptons, which is forbidden in the Standard Model and has not yet been observed, is very strong and robust. Its physics importance will persist in the future, whether the LHC finds evidence for the existence of SUSY or not. It is a way of searching for new physics beyond the Standard Model, in particular, supersymmetric extension with either the Grand Unification Theory (GUT) or the neutrino seesaw mechanism.

Among various LFV processes of charged leptons, the $\mu^- - e^-$ conversion in a muonic atom, $\mu^- + N \to e^- + N$, is the most promising for the future. The present published upper limit is $B(\mu^- + \mathrm{Ti} \to e^- + \mathrm{Ti}) < 6.1 \times 10^{-13}$. To proceed further in the search, we decided to take a staging approach, Phase 1 and Phase 2. The goal of the Phase 1 experiment is

$$B(\mu^- + \mathrm{Al} \to e^- + \mathrm{Al}) < 10^{-16}, \tag{11}$$

representing a factor of ~10,000 improvement over the present limit. In the second phase (the PRISM project), it is

$$B(\mu^- + \mathrm{Ti} \to e^- + \mathrm{Ti}) < 10^{-18}, \tag{12}$$

which corresponds to a factor of ~1,000,000 improvement over the present limits.

To improve the search for LFV with muons, a highly intense muon source is a requirement. On the basis of original ideas conceived at Osaka University, a highly intense muon beam source could be constructed if the necessary resources are made available. The target intensity of 10^{11}–$10^{12}\mu$/s is about 10^3–10^4 higher than the best currently available intensity. For the PRISM (Phase 2), not only the intensity but also the luminosity and purity of a muon beam can be improved by installing an additional muon storage ring (PRISM-FFAG ring).

References

1. Kuno Y., Okada, Y.: Rev. Mod. Phys. **73**, 151 (2001) [arXiv:hep-ph/9909265]
2. Brown, H.N., et al.: Phys. Rev. Lett. **92**, 161802(2004)
3. Feng, J.L., Matchev, K.T., Shadmi, Y.: Nucl. Phys. **B613**, 366 (2001)
4. Feng, J.L., Matchev, K.T., Shadmi, Y.: Phys. Lett. **B555**, 89 (2003)
5. A Letter of Intent to the J-PARC 50 GV Proton Synchrotron Experiments: Search for the Electric Dipole Moment of the Muon, January 1st, 2003
6. Hincks E.P., Pontecorvo, B.: Phys. Rev. Lett. **73**, 246 (1947)
7. Hall, L.J., Kostelecky V.A., Raby, S.: Nucl. Phys. B **267**, 415 (1986)
8. Barbieri, L., Hall, L., Strumia, A.: Nucl. Phys. B **445**, 219 (1955)
9. Hisano, J., Moroi, T., Tobe K., Yamaguchi, M.: Phys. Lett. B **391**, 341 (1997)
10. Hisano, J., Moroi, T., Tobe K., Yamaguchi, M.: Phys. Lett. B **397**, 357 Erratum (1997)

11. Hisano, J., Moroi, T., Tobe, T., Yamaguchi, M., Yanagida, T.: Phys. Lett. B **357**, 579 (1995)
12. Hisano, J., Nomura, D., Yanagida, T.: Phys. Lett. B **437**, 351 (1998)
13. Hisano, J., Nomura, D., Phys. Rev. D **59**, 116005 (1999)
14. Shanker, O.: Phys. Rev. D **25**, 1847 (1982)
15. Watanabe, R., Fukui, M., Ohtsubo H., Morita, M.: Prog. Theor. Phys. **78**, 114 (1987)
16. Watanabe, R., Muto, K., Oda, T., Niwa, T., Ohtsubo, H., Morita R., Morita, M.: At. Data Nucl. Data Tables **54**, 165 (1993)
17. Bryman, D.A., Blecher, M., Gotow K., Powers, R.J.: Phys. Rev. Lett. **28**, 1469 (1972)
18. Badertscher, A., et al.: Nucl. Phys. A **377**, 406 (1982)
19. Bryman, D.A., et al.: Phys. Rev. Lett. **55**, 465 (1985)
20. Ahmad, S., et al.: Phys. Rev. D **38**, 2102 (1988)
21. Dohmen C., et al. (SINDRUM II Collaboration): Phys. Lett. B **317**, 631 (1993)
22. Honecker W., et al. (SINDRUM II Collaboration): Phys. Rev. Lett. **76**, 200 (1996)
23. Wintz P.: In Klapdor-Kleingrothaus H.V., Krivosheina I.V. (eds.) Proceedings of the First International Symposium on Lepton and Baryon Number Violation, p. 534. Institute of Physics Publishing, Bristol and Philadelphia. Unpublished (1998)
24. Bertl, W., et al.: Eur. Phys. J. C **47**, 337 (2006)
25. Bachman, M., et al.: A research proposal to Brookhaven National Laboratory AGS "A Search for $\boldsymbol{\mu^- N \to e^- N}$ with Sensitivity Below $\mathbf{10^{-16}}$, Muon – Electron Conversion" (1997)
26. Brooks M., et al. (MEGA Collaboration): Phys. Rev. Lett. **83**, 1521 (1999)
27. Barkov L.M., et al.: A research proposal to PSI "Search for $\boldsymbol{\mu^+} \to e^+\gamma$ down to $\mathbf{10^{-14}}$ branching ratio" (1999)
28. A Letter of Intent to the J-PARC 50 GV Proton Synchrotron Experiments: An Experimental Search for a $\boldsymbol{\mu^- - e^-}$ Conversion at Sensitivity of the Order of $\mathbf{10^{-16}}$ With Slowly Extracted Pulsed Proton Beam, December 15th, 2006
29. A Letter of Intent to the J-PARC 50 GV Proton Synchrotron Experiments: The PRISM Project – A Muon Source o the World-Highest Brightness by Phase Rotation, January 1st, 2003
30. A Letter of Intent to the J-PARC 50 GV Proton Synchrotron Experiments: An Experimental Search for the $\boldsymbol{\mu^- - e^-}$ Conversion Process At An Ultimate Sensitivity Of The Order of $\mathbf{10^{-18}}$ With PRISM, January 1st, 2003
31. A Letter of Intent to the J-PARC 50 GV Proton Synchrotron Experiments: An Experimental Search for a $\boldsymbol{\mu^- - e^-}$ Conversion at Sensitivity of the Order of $\mathbf{10^{-18}}$ With A Highly Intense Muon Source, PRISM. April 28th, 2006
32. Palmer, R., Berg, J.S.: Proceedings of EPAC 2004. Lucerne, Switzerland (2004)

Index